Green Energy and Technology

For further volumes:
http://www.springer.com/series/8059

Avraam Karagiannidis
Editor

Waste to Energy

Opportunities and Challenges for Developing and Transition Economies

Springer

Assoc. Prof. Avraam Karagiannidis
Department of Mechanical Engineering
Aristotle University of Thessaloniki
Box 483, 54124 Thessaloniki
Greece
e-mail: akarag@auth.gr

ISSN 1865-3529 e-ISSN 1865-3537
ISBN 978-1-4471-2305-7 e-ISBN 978-1-4471-2306-4
DOI 10.1007/978-1-4471-2306-4
Springer London Dordrecht Heidelberg New York

British Library Cataloguing in Publication Data
A catalogue record for this book is available from the British Library

Library of Congress Control Number: 2011942904

Printed on acid-free paper

Springer is part of Springer Science+Business Media (www.springer.com)

Preface

Many countries, beyond those considered to be ‘developing’, have waste management systems with significant improvement margins to reach worldwide state-of-the-art engineering, health and safety standards. These are still characterized by a heavy dependence on landfilling (both engineered and non-engineered) and diverging, but generally low, levels of recycling and composting, as well as a general absence of waste-to-energy schemes and plants.

Waste-to-Energy (WtE) is understood in this edited volume to encompass a broad range of energy-from-waste approaches and techniques including incineration, pyrolysis, gasification, anaerobic digestion and co-combustion in existing industrial plants. This volume comprises of 15 chapters addressing different WtE aspects related to developing countries and transient waste management systems.

The editor would like to thank all authors for the highly professional work and output. Special thanks are due to Mrs Stamatia Kontogianni for her comprehensive support during the entire editing process.

Contents

Waste-to-Materials: The Longterm Option

Philip Nuss, Stefan Bringezu and Kevin H Gardner

Abstract Managing solid waste is one of the biggest challenges in urban areas around the world. Technologically advanced economies generate vast amounts of organic waste materials, many of which are disposed to landfills. In the future, efficient use of carbon containing waste and all other waste materials has to be increased to reduce the need for virgin raw materials acquisition, including biomass, and reduce carbon being emitted to the atmosphere therefore mitigating climate change. At end-of-life, carbon-containing waste should not only be treated for energy recovery (e.g. via incineration) but technologies should be applied to recycle the carbon for use as material feedstocks. Thermochemical and biochemical conversion technologies offer the option to utilize organic waste for the production of chemical feedstock and subsequent polymers. The routes towards synthetic materials allow a more closed cycle of materials and can help to reduce dependence on either fossil or biobased raw materials. This chapter summarizes carbon-recycling routes available and investigates how in the long-term they could be applied to enhance waste management in both industrial countries as well as developing and emerging economies. We conclude with a case study looking at the system-wide global warming potential (GWP) and cumulative energy demand (CED) of producing high-density polyethylene (HDPE) from organic waste feedstock via gasification followed by Fischer–Tropsch synthesis (FTS). Results of the analysis indicate that the use of organic waste feedstock is beneficial if greenhouse gas (GHG) emissions associated with landfill diversion are considered.

P. Nuss (✉) · K. H Gardner
Environmental Research Group, University of New Hampshire,
Durham, NH 03824, US
e-mail: philip.nuss@unh.edu

S. Bringezu
Research Group 3: Material Flows and Resource Management,
Wuppertal Institute for Climate, Environment and Energy,
Döppersberg 19, Wuppertal 42103, Germany

A. Karagiannidis (ed.), *Waste to Energy*, Green Energy and Technology,
DOI: 10.1007/978-1-4471-2306-4_1,

1 Evolution of Waste Management Practices and the Socio-Industrial Metabolism

1.1 Waste Generation and Management in a Development Perspective

Since prehistoric times, human activities generated waste materials that were discarded because they were considered of low-value or useless. In the early days, the disposal of wastes did not pose a significant problem, as the population was small and land for the assimilation of wastes was widely available. However, as the human population grew and began to settle in villages and communities, the accumulation of waste became a rogue consequence of life [62]. Since then, the turnover of materials has increased dramatically. This is not only due to global population growth but also due to the enormous growth of goods and assets used per person, in particular in affluent countries. Thus, along with the benefits of technology have also come the problems of disposal of resultant wastes.

Today, approximately 745 kg of municipal solid waste (MSW) are produced per capita per year in the United States [20] and an average of 522 kg MSW in the EU-27 [23]. Modern man consumes between 30 and 75 tons of material per person per year in their companies and households [7]. Of the materials consumed, an average of 90% of all biomass inputs and more than 90% of the non-renewable materials used are wasted on the way to making products available to the end-user [40]. Although materials are used more and more efficiently, there are no indications that overall material consumption will decline [8] and as a result it is expected that large amounts of waste will continue to be generated in the future [10]. Although in developing countries the quantity of solid waste generated in urban areas is low when compared to industrialized countries, waste management still remains inadequate [27]. Rapid economic growth and rise in community living standard in many of the low- or middle-income countries are likely to accelerate MSW generation as well as the complexity and variety in terms of substances present.

Managing these solid waste streams well and affordably is one of the key challenges of the Twenty first century [66]. Traditionally, municipal solid waste management encompasses the functions of collection, transport, resource recovery, recycling, and treatment. The primary goal of MSW management is to protect the health of the population, promote environmental quality, develop sustainability, and provide support to economic productivity [27]. In addition, climate change has drawn attention to the diversion of biodegradable municipal solid waste (BMSW), such as kitchen and garden waste, from landfills because it has the potential to form methane (a powerful greenhouse gas) under anaerobic conditions. According to the U.S. Environmental Protection Agency (EPA) the four basic options for integrated solid waste management include: (1) source reduction, including reuse, (2) recycling and composting, (3) combustion (waste-to-energy facilities), and (4) landfills [38]. Examples from e.g. Denmark and Japan suggest that a sustainable waste management system furthermore consists of a stable mixture of technologies and institutions which work flexibly under a defined policy umbrella [66]. Such sustainable waste management systems are designed to mimic an ecosystem that is robust and resilient. Taking a systems-perspective can help to e.g. determine whether materials currently regarded as wastes in one industrial sector could be viewed as raw materials by another sector.

1.2 Future Perspectives for Sustainable Waste Management

Ecosystems provide the best example of a system that works in a sustainable fashion [19]. One of the central principles in industrial ecology is the vision that industrial systems can use materials extracted and metabolized in a cyclical manner, driven by renewable energy which is used in a cascading manner [2]. One important measure relates to the systematic reuse of waste products in order to minimize the need to extract virgin raw materials and deplete environmental services [22]. However, to date recovery rates for materials such as metals, plastics, paper etc. from the municipal waste stream vary widely, even among industrialized countries. For example, in Germany in 2007 a total of 25% of all MSW generated was disposed to landfills and incinerators ([17]), while in the United States a total of 67% of all MSW generated in 2008 was discarded [20]. Furthermore, the EU landfill directive sets targets to progressively reduce the amount of BMSW disposed to landfills among the EU member states (including Germany), whereas in the United States large amounts of organic waste are sent to landfills. This happens despite the fact that organic waste, being rich in carbon, could serve increasingly as feedstock for thermal and biological processes recovering the carbon for further use as chemical feedstock ('carbon recycling').

The concept of *carbon recycling* is that, instead of releasing the carbon stored in biowaste into the atmosphere by applying conventional waste management practices such as incineration (for heat) or anaerobic digestion and landfilling (for biogas/landfill gas), the carbon inherent in the organic waste should be seen as a valuable feedstock resource [6]. Instead of carbon-capture and storage, which generally occurs at the beginning of the resource flows (e.g. at oil extraction sites to reduce fossil GHG emissions), the principle of carbon-capture and reuse could be further developed and applied throughout the whole socio-industrial metabolism. Specifically, technologies such as gasification, which allows the generation of a syngas, or anaerobic digestion, for the generation of an upgraded biogas (methane), could be applied.[1] Both syngas and biogas can then serve as feed

---

[1] In addition, in the future other carbon recycling technologies such as the synthetic tree air-capture unit, developed by Klaus Lackner of the Earth Institute at Columbia University, that stands in the open and captures CO_2 on its collector surfaces ("leaves") comprised of anionic resin [39], may serve as source of carbon for chemicals feedstock synthesis.

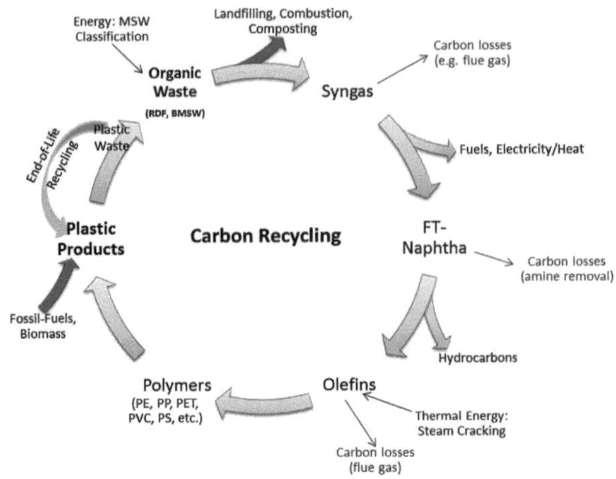

Fig. 1 Carbon recycling: making use of organic waste via the Fischer–Tropsch synthesis (FTS). Organic waste that could not be recovered via conventional waste recovery systems is gasified and transformed into a FT-naphtha (as well as by-products such as FT-diesel and electricity/heat). FT-naphtha is then transformed into olefins via conventional steam cracking. Polyolefins (PE, PP) and other polymers (PET, PVC, PS, etc) are generated via polymerization and used for the production of plastic products. At end-of-life these products can either be disassembled and the plastic parts be reused (preferred option if less energy and resource intensive than subsequent FTS) or the carbon and energy recovered via gasification producing a syngas and therefore closing the cycle

e.g. for the Fischer–Tropsch synthesis (FTS) to produce base compounds such as Fischer–Tropsch (FT)-naphtha and a number of subsequent chemical products and fuels. In addition, hydrolysis followed by fermentation can be applied to generate a variety of different base chemicals.

When fuels (e.g. FT-diesel, methanol, ethanol, etc.) are produced from organic waste and oxidized by use in combustion engines, the carbon (originally captured in the waste feedstock) is emitted back to the atmosphere. Assuming that the system-wide environmental burdens along this process route are lower than those of conventional fossil-based fuels production routes, this process route would lead to a mitigation of environmental burdens. However, this route of using carbon as fuel is still a linear process through the socio-industrial system which depends on significant amounts of waste feedstock being available [6].

If, in contrast, synthetic materials for the production of plastics could be synthesized, then the carbon would be kept longer in the use phase and add to the stock of durable goods in the technosphere. The plastic products could potentially be recycled at end-of-life to provide feedstock for either energy generation or as feedstock for the production of syngas in a cascading use scheme. Figure 1 exemplifies the concept of carbon recycling, making use of organic waste as feedstock for polyolefins production for the example of Fischer–Tropsch synthesis.

From an environmental perspective the use of waste would be advantageous as, in comparison to virgin greenwood biomass, it has no direct land-use requirement and collection and processing systems are oftentimes already in place. In addition, thermal treatment (i.e. gasification) has the further advantage of contributing to volume reduction, waste disinfection, and concentration of certain toxic elements (e.g. cadmium) in the gasification ash and slag produced.[2] A comprehensive systems analysis is required to assess the life-cycle wide performance of carbon recycling compared to conventional systems of waste management.

2 Carbon Recycling and Increased Resource Efficiency

2.1 Exemplary Routes of Carbon Recycling

Organic waste refers to all carbonaceous waste fractions that can potentially serve as feedstock for the thermochemical and biochemical platforms. These include:

- Biodegradable municipal solid waste (BMSW);
- Municipal plastic waste;
- Construction & demolition (C&D) derived biomass; and
- Liquid waste (e.g. sewage sludge).

BMSW includes all waste fractions of biological origin such as food wastes, paper, cardboard, yard wastes, and bulky wood waste. Of this the *cellulose and hemicelluloses fractions* can serve as feedstock for hydrolysis with fermentation or anaerobic digestion (kitchen organic waste, green organic waste and paper and cardboard). *Plastic waste* includes durable goods made from fossil-based[3] plastics such as PE, PP, PET, etc. *C&D derived biomass* originates from new construction sites and repairs and consists of treated and untreated wood fractions. *Organic liquid wastes* include municipal sludges such as sewage sludge and animal wastes that can be treated via anaerobic digestion or can be gasified after drying. In addition, industrial organic waste feedstock may be of interest as it often times is more homogeneous than waste from municipal sources.

---

[2] The removal of hazardous substances from the waste via thermal treatment leads to an ash or slag rich in hazardous substances, potentially enabling efficient recycling of metals from the waste stream in the future [10].

[3] In the beginning the thermochemical platform would, amongst other feedstocks, utilize conventional fossil-based plastics as feedstock for the production of syngas and subsequent plastics via the methanol to olefins (MTO) or Fischer–Tropsch synthesis (FTS). However, as this platform is continuously applied to recycle plastic waste by gasification and to produce new plastics from them, this implies that the feedstock origin will slowly shift from fossil- to waste based plastics (assuming that fossil-based feedstocks will become increasingly scarce over the course of the next decades). At the same time those plastics will slowly fade out that are less appropriate as feedstock or end-product of the recycling pathway.

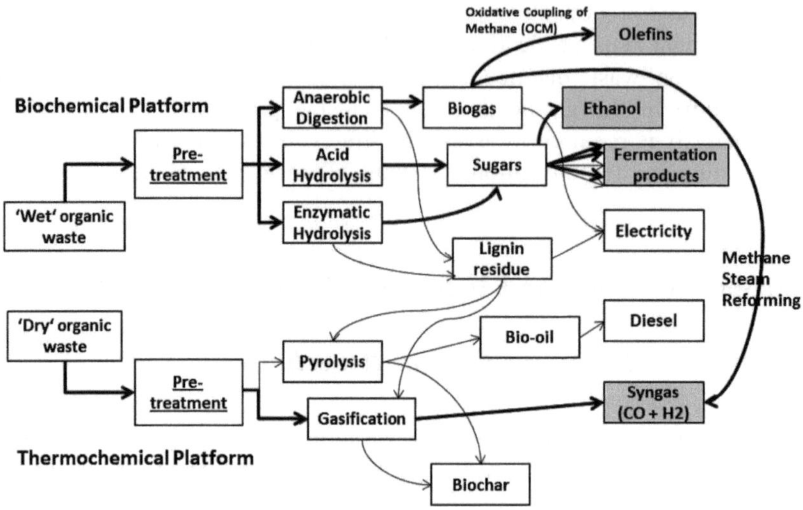

Fig. 2 The various conversion technologies possible for the treatment of organic waste (*Source* compilation adapted from [26]). Organic waste with high water content is treated in the biochemical platform in which either anaerobic digestion or acid/enzymatic hydrolysis are applied. Anaerobic digestion produces a biogas consisting mainly of CH_4 and CO_2 that can subsequently be converted into a syngas. Hydrolysis produces sugars which can be fermented into a variety of different base chemicals. Thermochemical processes apply gasification or pyrolysis of dry organic waste to derive at a syngas which serves as intermediate for the production of a potentially large number of chemicals (see Fig. 3). Please note that thermochemical processes, in contrast to anaerobic digestion and hydrolysis with fermentation, are able to utilize a large number of dry organic feedstock sources, including BMSW, plastic waste and C&D waste. Arrows in bold indicate routes of interest for the production of basic chemicals and polymers that would allow cascading use and carbon recycling

Organic waste can serve as feedstock for the production of transportation fuels, chemical feedstock and bio-energy using biochemical and thermochemical conversion routes. Current research with regards to biorefineries focuses mainly on the utilization of lignocellulosic materials, originating from agriculture and forestry, as second generation feedstock for the production of bio-fuels and chemicals. Interest in the use of organic waste residues as feedstock is growing. Biochemical processes will either employ anaerobic digestion or hydrolytic mechanisms to break apart the structural polysaccharides (lignocellulose) of the biomass. Alternatively thermochemical procedures can be used to dehydrate and volatilize the biomass feedstock. Research in bio-refining is proceeding quickly and commercial facilities are expected in the near-future [26]. Figure 2 provides an overview of the conversion technologies available for the treatment of organic waste. The bold arrows indicate pathways of interest for the synthesis of industrial feedstocks including plastic polymers.

Generally, the thermochemical platform, using gasification, will be superior to the biochemical platform if an organic waste fraction with low water content is used, whereas biochemical conversion generally works better if biomass with high water content is utilized [35].

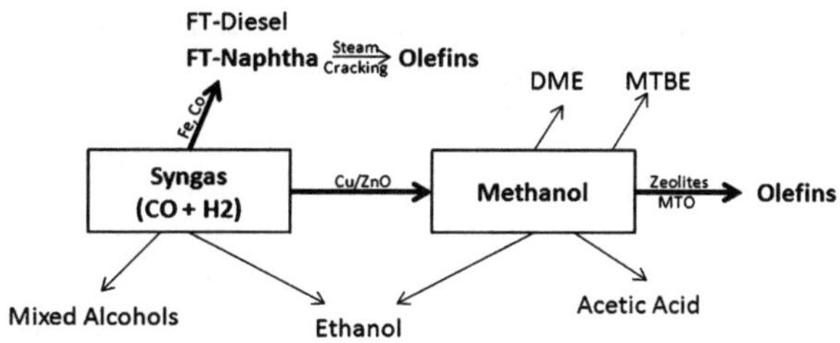

Fig. 3 The potential chemicals from syngas and some of the catalysts involved (*Source* compilation adapted from [60]). Syngas serves as industrial feedstock for the production of a variety of base chemicals. With respect to durable goods for cascading use, FTS and Methanol-to-Olefins (MTO) routes are of particular interest. Both allow the production of olefins which can subsequently be polymerized to derive at polyolefins

2.1.1 Thermochemical Platform ('Dry' Carbon Recovery)

Thermochemical conversion for the production of fuels and chemicals uses either *pyrolysis* or *gasification.* Pyrolysis is the thermal treatment of biomass in the absence of oxygen and results in the production of bio-oil, gases, or bio-char. Gasification occurs at higher temperatures and in less oxygen-restricted conditions than pyrolysis and leads to the formation of a synthesis gas (syngas) rich in hydrogen and carbon monoxide. The intermediate products of both processes have the potential as a feedstock for fuel and chemical synthesis via various catalytic pathways (e.g. Fischer–Tropsch synthesis) (Fig. 3).

2.1.2 Biochemical Platform ('Wet' Carbon Recovery)

Biochemical conversion either uses acids or enzymes to catalyze the conversion of the carbohydrate portion of the biomass (hemicelluloses and cellulose) into intermediate sugars that are then fermented to ethanol and other products. The remaining lignin residue that cannot be processed via the biochemical platform can be used for heat and power production, or alternatively used in the thermochemical conversion process to produce additional fuels and chemicals.

Anaerobic digestion is a fermentation technique that results in a biogas consisting mostly of CH_4 and CO_2 but generally carrying impurities such as H_2S, H_2O, NH_3, siloxane, and particulate matter. Anaerobic digestion is the principal process occurring in landfills (producing what is typically referred to as Land Fill Gas or LFG) and occurs naturally in marshes, wetlands and manure lagoons [69]. CH_4 for energy production can be obtained by upgrading the biogas. Syngas can be produced by steam reforming the upgraded syngas. Similar to the subsequent steps of the thermochemical platform, syngas can then be utilized for e.g. the production of methanol or FT Naphtha (Fig. 3). Direct olefin production from upgraded biogas is potentially possible via oxidative coupling (Fig. 2).

2.2 Chemicals and Polymer Production Pathways: What Is Potentially Possible?

Both thermochemical and biochemical conversion platforms allow the production of a variety of base chemicals and subsequent plastic polymers from organic waste. The reason for looking at base chemicals for the production of synthetic materials rather than fuels is the possibility of a more efficient cascading use in which a durable good (plastic polymer) is produced first and its energy content recovered at end-of-life. Gasification and fermentation both seem to be complementary to each other in terms of polymers they can produce. However, gasification has the clear advantage to be able to utilize a broader variety of waste feedstock (not only lignocellulosic waste but also plastics and C&D waste) and seems to have the advantage that possible toxic substances can be extracted directly from the syngas rather than the organic waste feedstock.

From a traditional base utilizing the natural complex macromolecules of e.g. starch and cellulose as raw materials for the production of biopolymers, the polymer industry is turning attention towards synthetic polymers based on renewable raw materials. Key polymer building blocks include e.g. alcohols such as methanol (C_1), ethanol (C_2) for the production of polyethylene and polypropylene polymers, glycerol (C_3) as a building block for the production of polyurethanes, C_3–C_6 carboxylic acids (e.g. lactic acid, succinic acid, and itaconic acid) as well as aromatic aldehydes (e.g. 5-hydroxy-methyl-furfural (HMF, C_6) and Fischer–Tropsch Naphtha (C_5–C_{12}). Strategies differ between replacement of conventional fossil-fuel derived plastics and the development of novel building blocks using biochemical and thermochemical conversion.

In terms of current production volume, ethylene and propylene as well as their derivatives dominate the plastics industry by feeding the polyethylene, polypropylene, ethylene oxide, styrene, polyvinylchloride, and a number of other supply chains [53, 59]. With a production volume of more than 150 million tons, light olefins (e.g. ethylene and propylene) are currently the most important basic petrochemicals to produce plastics, fibers and other chemicals [54]. In this regard, the methanol to olefins (MTO) route, Fischer–Tropsch Synthesis (FTS) towards FT-naphtha and biogas to olefins routes (either steam reforming or oxidative coupling) seem to provide interesting future pathways for olefins production from organic feedstocks. The MTO route as well as the route from methanol to acetic acid are well-established. FT-naphtha could play a key role as a base chemical, from which a variety of chemicals, including polymer building blocks, can be obtained.

2.3 MSW Feedstock Quality Issues

The quality of MSW as a feedstock for fermentation or gasification is important in terms of pre-treatment and conversion facility design. Barriers to fermentation and anaerobic digestion of MSW include the ability to effectively separate BMSW material from other wastes whereas gasification requires costly and possibly energy-intensive drying of moist feedstock as well as gas cleanup later in the process chain. Potential variations in feedstock quality and availability, as well as the cost of handling and competing uses such as recycling, compost, waste-to-energy (WtE) and landfill gas generation are further issues of concern.

The composition of MSW varies significantly among countries as well as among regions within individual countries (e.g. urban vs. rural areas). These variations are caused for example by differences in consumer habits, diet and disposal patterns and relate furthermore to the level of affluence and development of the country [33]. *One of the biggest challenges faced by developers of waste conversion facilities is the heterogeneity of the feedstock.* Varying MSW composition over time[4] is a challenge for most conversion facilities and performance will depend on their flexibility to cope with these changes and to be able to process a number of alternative feedstocks. Methods that can be applied to deal with these issues are: on-site storage and blending, mixing and shredding of the waste, compression and baling of the input, and integration with a Materials Recovery Facility (MRF) to obtain a more homogeneous waste feedstock [33].

A number of studies on the use of MSW as raw material for the production of fuels and chemicals has been published to date [1, 11, 25, 29, 30, 41, 42, 44, 47, 48, 56–58, 67], [69, 70]. With respect to biochemical conversion (hydrolysis + fermentation), these studies indicate that, by optimizing BMSW pre-treatment and hydrolysis procedures, more than 85% of the waste cellulose fraction can be converted into glucose [43] which could be converted to fermentation products such as ethanol and other platform chemicals. Depending on the waste composition, pretreatment methods using dilute sulfuric acid or hydrochloric acid followed by enzymatic hydrolysis and steam and pressure pretreatment have been investigated. However, chemicals (e.g. biosurfactants and antimicrobials) present in the feedstock have the potential, if not removed, to inhibit enzymatic hydrolysis or fermentation resulting in lower yields of the intermediate-products [44]. Most studies looking at fermentation of BMSW have focused on the production of ethanol as biofuels for transportation purposes. All of these studies looked at the conversion of MSW on the lab- or pilot-scale. *So far, no commercial plants applying hydrolysis followed by fermentation of the sugars are operating.*

In contrast to fermentation, *pyrolysis and gasification techniques are widely used for the processing of waste feedstock.* As of 2001, there were 110 plants operating in 22 countries processing over 5 million tons of waste per year applying gasification and/or pyrolysis [33]. The majority of these efforts focus on the utilization of MSW and other dry waste fractions to recover energy. However, a wide range of technologies is emerging for the conversion of organic waste to biofuels. These technologies are able to use a wide variety of waste feedstocks, including C&D derived biomass as well as plastics waste.

---

[4] This includes variations caused by e.g. changing houshold patterns due to the season (e.g. more garden waste in summer); unusual events such as Christmas, holidays; etc.

Anaerobic digestion (AD) of BMSW has advanced mainly in Europe but facilities were recently also built in Canada, Japan, Australia and several other countries [52]. AD systems are applied in many wastewater treatment facilities for sludge degradation and stabilization and are used to treat those wastewaters prior to discharge. Some facilities are also employed at animal feeding operations to reduce the impacts of manure and to use it as a feed for energy production from biogas. Of the organic waste fraction of MSW, wet BMSW such as food and yard wastes can be treated in anaerobic digesters. AD therefore represents a commercially available alternative to fermentation techniques. European technologies all use extensive pre- and post-digestion processing units. These include visual manual or robotic sorting and removal of bulky or potentially harmful items, particle size reduction and separation (see [52] for further details) adding to the cost of these technologies.

Gasification seems to be favored over biochemical conversion due to the fact that contaminants (alkali metals, halides, sulfur gases, and tars) present in the biodegradable fraction of the waste can (in comparison to pre-sorting and steam-cleaning the biomass itself) be removed from the produced syngas before catalytic conversion (e.g. Fischer–Tropsch) to the intermediate products takes place. If not removed, these contaminants can poison the noble metal catalysts. In addition, gasification utilizes both the lignin as well as the cellulose and hemicelluloses fractions of the BMSW feedstock and has the potential to utilize additional waste fractions such as plastics and C&D waste.

Among the advantages of using organic waste as a primary feedstock for biofuels and bio-materials are that unlike other lignocellulosic feedstocks, MSW has an already well-established collection system and processing infrastructure and is generally available at a negative cost. In contrast to agricultural waste and energy crops which are harvested on a seasonal basis, BMSW provides a year-long supply of feedstock for the biochemical and thermochemical platform. Since the major fraction of MSW consists of organic waste, utilization of MSW provides environmental benefits, as for instance reduction of GHG emissions (CO_2, CH_4) and landfill space (landfill diversion).

2.4 Potentials for Developing and Emerging Countries

Waste gasification, anaerobic digestion and fermentation are technologies still under development. Implementation will require significant investments and initial investors will have to carry the risk of whether they are able to successfully introduce these technologies to the market. While gasification systems may be affordable in affluent countries such as Germany or the United States, they are unlikely to be either appropriate or financially affordable in developing countries in the short-term, simply because citizens have lower incomes and are therefore not be able to pay as much for waste management and carbon recycling. A modern gasifier designed for high-heating value European wastes is likely to require additional fuel inputs to gasify a typical high-organic and relatively wet waste in a developing country. Furthermore, the costs and expertise required to operate and maintain the system in a continuous manner is likely to restrict it to a few cities with most advanced waste collection and separation systems in place in developing or transitional countries. In addition, a novel conversion technology that has not yet been introduced to e.g. the European market is a risky choice for the developing world which requires systems that are guaranteed to be reliable in collecting, treating and disposing of the waste, all year around. Therefore, we envision these technologies to be first introduced in industrialized nations and mega-cities of emerging economies with high volume generation rates of organic waste feedstock.

However, it should be pointed out that in particular the thermochemical platform has the capability of combining safe waste handling of organic waste with the production of energy, fuels and chemical feedstock. According to UN-Habitat data, significant increases in the occurrence of sickness among children living in households where (organic) waste is dumped or burned in the yard can be observed [65]. Organic waste materials can pollute surface and groundwater and therefore pose a threat to the health of people who depend on these water resources for drinking water. The potential of gasification technologies to destroy harmful microorganisms at high temperatures and concentrate hazardous metals in the slag and ash could become of increasing interest for developing countries in the future. In addition, feedstock flexibility would potentially allow utilizing both, organic waste as well as virgin green wood biomass as gasification feed. Finally, operating smaller decentralized conversion facilities would allow the production of energy, fuels and chemicals without having to build large refineries and power plants.

3 Status of Knowledge: Waste as Feedstock for Thermo- and Bio-chemical Conversion

3.1 Resource Potentials with a Focus on Developing Countries

The data on solid waste generation and recovery rates in developing countries is scarce [63]. Even a rough estimate of waste amounts and composition as well as recovery and recycling rates is often not possible. When data exists it is difficult to do comparisons even within a city because of inconsistencies in data recording, collection methods and seasonal variations. However, a recent overview of a number of reference cities in developing and emerging countries is given in [66]. According to this study, in low gross domestic product (GDP) cities, waste density can be as high as 400 kg per cubic meter due to high fractions of wet organic waste. A comparison of all reference cities indicates that organic waste is a very large part of the waste stream in all cities investigated. The organic fraction is often between 50 and 70 weight-% of MSW in developing countries. Low- and middle-income countries were found to have relatively high percentages of organic waste (above 45 weight-%) in cities such as Cairo, Cluj, Lima, Pune, and Quezon City.[5]

While in industrialized countries the value of organic waste often times is due to composting or incineration and anaerobic digestion for energy production, in cities of the developing world organic waste is used mainly to feed livestock (especially swine feeding) and to generate compost for land application [66]. It is important to note that the informal sector does most of the recycling related to organic waste in developing countries. This includes street pickers, dump-pickers, itinerant waste buyers and junk shops that collect and deal with the waste feedstock as long as a market for the product exists. This is, however, only partially true for organic wastes. While food waste may have a market value as animal feed, products made from compost are increasingly being replaced e.g. by chemical fertilizers.

Figure 4 shows the amounts of organic waste going to animal feeding, composting or land application in a number of cities around the world. As can be seen from the figure, still large amounts of organic waste feedstock remain unutilized. This fraction could potentially serve as feedstock for thermochemical or biochemical conversion technologies.

3.2 Environmental Performance Evaluation: Waste-to-Chemicals

A limited number of studies looking at the life-cycle-wide environmental implication of the route MSW to fuels/chemicals have been carried out to date. These studies focus on the production of heat, electricity and fuel, including methanol [5], ethanol [12, 34, 61] and synthesis gas [36] from MSW. All of these products are interesting intermediates on the way to synthetic materials. Methanol could serve as base chemical for the methanol-to-olefins (MTO) route, whereas ethanol can be transformed into ethylene and subsequent polypropylenes. Syngas acts as a base chemical for a variety of routes including MTO and the FTS and can be used for the production of electricity.

These studies indicate that utilizing the organic fraction of MSW for energy recovery or material recycling may have advantages in terms of GHG emissions savings and to reduce fossil-energy consumption when compared to conventional use including current waste management practices (e.g. landfilling and incineration) and transportation purposes to replace fossil-based petrol. Material recycling through the provision of base chemicals (syngas, methanol, and ethanol) via fermentation and gasification seems to be possible. The overall environmental performance will largely depend on the choice of assumptions made and

[5] See Key Sheet II in Chap. 4 of the [66] report.

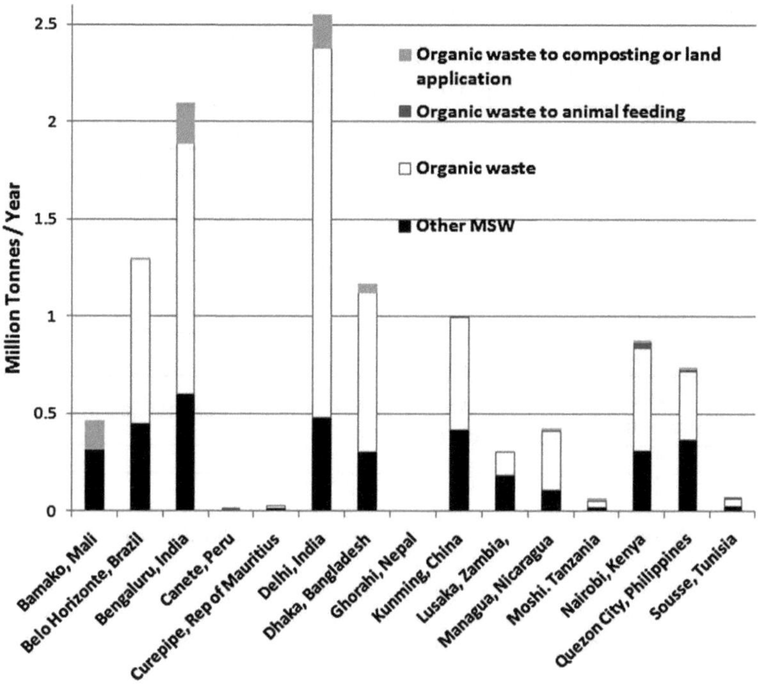

Fig. 4 Destination of organic waste generated in MSW per year in a number of cities around the world (*Source* compilation using data from [66]). For example, in Delhi, India, a total of 2.55 million tones of MSW are generated per year. Of this roughly 2.10 million tons consist of organic waste of which 8% (165,565 t) is diverted to composting or land application

comparisons applicable (e.g. landfilling with or without landfill gas recovery, inclusion of MSW collection and classification, etc.). Key processes and their performance will be exemplified in the following.

4 MSW Processes to High-Value Products

This section presents results from one case-study carried out on gasification routes from organic waste to chemical feedstock. We assessed the system-wide global warming potential (GWP) and cumulative energy demand (CED) associated with these routes using attributional life-cycle assessment (LCA). Data collection for the foreground system as shown in the case study below was gathered from available literature. All supplies of materials, electricity, energy carriers, services, etc. were modeled with best available background data from the ecoinvent database [18], the U.S. LCI database [50] and other published LCI data sources. SimaPro LCA software was used to calculate the life cycle inventory and carry out the impact assessment.

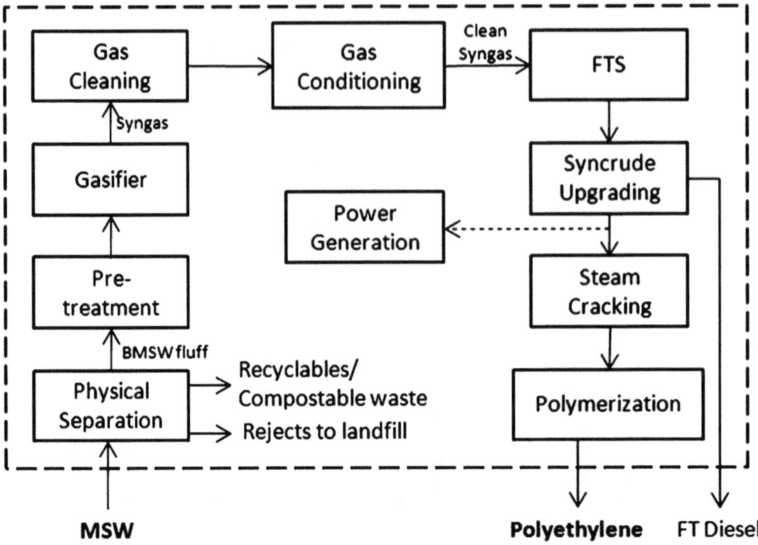

Fig. 5 Process diagram of the MSW to polyolefin process (Source: own compilation). Mixed MSW enters the physical separation facility (MRF) in which the recyclable fractions are separated and a portion of the biodegradable fraction (BMSW) is sent to composting facilities (Waste Classification). The remaining fraction consisting of BMSW is further pre-treated and then converted into syngas via gasification. Additional steps include gas cleaning and conditioning followed by FTS. The products of FTS consist of hydrocarbons of various chain length (syncrude) of which the naphtha fraction (C_5–C_8) is converted into polyolefins using syncrude upgrading, steam cracking and polymerization

4.1 Case Study: Waste-to-Olefins via Fischer–Tropsch Synthesis (FTS)

4.1.1 Methodology

This analysis compares the use of organic waste for polyethylene (PE) production with the production process using crude oil in a conventional refinery. The goal is to estimate the life-cycle environmental burdens with regards to GWP[6] and CED associated with the production of 1 kg of PE at the factory gate. This analysis is based primarily on U.S. waste collection practices, technological parameters and background data. Electricity inputs to the foreground system (Fig. 5) are assumed to come from the U.S. power grid.[7] An LCA model is developed following the ISO 14040 standards. It is assumed that organic waste needs to be disposed of and the

[6] Biogenic carbon present in the BMSW feedstock has been excluded from the analysis

[7] The process 'electricity, medium voltage, at grid from the ecoinvent database is used.

environmental implications of the processes that generated the waste are therefore excluded. Utilizing MSW as feedstock implies a diversion of the waste, as opposed to the cultivation of additional feedstock (e.g. biomass).

The need to collect MSW regardless of its end-use implies no significant changes to the collection process and environmental burdens associated with collection are therefore excluded from the LCA. However, in order to process the waste and separate the organic fraction from the remaining waste stream, the waste feedstock needs to be pre-sorted and separated in a Materials Recovery Facility (MRF) (MSW classification).[8] Further pre-treatment steps for comminution and drying are required (Fig. 5).

In this system, MSW is first separated to remove recyclables and shredded and milled to reduce size. It is then dried prior to gasification. The syngas is then cleaned to remove tars, dust, alkali, BTX (benzene, toluene and xylenes) and halons. The cleaning stages envisaged are suitable for subsequent FTS. The six main stages of the life cycle considered are: Classification (sorting), Pre-treatment (Fluff shredding/Drying), Gasification/FTS, FT Syncrude upgrading, Steam Cracking, and Polymerization. Technologies included represent existing processes that are available on pilot or demonstration scale (e.g. gasification system) as well as currently operated processes (i.e. naphtha steam cracker, etc.). The transport from the MSW classification plant to the conversion plant is taken as 50 km. The transport is performed by a 28 t truck.

The analysis exclusively considers MSW destined for landfills and incineration plants. This excludes recyclables which are reused as well as agricultural and forestry residues. Commercial scale FTS plants utilizing organic waste as feedstock do not yet exist, but mass and energy balances on syngas generation from waste feedstock including raw MSW, BMSW and refuse derived fuel (RDF) [28, 30, 33, 49, 51] as well as data for subsequent FT syncrude production [3, 13, 30–32, 46, 68] are available from the literature. In fact, several studies indicate that in particular RDF would be a suitable feedstock for gasification-based FTS [30, 51].

We assume that all BMSW destined for landfills and incinerators can be separated from the remaining waste either at the source or during the classification process (during which marketable aluminum, glass, steel and plastic material are recovered). The waste composition and energy content of the BMSW fluff diverted to the gasification plant is assumed to be similar to the U.S. average and is taken from [20, 62]. The wet tones of MSW constitute the mass that must be treated in the classification plant. The classification process is modeled based on [9] assuming that electricity is used for meeting all of the energy requirements in the

[8] Kalogo et al. [34] reported that there is some discussion as to whether MSW classification should be included in the analysis. Some authors share the opinion that this step does not need to be included in an LCA. They state the fact that MSW is anyways classified into the different waste fractions because it is economically feasible due to the value of recovered material and because of legal mandates for prior separation.

classification process.[9] The analysis assumes average recycling and recovery rates as given in [20]. Energy use is allocated as follows: BMSW fluff (37%), recyclables (i.e. glass, ferrous, non-ferrous, etc. 24%), compostable waste (9%), and scraps (30%). The input of BMSW fluff to the conversion plant is calculated based on the average energy content of the waste fluff after classification (11.588 MJ per kg wet BMSW fluff). The BMSW composition is: paper & paperboard (38%), wood (16%), food scraps (34%), and yard trimmings (13%).

Three technologies for converting BMSW fluff into synthetic gas ($CO + H_2$) are selected and compiled from [14, 30–33, 36, 49, 51]. Low-temperature wet gas cleaning is envisaged as cleaning process after gasification. Various reports are available describing the in-depth technical details of those technologies [4, 33, 37, 45] and therefore they are not explained in detail. Table 1 gives an overview of the conversion technologies selected.

Fischer–Tropsch synthesis followed by upgrading of the FT raw liquid yields mainly naphtha and distillate as well as electricity. The FT product of primary interest to this study is naphtha that can be sent to a petroleum refinery. For this study, a *combined credit/allocation approach* is used for allocation. The environmental burdens from conversion and hydrocarbon recovery of the syngas-based FT plants are allocated based on the ratio of the energy content (Lower Heating Value) of the specific fuel relative to the total product. However, electricity co-produced is sold to the grid and can therefore be considered an end-use for FT-liquids and syngas. In order to compensate for this, excess electricity is treated with the credit approach, whereby electricity is assumed to come from the U.S. medium voltage grid.

The Fischer–Tropsch synthesis (FTS) is based on modeling results from [32, 68]. Two different FT systems are investigated as part of this study. On the one hand, clean syngas generated by the BHTGS and MTCI gasification units is fed into a slurry-bed, iron-based catalyst FT-reactor system based on a model developed from public information and published in [68]. The FT-model used in their study is based on data originally published by Bechtel/Amoco in 1993.while for cobalt catalysts a H2/[10] On the other hand, syngas generated from the Choren Carbo-V process is converted into FT syncrude using a cobalt catalyst in a tubular-fixed-bed reactor (TFBR). This process is based on aggregated inventory data (due to confidentiality issues) directly taken from [32, 55]. Although assumptions with regards to allocation and emissions profiles may vary somewhat from our LCA model, it was decided to use the aggregated dataset to cross-check results of the other two conversion systems investigated in this paper. Syngas characteristics and conditioning are critical for fuels and

[9] The study by Broder et al. [9] looked specifically at classification processes that would be able to generate a clean RDF suitable for biochemical ethanol synthesis. We assume that this sort of classification system will produce a pure organic feedstock that would be suitable for subsequent conversion towards chemicals via gasification and AD.

[10] Baseline Design/Economics for Advanced Fischer–Tropsch Technology, DOE Contract No. DE-AC22-91PC90027, Topical Report Volume 1, Process Design – Illinois No. 6 Coal Case with Conventional Refining, October, 1994.

Table 1 Technologies chosen for syngas production using gasification of RDF

Data Source	Niessen et al. [49], Paisley et al. [51]	Juniper Consultancy Services [33], Niessen et al. [49]	Jungbluth et al. [32]
Manufacturer	Batelle (BHTGS)[a]	MTCI ThermoChem[b]	Choren[c]
Type	Circulating fluidized bed gasifier (CFB)	Bubbling fluidized bed gasifier (BFB)	Two-stage entrained flow gasification (Carbo-V process)
Direct/Indirect heating	Indirect	Indirect	Indirect
Pressurized/ Atmospheric	Atmospheric	Atmospheric	Pressurized
Air/Oxygen/ Steam-blown	Steam/Air	Steam	Oxygen
Temperature (°C)	766	843	400–600 (1st step) 1,300–1,500 (2nd step)
Feedstock	RDF	RDF	RDF[d]
Water content (%)	20	20	20
Scale	Demonstration	Semi-commercial	Semi-commercial

[a] Batelle High Throughput Gasification System (BHTGS)
[b] Manufacturing and Technology Conversion International, Inc (MTCI)
[c] Only aggregated datasets for the generation of FT-liquids were available
[d] The study by Jungbluth et al. [31, 32] looks at woody biomass (willow-salix) for FT Diesel production. We assume that pre-plant classification produces an organic feedstock (RDF) that would be acceptable for gasification and subsequent FT liquids production using the Carbo-V process

chemicals synthesis. High purity syngas (with low quantities of inert gas such as N_2) is beneficial as it substantially reduces the size and cost of downstream equipment. Supporting process equipment (e.g. scrubbers, compressors, coolers, Water–Gas-Shift, etc.) can be applied to adjust the conditioning of the product gas. When using an iron catalyst the H_2/CO ratio of the syngas should be adjusted to approximately 0.6,[11] while for cobalt catalysts a H_2/CO ratio near 2.0 should be used [14]. An authothermal reformer (ATR) using steam and enriched air/oxygen with partial CO_2 recycle is used for syngas preparation. It is important to point out, that varying calorific values of the product gas do not affect subsequent FTS as long as H_2/CO and impurity levels are met [14]. No transportation is accounted for as it is assumed that the gasifier, syngas cleaning and FTS platforms are integrated and located within one conversion plant. The full amount of heat and the main part of electricity is used inside the conversion plant (note that excess electricity generated in the FTS platform is delivered to the gasifier to meet some or all of the energy requirements).

[11] The iron-based F-T catalyst promotes the water–gas shift reaction which produces hydrogen for the F-T synthesis reaction ($CO + H_2O = CO_2 + H_2$).

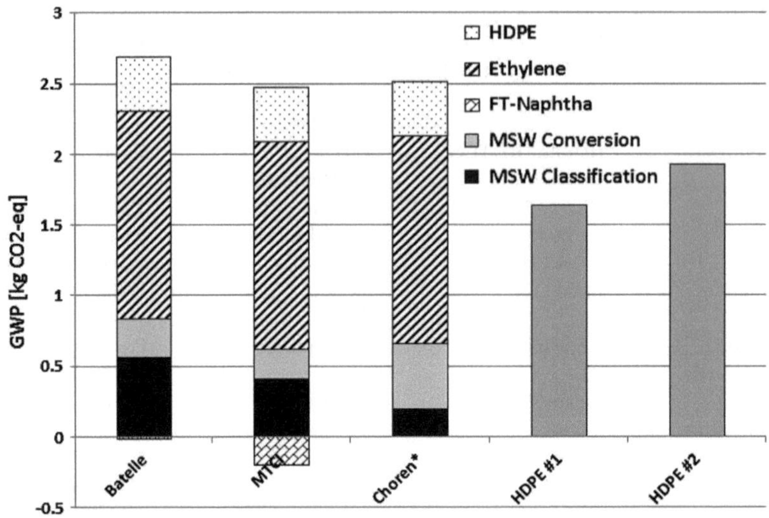

Fig. 6 Comparison of the system-wide global warming potential (GWP) of producing 1 kg of HDPE from MSW with its fossil-based counterpart. Data for fossil-based HDPE comes from the US LCI database (HDPE #1) and ecoinvent (HDPE #2). *The process 'MSW Conversion' for the Choren plant includes gasification and FT-naphtha production (aggregated dataset)

Dancuar et al. [16] investigated the suitability of FT naphtha for use as a steam cracker feedstock and found that the substance mix was extremely well suited for the production of olefins (ethylene and propylene) by steam cracking. Accordingly, this study assumes the use of conventional naphtha steam cracking for the generation of ethylene. Data from the CPM database [15] and the ProBas database [64] is used to model the FT naphtha steam cracking process.

The life cycle inventory for high density polyethylene (HDPE) resin production from FT-derived ethylene is based on data from the U.S. life cycle inventory database [24, 50].

4.1.2 Results

The results (see Figs. 6, 7, 8) summarize the system-wide GWP (Fig. 6) and CED (Fig. 7) that we estimate would occur if BMSW from the MSW stream were used as feedstock for HDPE production. Figure 8 considers the essential fact that the use of BMSW for chemical supply derives waste from landfills and thus may relieve the overall GHG balance. Results are shown for the *functional unit* of 1 kg HDPE at the factory gate and are compared to conventional (fossil-based) HDPE production routes. Data for these comes from the U.S. LCI database (HDPE #1) and ecoinvent (HDPE #2).

The comparison shows that GWP associated with the waste-derived polymers is with 2.7–2.3 kg CO_2-eq slightly higher than their fossil-based counterparts.

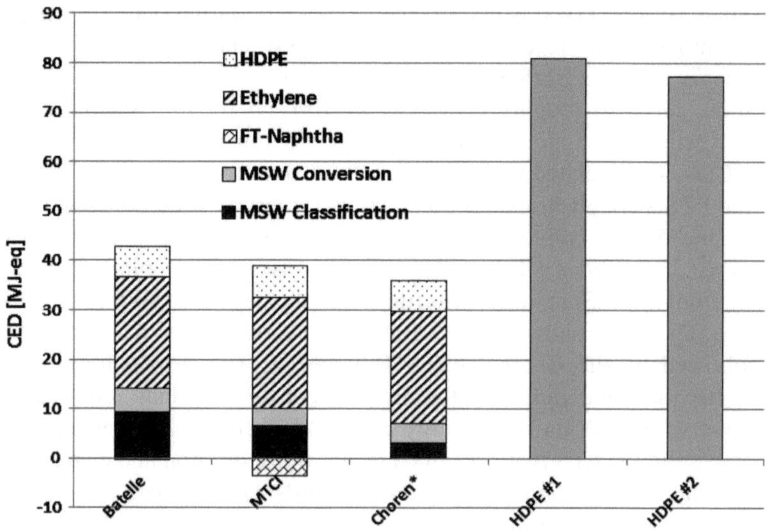

Fig. 7 Comparison of cumulative energy demand (CED) of producing 1 kg of HDPE from MSW with its fossil-based counterpart. Data for fossil-based HDPE comes from the US LCI database (HDPE #1) and ecoinvent (HDPE #2). *The process 'MSW Conversion' for the Choren plant includes gasification and FT-naphtha production (aggregated dataset)

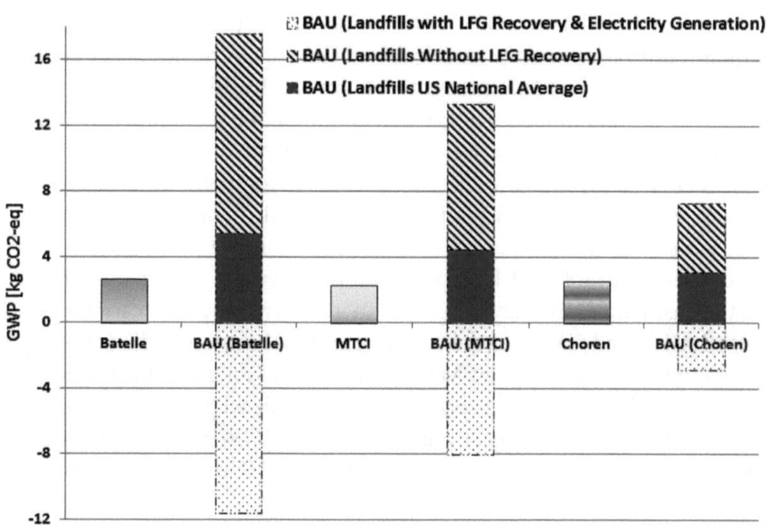

Fig. 8 System expansion accounting for the fact that in a business-as-usual (BAU) case HDPE is produced from petroleum and BMSW is landfilled. The amount of BMSW going to landfills depends on the feedstock requirements of the carbon-recycling systems (Batelle, MTCI and Choren). U.S. landfill net emission factors from the WARM model are used. Data for fossil-based HDPE comes from the US LCI database (HDPE #1) and ecoinvent (HDPE #2)

A large share of total GWP is due to steam cracking, during which FT-naphtha is converted into ethylene feedstock, as well as MSW classification where the raw MSW is separated into BMSW fluff and other waste fractions. MSW conversion (gasification) leads to roughly 0.21–0.27 kg CO_2-eq for the Batelle and MTCI systems investigated. Aggregated data for the Choren plant indicates a GWP of 0.463 kg CO_2-eq associated with the conversion step from BMSW to FT-naphtha. During the FTS step, electricity is co-generated, most of which is used internally. However, the MTCI conversion system generates a small amount of excess electricity (0.23 kWh/kg FT liquids) which is assumed to offset conventional electricity from the U.S. national grid (therefore the negative GWP for FT-naphtha generation). The figure shows that the Batelle conversion-plant leads to the highest GWP, followed by the Choren and MTCI design. The reason for this is that, according to the data gathered, the Batelle conversion-plant requires slightly higher inputs of BMSW fluff (by energy content) and electricity to generate a clean syn gas for use in the FTS platform. As a result, transportation and energy required for MSW classification contribute more towards GWP and excess electricity exported to the grid is minimal.

In contrast, results for CED are highest for fossil-based HDPE (77.3–81.0 MJ-eq). This is followed by 42.61 MJ-eq for the Batelle conversion facility and 35.97 and 35.57 MJ-eq for the Choren and MTCI systems, respectively. CED of the MSW-based routes is about half that of the conventional fossil-based routes. The CED indicator encompasses non-renewable (i.e. fossil and nuclear) as well as renewable (i.e. biomass, wind, solar, etc.) energy demand. However, renewables account for less than 1% of total CED. The reason that CED for the waste-derived polymers is lower than for their fossil-based counterparts is the fact that the intrinsic energy content of the waste feedstock is not accounted for (cut-off approach). In contrast, for fossil-based polymers the direct and indirect energy consumption of e.g. natural gas and crude oil resources used to synthesize the HDPE polymer (some of which is later present as 'feedstock energy' in the final product) are accounted for in the CED values. Similar to GWP, steam cracking and MSW classification, both being very energy intensive processes, account for a large share of CED. The magnitude to which MSW conversion and FTS contribute to total CED depends on the amounts of waste feedstock transported to the gasifier and further energy and materials requirements for the conversion facility. Both CED and GWP for the MSW classification step of the Choren plant are small compared to the Batelle and MTCI conversion systems. This is due to the fact that in the Choren design, which is optimized for FT diesel production, less naphtha is produced and therefore the largest part of CED and GWP associated with BMSW provision to the conversion system is allocated to the FT-distillate (for diesel).

4.1.3 System Expansion: Avoided Landfilling

When paper, wood, food scraps and yard trimmings are landfilled, anaerobic bacteria degrade the materials, producing methane and carbon dioxide. Although landfills during use operate as net-carbon sink (and as a source afterwards), methane generated is counted as an anthropogenic GHG because degradation would not take place if the BMSW were not landfilled. The impact of waste diversion from landfills is significant for landfills with no recovery equipment (i.e. landfill gas (LFG) recovery for flaring or electricity generation). In contrast to many countries in Europe, in the United States and many developing countries significant amounts of BMSW are sent to landfills. We use system expansion to compare: (1) GHG emissions associated with the production of 1 kg HDPE from BMSW (*'carbon recycling'*) with (2). GHG emissions associated with landfilling the BMSW and the production of 1 kg of fossil-based HDPE [*'business-as-usual (BAU)'*]. We use emission factors from the WARM model for the United States to estimate the GHG emissions from landfilling the BMSW fluff [21].

Given the U.S. national landfills average from the WARM model, the emissions avoided of removing 1 kg of wet BMSW (with the average waste composition mentioned above) from landfills equals 0.167 kg CO_2-eq. As a result, the 'BAU—U.S. National Average' case would lead to a higher system-wide GWP of 5.5–3.1 kg CO_2-eq per kg of HDPE produced when compared to the production of 1 kg of waste-derived HDPE (carbon recycling). With 3.1 kg CO_2-eq, GHG emissions are lowest for the BAU case in which BMSW, otherwise used as feedstock in the Choren plant, is landfilled and HDPE is produced from fossil-fuels. This is due to the fact that most of the BMSW is used to produce distillate (for fuels) and less for naphtha (for HDPE), and thus a major part of landfill emissions in the BAU scenario are allocated to the distillate. The results indicate that, accounting for average landfills emissions in the U.S., carbon recycling may have the potential to lead to an overall reduction in GWP when compared to current (BAU) waste management and HDPE production practices (Fig. 8).

However, the magnitude to which landfill diversion results in net GWP reduction depends significantly on whether landfill gas (LFG) recovery and energy recovery equipment is deployed and how effectively it is operated. Flares and generators on-site have the potential to convert methane into CO_2, therefore reducing GWP. If only landfill systems without LFG recovery equipment are considered, system-wide emissions of the BAU case would amount to 17.6–7.3 kg CO_2-eq per kg HDPE (BAU Landfills without LFG Recovery), while only considering landfills with LFG recovery equipment and electricity generation would amount to -11.6 to -2.9 kg CO_2-eq in GHG savings (BAU Landfills with LFG Recovery & Electricity Generation), therefore competing with polymer production about the most beneficial use of BMSW to reduce GHG emissions (Fig. 8). The difference between landfills without LFG recovery equipment and those with LFG recovery equipment illustrates the impact that assumptions on waste diversion can have on the net GWP of the expanded system.

4.1.4 Discussion

Ignoring the fact that waste needs a safe final disposal, the use of BMSW for the production of polyethylene seems to result in only slightly higher GHG emissions as compared to conventional fossil-based routes. Under the same assumption, CED of the MSW-based polymer production routes was found to be roughly half that of conventional HDPE production. The impact of BMSW landfilling presents the greatest system uncertainty. Depending on the landfill system chosen, these assumptions can change GHG emissions for the BAU scenarios from positive to negative. When using U.S. national landfills average data, carbon recycling systems investigated in this paper may have the potential to significantly reduce GHG emissions.

However, the required capital for methane recovery installations particularly in developing countries may be lacking, and the low price of commercially produced gas may not make methane recovery an economically viable option. In addition, landfill space may be limited, in particular in urban areas. These conditions could make carbon recycling technologies an attractive option for developing countries and emerging economies in the future. Furthermore, the WARM model makes key assumptions that are critical for the interpretation of our results. For instance, when LFG is recovered for energy production, co-product credits for the displacement of an equivalent amount of energy from the U.S. electricity grid, which is dominated by coal with high GHG emissions, are applied by the model. Therefore, considering a less carbon-intensive electricity mix (e.g. in a future scenario with larger shares of electricity being supplied by renewable energy systems) could change the balance more in favor of carbon recycling systems. Further investigations should also consider other end-of-life waste management techniques such as combustion and composting and include an uncertainty and sensitivity analysis (using e.g. economic allocation). Furthermore, the analysis should be expanded to other impact categories, including total material requirement (TMR), acidification, eutrophication and health impacts, as well as cost. Finally, we assumed that MSW classification is required to obtain a clean gasification feedstock. MSW classification leads to roughly one fifth of these impacts and therefore excluding this process step from the LCA would result in further lowering of life-cycle wide impacts. This may be justified in circumstances where classification takes place solely due to the value of recovered material (e.g. plastics, metals, etc.) or because of legal mandates prior to separation.

5 Conclusion

Carbon recycling, in which organic waste is recycled into chemical feedstock for material and fuel production, may have the potential to provide benefits in resource efficiency and a more cyclical economy—but may also create ‘trade-offs’ in increased impacts elsewhere. Preliminary LCA model results derived from the combination of various existing technologies (i.e. MSW classification, gasification, FTS, steam cracking, etc.) and considering landfill diversion, indicate that the use of biodegradable waste for HDPE production could lead to a reduction in system-wide GHG emissions when compared to conventional fossil-based production routes. However, as yet the conversion technologies assessed do not work in the integrated fashion modelled in this paper.

Developing pilot plants for the conversion of BMSW into base chemicals such as HDPE could be a future option in particular for mega-cities in developing countries and emerging economies with high-volume generation rate of organic waste and a lack of landfill gas recovery equipment and landfill space. Varying the process parameters of technologies such as FTS could allow the generation of both base chemicals (such as naphtha for polymers) and liquid fuels. The potential of gasification technologies to destroy harmful microorganisms at high temperatures and concentrate hazardous metals in the slag and ash could become of increasing interest for developing countries in the future.

In summary, the described technologies of carbon recycling may contribute to further develop the waste management sector towards a more sustainable resource management. Besides the recycling of carbon flows, also the extraction, use, recycling and disposal of all material resources should be considered when developing the physical basis of society and economy towards increased sustainable resource management.

References

1. Aiello-Mazzarri C, Coward-Kelly G, Agbogbo F, Holtzapple M (2005) Conversion of municipal solid waste into carboxylic acids by anaerobic countercurrent fermentation. Appl Biochem Biotechnol 127(2):79–93
2. Ayres RU, Ayres L (2002) A handbook of industrial ecology. Edward Elgar Publishing, Cheltenham
3. Bechtel (1998) Baseline design/economics for advanced Fischer-Tropsch technology. U.S. Department of Energy Technology Center, US
4. Belgiorno V, De Feo G, Della Rocca C, Napoli RMA (2003) Energy from gasification of solid wastes. Waste Manage 23(1):1–15
5. Bez J, Goldhan G, Buttker B (2001) Methanol aus Abfall–Ökobilanz bescheinigt gute Noten. Müll und Abfall 33(3):158–162
6. Bringezu S (2009) Chapter 4: Visions of a sustainable resource use. In: Bringezu S, Bleischwitz R (eds) Sustainable resource management: global trends, visions and policies. Greenleaf Publishing, Sheffield, pp 155–215
7. Bringezu S, Schütz H, Moll S (2003) Rationale for and interpretation of economy-wide materials flow analysis and derived indicators. J Ind Ecol 7:43–64
8. Bringezu S, Schütz H, Steger S, Baudisch J (2004) International comparison of resource use and its relation to economic growth: The development of total material requirement, direct material inputs and hidden flows and the structure of TMR. Ecol Econ 51(1–2):97–124
9. Broder J, Eley M, Barrier J (1993) Municipal solid waste and waste cellulosics conversion to fuels and chemicals-vol 2: Front-end classification systems. Tennessee Valley Authority, Muscle Shoals
10. Brunner PH, Morf L, Rechberger H (2004) VI.3 Thermal waste treatment–a necessary element for sustainable waste management (Internet). In: Irena Twardowska HEA (ed) Solid waste: assessment, monitoring and remediation. Elsevier, (cited 2010 Dec 3), pp 783–806
11. Champagne P (2007) Feasibility of producing bio-ethanol from waste residues: a Canadian perspective: feasibility of producing bio-ethanol from waste residues in Canada. Resour Conserv Recycl 50(3):211–230
12. Chester M, Martin E (2009) Cellulosic ethanol from municipal solid waste: a case study of the economic, energy, and greenhouse gas impacts in California. Environ Sci Technol 43(14):5183–5189

13. Choi G, Kramer S, Tam S, Fox J (1997) Design/Economics of a once-through natural gas Fischer-Tropsch plant with power co-production (Internet). 1997 (cited 2010 Oct 25); Available from: http://www.fischer-tropsch.org/DOE/_conf_proc/Coal%20Conferences/coal_liq_conf97/choi.pdf
14. Ciferno J, Marano J (2002) Benchmarking biomass gasification technologies for fuels, chemical and hydrogen production (Internet). U.S. Department of Energy National Renewable Energy Technology Laboratory; (cited 2010 Nov 4). Available from: http://www.netl.doe.gov/technologies/coalpower/gasification/pubs/pdf/BMassGasFinal.pdf
15. CPM (2010) CPM LCA database (Internet). Center for Environmental Assessment of Product and Material Systems (CPM), Chalmers University of Technology, Goteborg, Sweden (cited 2010 Nov 22). Available from: http://www.cpm.chalmers.se/CPMDatabase/
16. Dancuar L, Mayer J, Tallman M, Adams J (2003) Performance of the SASOL SPD naphtha as steam cracking feedstock. Prepr Am Chem Soc (a division of Petroleum Chemistry) 48:132–138
17. DESTATIS (2009) Environment Waste Balance 2007. German Federal Statistical Office, Wiesbaden
18. Ecoinvent (2010) Ecoinvent life cycle inventory database v2.2 (Internet). Swiss Centre for Life Cycle Inventories; (cited 2010 Nov 7). Available from: http://www.ecoinvent.ch/
19. Ehrenfeld JR (2000) Industrial ecology: paradigm shift or normal science? Am Behav Scientist 44(2):229–244
20. EPA (2009) Municipal solid waste generation, recycling and disposal in the United States: facts and figures for 2008 (Internet). U.S. Environmental Protection Agency; 2009 (cited 2009 Dec 20). Available from: http://www.epa.gov/waste/nonhaz/municipal/pubs/msw2008rpt.pdf
21. EPA (2010) Solid waste management and greenhouse gases: documentation for greenhouse gas emission and energy factors used in the waste reduction model (WARM) (Internet). U.S. Environmental Protection Agency; 2010 (cited 2010 Dec 11). Available from: http://epa.gov/climatechange/wycd/waste/SWMGHGreport.html#documentation
22. Erkman S (1997) Industrial ecology: an historical view. J Cleaner Prod 5(1–2):1–10
23. Eurostat (2009) Environmental data centre on waste (Internet). 2009 (cited 2009 Dec 21); Available from: http://epp.eurostat.ec.europa.eu/portal/page/portal/waste/introduction
24. Franklin Associates (2007) Cradle-to-gate life cycle inventory of nine plastic resins and four polyurethane precursors (Internet). The Plastics Division of the American Chemistry Council, Prairie Village; 2007 (cited 2010 Nov 26). Available from: http://www.americanchemistry.com
25. Green M, Shelef G (1989) Ethanol fermentation of acid hydrolysate of municipal solid waste. Chem Eng J 40(3):B25–B28
26. Hayes DJ (2009) An examination of biorefining processes, catalysts and challenges. Catal Today 145(1–2):138–151
27. Henry RK, Yongsheng Z, Jun D (2006) Municipal solid waste management challenges in developing countries-Kenyan case study. Waste Manage 26(1):92–100
28. Higham I, Palacios I, Barker N (2001) Review of BAT for new waste incineration issues - part 1 waste pyrolysis & gasification activities. Enviroment Agency, Bristol
29. Jones A, O'Hare M, Farrel A (2007) Biofuel boundaries: estimating the medium-term supply potential of domestic biofuels (Internet). UC Berkeley Transportation Sustainability Research Center, Berkely, Working Paper; 2007 (cited 2009 Dec 23). Available from: http://escholarship.org/uc/item/950662sc
30. Jones S, Zhu Y, Valkenburg C (2009) Municipal solid waste (MSW) to liquid fuels synthesis, vol 2: A techno-economic evaluation of the production of mixed alcohols (Internet). Pacific Northwest National Laboratory, Richland; 2009 (cited 2009 Oct 30). Available from: http://www.pnl.gov/main/publications/external/technical_reports/PNNL-18482.pdf
31. Jungbluth N, Chudacoff M, Dauriat A, Dinkel F, Doka G, Faist Emmenegger M et al. (2007) Life cycle inventories of bioenergy (Internet). Swiss Centre for Life Cycle Inventories, Dübendorf, CH
32. Jungbluth N, Frischknecht R, Emmenegger MF, Tuchschmid M (2007) RENEW: Renewable fuels for advanced powertrains - life cycle assessment of BTL-fuel production: inventory analysis (Internet). ESU-Services Ltd.; 2007 (cited 2010 Jan 11). Available from: http://www.renew-fuel.com/fs_documents.php

33. Juniper Consultancy Services (2001) Pyrolysis and gasification of waste: a worldwide business and technology review, vols 1 and 2 (Internet). Juniper Consultancy Services Ltd, England. Available from: http://www.wastereports.com/free_downloads/pyrolysis-and-gasification.html
34. Kalogo Y, Habibi S, MacLean HL, Joshi SV (2007) Environmental implications of municipal solid waste-derived ethanol. Environ Sci Technol 41(1):35–41
35. Kamm B, Gruber PR, Kamm M (2006) Biorefineries - industrial processes and products: status quo and future directions, vol 2. Wiley-VCH, London
36. Khoo HH (2009) Life cycle impact assessment of various waste conversion technologies. Waste Manage 29(6):1892–1900 Jun
37. Klein A (2002) Gasification: an alternative process for energy recovery and disposal of municipal solid waste (Internet). 2002 May (cited 2010 Aug 8); Available from: http://www.seas.columbia.edu/earth/wtert/sofos/klein_thesis.pdf
38. Kreith F, Tchobanoglous G (2002) Handbook of solid waste management. McGraw Hill Professional, New York
39. Lackner KS, Brennan S (2009) Envisioning carbon capture and storage: expanded possibilities due to air capture, leakage insurance, and C-14 monitoring. Clim Change 96(3):357–378
40. Lettenmeier M, Rohn H, Liedtke C, Schmidt-Bleek F (2009) Resource productivity in 7 steps : how to develop eco-innovative products and services and improve their material footprint (Internet). Wuppertal, Germany: Wuppertal Institut für Klima, Umwelt, Energie GmbH; 2009 http://www.wupperinst.org/uploads/tx_wibeitrag/ws41.pdf
41. Li A, Khraisheh M (2008) Municipal solid waste used as bioethanol sources and its related environmental impacts. Int J Soil, Sediment Water 1(1):5–10
42. Li A, Khraisheh M (2008) Rubbish or resources: an investigation of converting municipal solid waste (MSW) to bio-ethanol production. WIT Trans Ecol Environ 109:115–122
43. Li A, Khraisheh M (2009) Bioenergy II: bio-ethanol from municipal solid waste (MSW): The UK potential and implication for sustainable energy and waste management. Int J Chem Reactor Eng (Internet). 2009; 7. Available from: http://www.bepress.com/ijcre/vol7/A78
44. Li A, Antizar-Ladislao B, Khraisheh M (2007) Bioconversion of municipal solid waste to glucose for bio-ethanol production. Bioprocess Biosyst Eng 30(3):189–196
45. Malkow T (2004) Novel and innovative pyrolysis and gasification technologies for energy efficient and environmentally sound MSW disposal. Waste Manage 24(1):53–79
46. Marano JJ, Ciferno JP (2001) Life-cycle greenhouse-gas emissions inventory for Fischer-Tropsch fuels (Internet). U.S. Department of Energy National Energy Technology Laboratory; 2001 (cited 2009 May 21). Available from: http://www.nrel.gov/docs/legosti/fy98/23076.pdf
47. McCaskey T, Zhou S, Britt S, Strickland R (1994) Bioconversion of municipal solid waste to lactic acid by Lactobacillus species. Appl Biochem Biotechnol 45–46(1):555–568
48. Mtui G, Nakamura Y (2005) Bioconversion of lignocellulosic waste from selected dumping sites in Dar es Salaam, Tanzania. Biodegradation 16(6):493–499
49. Niessen W, Marks C, Sommerlad R, Niessen WR, Marks CH, Sommerlad RE (1996) Evaluation of gasification and novel thermal processes for the treatment of municipal solid waste. National Renewable Energy Laboratory (NREL), Golden
50. NREL (2008) U.S. life cycle inventory database (U.S. LCI), v1.6.0 (Internet). National Renewable Energy Laboratory (NREL); 2008 (cited 2010 Jan 26). Available from: http://www.nrel.gov/lci/database/
51. Paisley M, Creamer K, Tweksbury T, Taylor D (1989) Gasification of refuse derived fuel in the Battelle high throughput gasification system (Internet). 1989 (cited 2010 Oct 25). Available from: http://www.osti.gov/bridge/servlets/purl/5653025-QRQFYH/
52. Rapport J, Zhang R, Jenkins B, Williams R (2008) Current anaerobic digestion technologies used for treatment of muncipal organic solid waste (Internet). California Integrated Waste Management Board, California; 2008 (cited 2010 Jan 2). Available from: http://www.calrecycle.ca.gov/publications/Organics/2008011.pdf
53. Ren T (2009) Petrochemicals from oil, natural gas, coal and biomass: energy use, economics and innovation (Internet). 2009 (cited 2009 Aug 22); Available from: http://igitur-archive.library.uu.nl/dissertations/2009-0212-200641/UUindex.html

54. Ren T, Patel MK, Blok K (2008) Steam cracking and methane to olefins: energy use, CO_2 emissions and production costs. Energy 33(5):817–833 May
55. RENEW (2006) RENEW–Renewable fuels for advanced powertrains: WP5.4 technical assessment, Europäisches Zentrum für erneuerbare Energie Güssing GmbH [cited 2010 Nov 5]. Available from: http://www.renewfuel.com/fs_documents.php
56. Sakai K, Taniguchi M, Miura S, Ohara H, Matsumoto T, Shirai Y (2003) Making plastics from garbage. J Ind Ecol 7(3–4):63–74
57. Shi AZ, Koh LP, Tan HT (2009) The biofuel potential of municipal solid waste. GCB Bioenergy 1(5):317–320
58. Shi J, Ebrik M, Yang B, Wyman CE (2009) The potential of cellulosic ethanol production from municipal solid waste: a technical and economic evaluation. University of California Energy Institute, Berkely
59. Skibar W, Grogan G, McDonald J, Pitts M (2009) UK expertise for exploitation of biomass-based platform chemicals—a white paper by the FROPTOP Group (Internet). FROPTOP (From Renewable Platform Chemicals to Value Added Products); 2009 (cited 2009 Jun 9). Available from: www.chemistryinnovation.co.uk/FROPTOP
60. Spath P, Dayton D (2003) Preliminary screening-technical and economic assessment of synthesis gas to fuels and chemicals with emphasis on the potential for biomass-derived syngas (Internet). U.S. Department of Energy National Renewable Energy Technology Laboratory; 2003. Available from: http://www.nrel.gov/docs/fy04osti/34929.pdf
61. Stichnothe H, Azapagic A (2009) Bioethanol from waste: life cycle estimation of the greenhouse gas saving potential. Resour Conserv Recycling 53(11):624–630
62. Tchobanoglous G, Theisen H, Vigil S (1993) Integrated solid waste management: engineering principles and management issues. McGraw-Hill, New York
63. Twardowska I (2004) I.1 Solid waste: what is it? In: Solid waste: assessment, monitoring and remediation. Elsevier, pp 3–32
64. UBA (2010) ProBas-Lebenszyklusdatenbank (Internet). Dessau (Germany); Freiburg (Germany): Umweltbundesamt (German Federal Environmental Agency) and Öko-Institut; 2010 (cited 2010 Nov 22). Available from: http://www.probas.umweltbundesamt.de/php/index.php
65. UN-HABITAT (2008) State of the world's cities 2008/2009: harmonious cities (Internet). UN-HABITAT; 2008 (cited 2010 Dec 5). Available from: http://www.unhabitat.org/pmss/listItemDetails.aspx?publicationID=2562
66. UN-HABITAT (2010) Solid waste management in the world's cities: water and sanitation in the world's cities 2010. Earthscan Publications Ltd, London
67. Valkenburg C, Gerber M, Walton C, Jones S, Thompson B, Stevens D (2008) Municipal solid waste (MSW) to liquid fuels synthesis, vol 1: Availability of feedstock and technology (Internet). Pacific Northwest National Laboratory, Richland; 2008 (cited 2009 Oct 30). Available from: http://www.pnl.gov/main/publications/external/technical_reports/PNNL-18144.pdf
68. Van Bibber L, Shuster E, Haslbeck J, Rutkowski M, Olson S, Kramer S (2007) Technical and economic assessment of small-scale fischer-tropsch liquids facilities [Internet]. U.S. Department of Energy/National Energy Technology Laboratory; 2007 [cited 2010 Oct 25]. Available from: http://www.purdue.edu/discoverypark/energy/pdfs/cctr/DOE-NETL-F-T-2007.pdf
69. Williams RB (2007) Biofuels from Municipal solid wastes - background discussion paper (Internet). University of California, Davis and California Biomass Collaborative. Available from: http://biomass.ucdavis.edu/materials/reports%20and%20publications/2007/2007_Annual_Forum_Background_Paper.pdf
70. Zheng Y, Pan Z, Zhang R, Labavitch J, Wang D, Teter S et al (2007) Evaluation of different biomass materials as feedstock for fermentable sugar production. Appl Biochem Biotechnol 137–140(1):423–435

Planning Tools and Procedures for Rational Municipal Solid Wastes Management

Alexander P. Economopoulos

Abstract A rational approach for developing optimal municipal solid wastes (MSW) management plans comprises two steps, the strategic and the detailed planning one. The objectives of the strategic planning are the screening of alternative technologies, the definition of the number and approximate location of sites for treatment and/or disposal installations, the formulation of alternative management plans and the selection of the most prominent ones. The detailed optimal planning, which normally follows, focuses on the latter and develops the plan that meets all legal and other requirements with the least cost. In addition, it performs sensitivity analysis involving the development of alternative optimal plans, each of which meets a set of constraints. The latter reflects the preferences and/or objections of the communities concerned (e.g. the exclusion or imposition of sites and/or technologies, the application of capacity limits etc.). The local authorities can then select among the alternative plans the one, which balances best their preferences and objections against the associated costs.

1 Introduction to MSW Management

The diagrams in Fig. 1 illustrate the evolution of the waste management schemes, from a typical initial to a more advanced stage. The management in its initial stage is often limited to the source separation of some recyclable materials and to the landfilling of the remaining wastes, Fig. 1a. In more advanced stages, a portion of

A. P. Economopoulos (✉)
Environmental Engineering Department,
Technical University of Crete,
731 00 Chania, Greece
e-mail: eco@otenet.gr

A. Karagiannidis (ed.), *Waste to Energy*, Green Energy and Technology,
DOI: 10.1007/978-1-4471-2306-4_2,

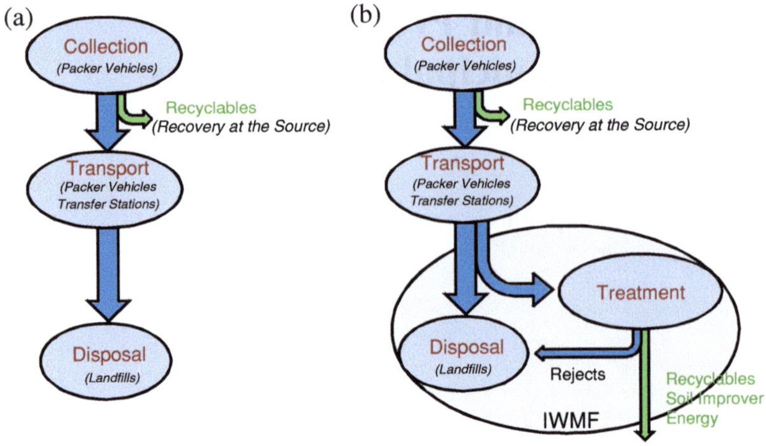

Fig. 1 Evolution of the MSW management system from an initial to a more advanced stage

the commingle MSW is treated so as to recover recyclable materials, to produce biostabilized organics (used as soil improver or landfill cover material) and/or to generate energy. From the treated MSW only the rejects are landfilled, Fig. 1b.

In the most advanced management stages, emphasis is given on the waste prevention, reuse and source separation into selected streams (e.g. recyclable material and/or kitchen waste streams). This improves the yield and the quality of the waste products and reduces the treatment and final disposal requirements.

The treatment and final disposal of the wastes take place in integrated waste management facilities (IWMFs). These, usually serve areas with large populations so as to achieve the desirable economy of scale. The collection of the wastes to central IWMFs often involves distant transportation of the MSW and this is economically achieved through transfer stations than make optimal use of road, railroad and sea transportation media.

In the general case, treatment of the MSW through successive steps at different installations may be involved and final disposal of the rejects at different sites may be required, depending on the nature of the rejects. Moreover, the IWMF can be physically distributed at different sites, each hosting one or more of the required installations. Some of the above possibilities are discussed in Sect. 4.1.

2 Overview of MSW Management Technologies

The purpose of this section is to outline the technologies often used in the transportation, treatment and final disposal of MSW and to provide relevant data and information required for the formulation and evaluation of alternative management plans.

Fig. 2 Transfer station with a ramp, a hopper and a semitrailer under the hopper in loading position

2.1 Transfer Stations

As the quantities of the MSW grow and the travel distances to their management facilities increase, the cost of direct transportation by the collection vehicles becomes increasingly expensive. From a point onward, the transportation becomes more economic though the use of transfer stations. The latter receive the waste of collection vehicles and transfer it to large trucks, tractor-trailers, semi-trailers, railroad cars and/or barges for economic long distance transportation.

Transfer stations without waste compression are the simplest to construct and require limited investment. However, most modern transfer stations provide waste compression, as this reduces the transportation costs enabling the howl vehicles to transport heavier net pay loads. The latter are usually of the order of 19.5 t, but this depends on the applicable gross weight limits of the roads and on the vehicle design and configuration. Two of the most frequently used transfer station technologies are described in the Sects. 2.1.1 and 2.1.2 that follow.

2.1.1 Transfer Station with Mobile Compactor Units

In a typical station of this type, the packer vehicle ascent to an elevated platform through a ramp, so as to discharge their wastes into a hopper. From there, the wastes are fed to the front end of a container fixed on a semitrailer, Fig. 2. The container is equipped with a hydraulic pusher mechanism, which facilitates (a) the waste loading by pushing periodically the wastes towards the back-side of the container, providing at the same time a degree of compaction, and (b) the waste unloading at the waste reception site.

Fig. 3 Transfer station with a hopper, a compactor unit and semitrailers for the transportation of containers

2.1.2 Transfer Stations with Compactor Units

As in the previous case, the packer vehicles ascent to an elevated platform through a ramp so as to discharge their wastes into hoppers for temporary storage. From there the wastes are fed into compactors and are compressed into containers. The latter are loaded into large trucks or semitrailers for transportation to the reception site, Fig. 3. When large quantities of wastes are to be transported to remote IWMFs, the transportation of containers through railroad or sea can be more economic. In these cases, the direct loading of the containers from the transfer station into railroad cars, as in Fig. 4, or into barges is particularly cost effective.

2.2 Treatment Methods

The technologies considered in the present section are suitable for the treatment of commingled MSW, as their use is expected to remain predominant for several years in countries with a short history in material separation at the source programs. Four such commonly used technologies are briefly described below.

The material and energy balances presented in the diagrams of Figs. 5, 6, 7 and 8 and Table 1 are indicative and correspond to the anticipated mean composition of the waste feedstock (the wastes that remain after the application of material recovery at the source programs) in Greece.

2.2.1 Aerobic Mechanical:Biological Treatment

The Aerobic Mechanical-Biological Treatment (MBT) comprises a material separation unit followed by an aerobic composting one, Fig. 5. The former can

Fig. 4 Transfer station with direct loading of containers to train

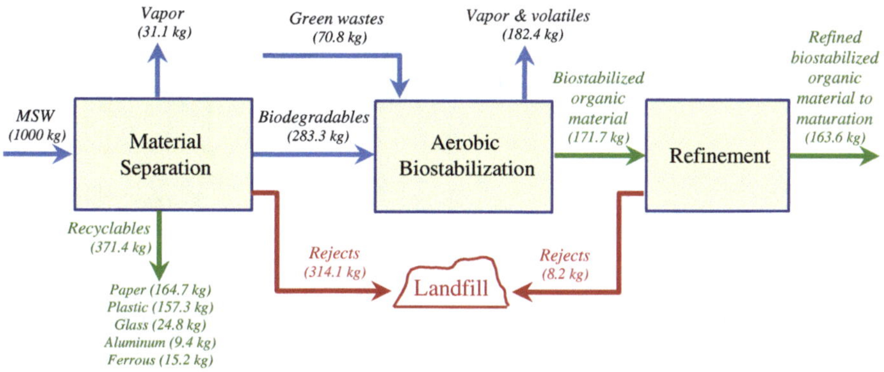

Fig. 5 Typical flow diagram of Aerobic MBT plants with indicative material balances (*Source* Economopoulos [8])

recover recyclable materials (paper, plastic, glass, metal etc.) and/or RDF (Refuse Derived Fuel) and separate the remaining wastes into organics and rejects. The organics, mixed in proper proportions with green wastes (and optionally with wastewater sludge), are fed into the aerobic composting unit, in which they are adequately aerated and biostabilized. The product can be used directly as disturbed soil improver (eg. in quarries) or as landfill cover material.

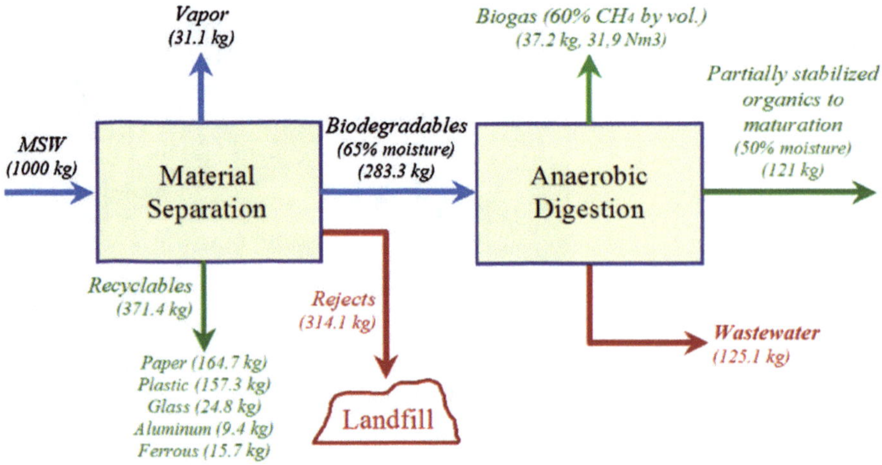

Fig. 6 Typical flow diagram of Anaerobic MBT plants with indicative material balances (*Source* Economopoulos [8])

Fig. 7 Typical flow diagram of biological drying plants with indicative material balances (*Source* Economopoulos [8])

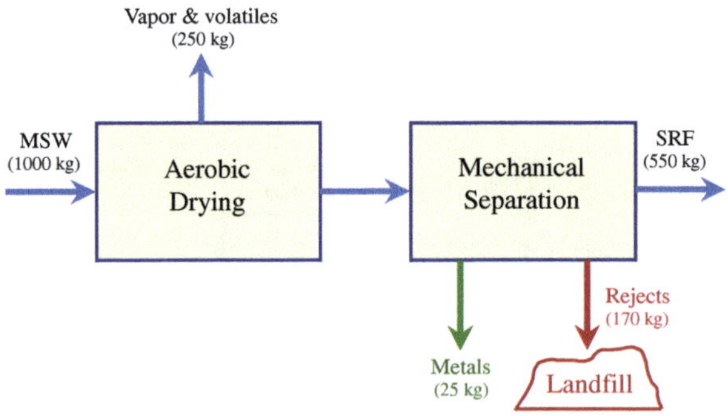

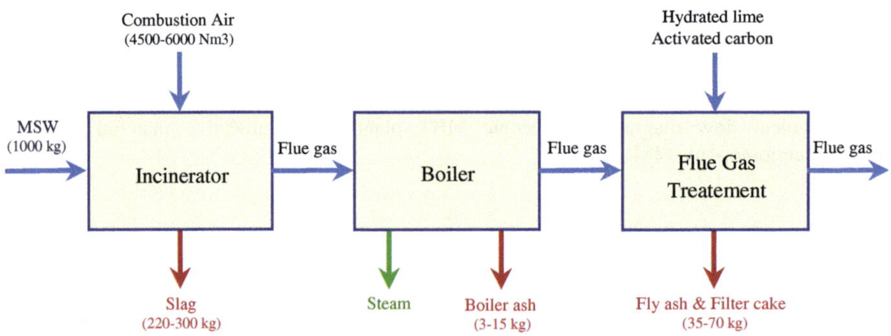

Fig. 8 Typical flow diagram of waste incineration plants with indicative material balances (*Source* Economopoulos [8])

Table 1 Electric energy export potential of alternative treatment methods [8]

Aerobic MBT with materials recovery		
Internal energy consumption	kWh/1,000 kg MSW	**32,0**
Excess electricity to be exported	kWh/1,000 kg MSW	**−32,0**
Biological drying and incineration of SRF		
Biological drying:		
Secondary fuel production	kg SRF/1,000 kg MSW	550,0
LCV of fuel	MJ/kg SRF	17,5
Internal energy consumption	kWh/1,000 kg MSW	**140**
SRF incineration:		
Overall conversion efficiency	%	27,0
Electricity generation	kWh/1,000 kg MSW	**721,9**
Internal electricity consumption	kWh/1,000 kg MSW	**197,8**
Excess electricity to be exported	kWh/1,000 kg MSW	**384,1**
Mass incineration		
Feedstock	kg MSW	1000,0
LCV	MJ/kg MSW	11,5
Overall conversion efficiency	%	20,0
Electricity generation	kWh/1,000 kg MSW	**640,0**
Internal electricity consumption	kWh/1,000 kg MSW	**175,4**
Excess electricity to be exported	kWh/1,000 kg MSW	**464,6**

2.2.2 Anaerobic Mechanical: Biological Treatment

The Anaerobic MBT method, illustrated by the flow diagram of Fig. 6, differs from the Aerobic MBT one in that the organic fraction undergoes anaerobic (in enclosed reactors, without oxygen), instead of aerobic, decomposition.

Most of the anaerobic systems today are of the high solids (with the feedstock diluted with water to a total solids content of around 25%), thermophilic and single-stage type, with retention times ranging from 14 to 20 days.

The stabilized residues contain large amounts of water, most of which is removed by filtration. Only a portion of the removed water can be reused so as to maintain the electrical conductivity within proper limits in order to protect the microbial activity in the reactors. The excess water forms a strong effluent that requires advanced treatment prior to disposal. The filter cake can be matured under aerobic conditions for a period of 2–4 weeks.

2.2.3 Biological Drying

The biological drying, a pretreatment process that converts the MSW into SRF (Solid Refuse Fuel—a secondary fuel), comprises waste shredding, aerobic drying for reducing the moisture of the wastes to less than 20%, metal recovery, and separation of rejects, Fig. 7.

The energy required for drying the wastes in the bioreactor is generated by the exothermic aerobic decomposition of a limited fraction of easily biodegradable organics. This process does not biostabilize the wastes and is in fact designed to minimize the bio-decomposition so as to preserve the energy content of the wastes.

2.2.4 Incineration

The waste incinerator plants comprise the incinerator, usually of the inclined moving or roller grate type, the boiler for energy recovery and electricity production, and the flue gas treatment system for the removal of particles, HCl, HF, SO_2 dioxins, furans and heavy metals, including Hg, Fig. 8.

Most of the solid residues are generated from the combustion chamber and the boiler. Smaller, but more toxic, quantities are generated from the flue gas treatment system. In EU, the control of emissions is subject to Directive 2000/74. Guidance on pollution prevention and control is given by the BREF report, European Commission [14].

2.2.5 Electric Energy Balances

Table 1 presents typical electric energy balances for the above technologies, assessing their net electric energy export potential on the basis of their electric energy generation and internal consumption potential, Economopoulos [8].

2.2.6 Treatment Products and Possible Uses

All products from the treatment methods considered above can find a market (some, such as glass and biostabilized organics, at nearly zero or even slightly negative prices) with the exception of the SRF from the biological drying plants and the RDF from the Mechanical Separations Units, the alternative uses of which are illustrated in the diagram of Fig. 9.

The use of the SRF in the clinker kilns of the cement industry is a rational option, as no new installation is involved and little pollution is generated; The cement industry however, can accommodate small quantities of this fuel, e.g. Juniper Consultancy Services Ltd. [16]. Co-firing of SRF with fossil fuels in utility boilers is feasible, but in limited proportions due to operating problems. This however is not a practical option as the flue gas control system of the power plant needs to be upgraded to waste incineration standards and the toxic residues from the SRF incineration are mixed with large quantities of fly ash making the management of the latter expensive. As a result, incineration is the only practical SRF utilization method, Economopoulos [8]. Similar is the situation with the RDF, a secondary fuel, which can be optionally produced by aerobic or anaerobic MBT plants (see Sect. 2.2.1).

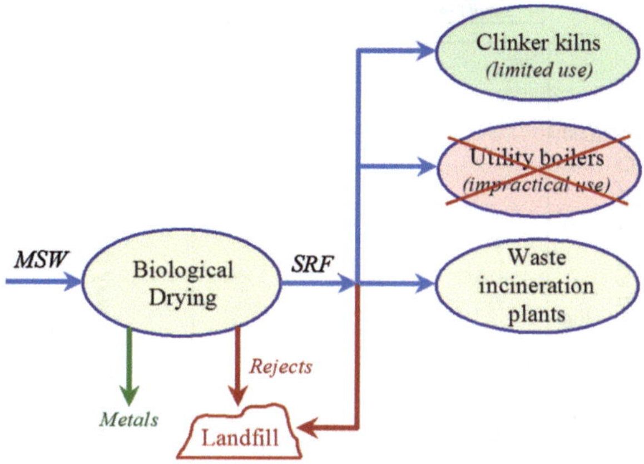

Fig. 9 Possible uses of the SRF produced from biological drying plants (adapted from [8])

2.3 Final Disposal Sites

In modern landfills the design comprises a low permeability liner to restrict leachates, a strong effluent, from percolating through the base of the landfill and a pipe system enabling collection of leachates for proper treatment and disposal. It comprises also a pipe system for collecting most of the biogas generated by the anaerobic decomposition of the organic materials so as to be flared or used for energy generation. Thus, modern landfills are not very simple facilities.

3 Planning Tools

The present section describes a graphical method that allows convenient determination of the approximate number and location of the required IWMFs in the study area. It provides also cost functions that allow economic analysis of management plans. The above are essential tools used in Sects. 5 and 6 for the formulation, evaluation and optimization of management plans.

3.1 Optimal Number and Location of IWMFs

For the development of cost-effective plans, planners need to define the near optimal number and the approximate location of sites for building the required waste treatment and final disposal installations. The above reflect the optimal balance between the economy of scale offered by large central installations and the increased cost of waste transportation required for collecting the wastes to these installations.

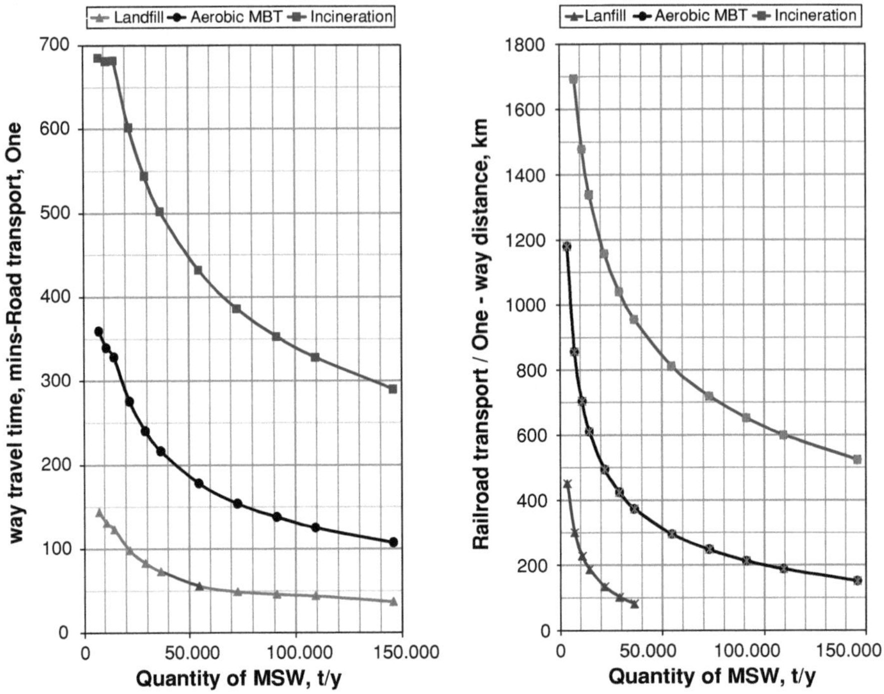

Fig. 10 Maximum time of road and maximum distance of rail transport to central IWMF as function of the quantity of local wastes, in case where a transfer station is already used

To cope with this requirement, an easy to use methodology is presented, which is based on the graphs of Figs. 10 and 11, Economopoulou and Economopoulos [11]. These graphs provide the maximum distance that is profitable to transport the locally produced wastes for treatment and/or disposal at a large central rather than at a small local IWMF. At this maximum distance, the extra cost of transportation becomes equal to the cost savings realized by the increased economy of scale of the large central over the small local IWMF.

The development of the graphs in Figs. 10 and 11 is based on the combined use of cost functions that yield the capital investment and annual operating costs of collection vehicles, transfer stations and their haul vehicles, treatment plants and final disposal installations. Updated graphs will soon be available based on the use of more recent cost functions, Economopoulou and Economopoulos [12].

As the graphs in Figs. 10 and 11 show, the maximum distance depends on the kind of treatment method considered, the quantity of local wastes, the transportation media (road or railway) used, as well as on the availability of local transfer stations.

In relation to the latter, two alternatives are considered: (a) Local transfer stations do not exist and are not required for transporting the wastes to the local IWMF. In this case, transfer stations may have to be constructed for the transportation of the

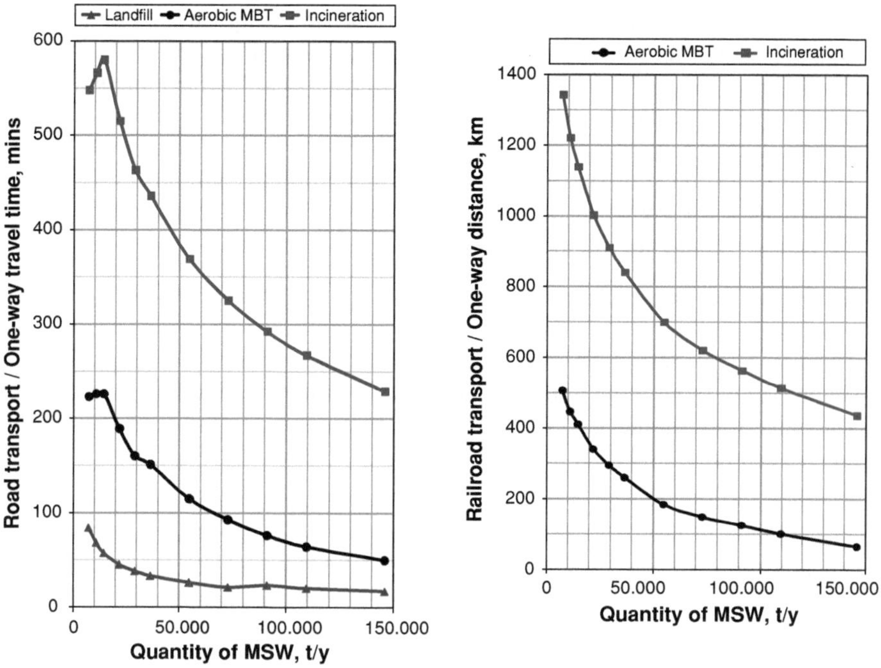

Fig. 11 Maximum time of road and maximum distance of railroad transport to a central IWMF as function of the quantity of wastes produced locally, in case where a transfer station is not used

wastes to the central IWMF and, if this is the case, the transportation cost must include the annual capital investment and operating cost of the transfer station, along with its temporary storage facilities and haul vehicles, and (b) Local transfer stations already exist and/or are required for the transportation of the wastes to the local IWMF. In this case, the same transfer stations can be used for transporting wastes to the central IWMF and hence, only the annual capital investment and operating cost of the haul vehicles for the extra transportation time need to be considered.

For road transportation, the graphs yield the maximum one-way driving time. This needs to be multiplied by the mean vehicle velocity for obtaining the one-way transportation distance. For railroad transportation the graphs yield directly the maximum one-way distance.

With the use of graphs in Figs. 10 and 11, planners can assess the approximate number and location of the required IWMF sites though the following steps:

1. Consider the administrative levels of the study area (e.g. municipality, prefecture) and select the one, which divides the study area in a reasonable number of sectors, e.g. 10–50.
2. For each administrative sector calculate the annual waste load generated (see Sect. 5.1), define its barycentric population point, use the graphs in Figs. 10 or 11 for estimating the maximum transport distance for the desirable

treatment method, and draw a circle around the barycentric point with a diameter equal to 0.8–0.9 of the maximum transport distance.

3. The overlay areas of surrounding circles define the possible location of a central IWMF that can serve the corresponding administrative sectors. As most administrative sectors can be served by a number of alternative central IWMFs, planners can define the location of IWMFs and group the administrative sectors served by each IWMF in many different ways. The following provide some guidance for making rational selections:

- Administrative sectors are served best by central IWMFs with good road connection.
- The use of overlay areas with sites- or near sites- known to be particularly suitable (e.g. large mine fields or sites already in use) should be given priority.
- The selections should be in the direction that equalizes the load distribution among central IMWFs.

In some cases, the same exercise could be repeated with the exclusion of congested administrative sectors with very large populations. This exclusion is of practical interest in cases where it is difficult to find proper sites within- or near- the congested administrative sector and it is justified by the fact that transportation of the wastes from the congested sectors to sites at reasonably long distance is not prohibitively expensive.

It is interesting to note, that alternative sets of sites, defined through different selections in the above procedure, are likely to lead into optimal solutions with nearly identical total management costs.

3.2 Normalized Cost Elements

In the present section, typical cost data are given for waste treatment, transportation and final disposal installations, so as to enable the economic analysis of management plans.

3.2.1 Cost of Treatment

For plants with configurations similar to these in the diagrams of Figs. 5, 6, 7 and 8, cost functions allowing estimation of the initial capital investment and annual operating cost versus the annual quantity of MSW processed, have been developed and presented in graphical and mathematical form, Economopoulos [8].

For the Aerobic MBT plants, the cost data have been generated by an advanced plant design and cost estimation model and have been found in reasonable agreement with cost data reported in the literature from European plants. For the remaining processes, data were collected from recent EU literature sources.

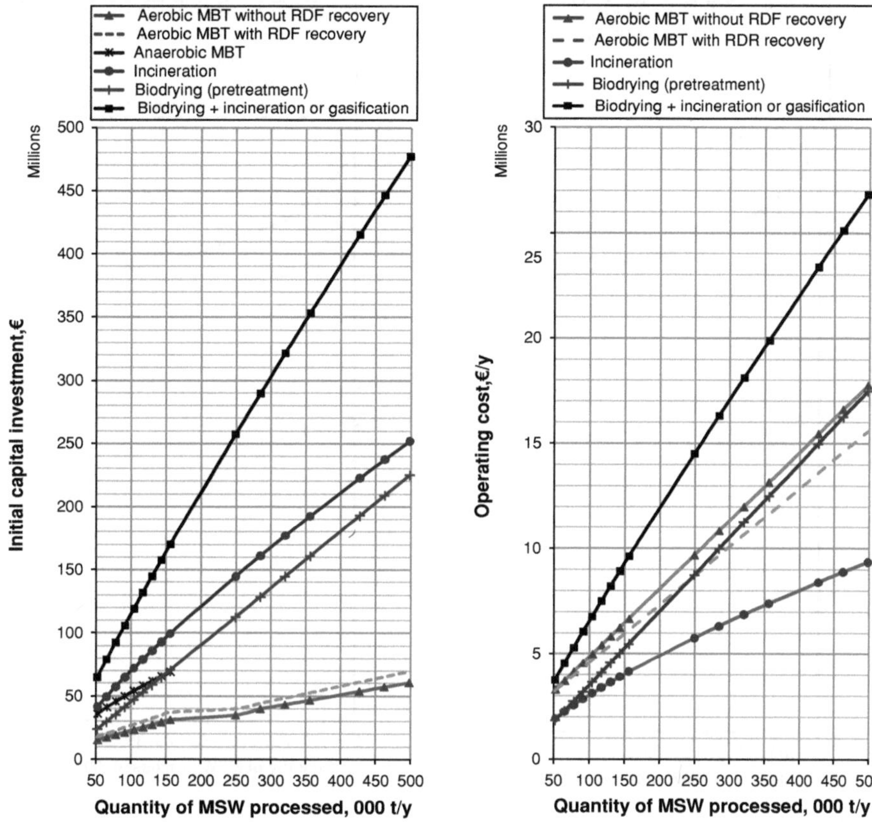

Fig. 12 Initial capital investment and annual operating cost of alternative waste treatment technologies (*Source* Economopoulos [8])

The datasets thus established were fit so as to derive cost functions for the initial capital investment and the annual operating cost of alternative treatment technologies and the results are presented in the diagrams of Fig. 12.

As it can be seen from the above diagrams, the biological drying plans do not offer any significant economy of scale; this is due to their modular construction and the use of multiple parallel units in large installations. It can be also noticed that the total cost for the biological drying of the wastes and the SRF incineration of the SRF produced, is considerably higher than that of the direct mass incineration of MSW.

The capital investment and operating cost functions of Fig. 12 have been used for deriving the normalized treatment cost functions presented in the graphs of Fig. 13. The latter are based on the following two typical scenarios:

(a) The plant is owned and operated by a municipality association. In this case the annual cost of the invested capital is assumed 5.5% and the average life of the installation 20 years (zero salvage value). As the plant provides service to its Municipality members, VAT is not charged.

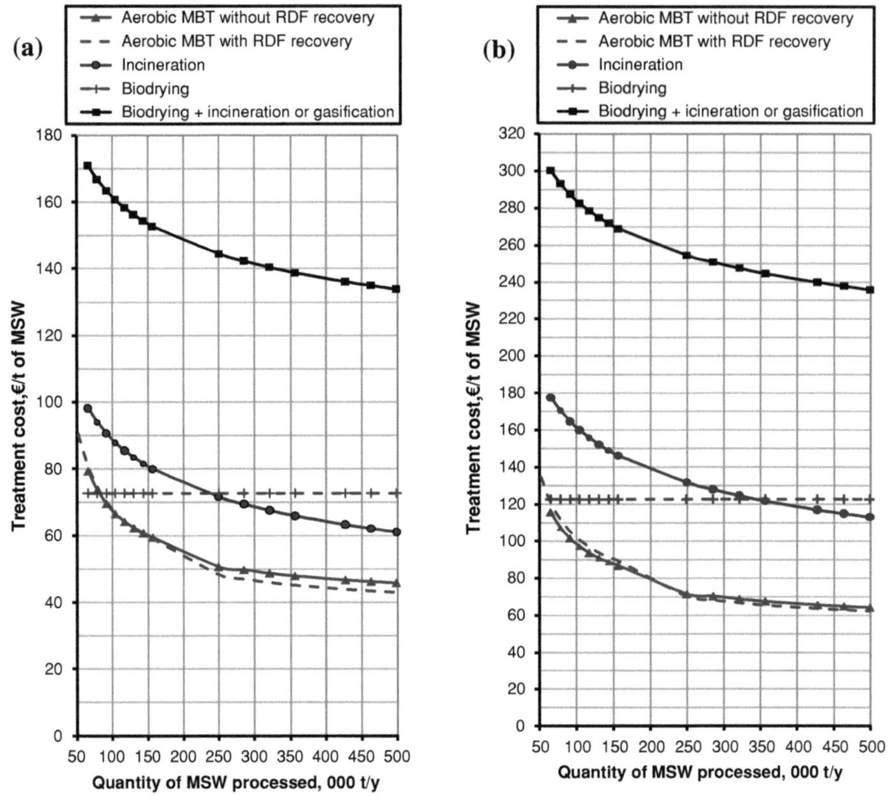

Fig. 13 Normalized treatment costs of alternative technologies with plants built and operated by **a** municipal associations, and **b** private enterprises (Economopoulos) [8]

(b) The plant is built and operated by private entrepreneurs. The annual internal return on investment is assumed 14% and the average life of the installation 20 years (zero salvage value). In this case, the gate fees include 19% VAT.

The graphs of Fig. 13 yield the normalized treatment cost (in €/t of MSW processed) for the alternative technologies considered in Sect. 2, as well as for the biological drying of MSW and the gasification or incineration of the SRF produced.

The treatment costs in Fig. 13 refer to year 2009 and do not include the revenue from the sale of products (recyclable materials and/or energy), nor the expenditures for the transportation and management of unused products and residues. However, the material and energy balances in the diagrams of Figs. 5, 6, 7 and 8 provide data about the quantities of products and rejects, facilitating the estimation of the relevant annual revenues and expenditures.

It should be noted that the cost functions of Fig. 12 are based on data obtained, mostly, from literature. As such they are approximate and do not take into consideration the particular conditions that influence the actual cost of each installation. Moreover, the treatment cost functions of Fig. 13 are based on the specific assumptions associated with each scenario considered. Different assumptions result to different treatment cost functions.

3.2.2 Cost of Transportation and Final Disposal

Cost functions for assessing the capital investment and annual operating costs of transfer stations and final disposal sites are not yet available. They are getting however developed and will be presented elsewhere, Economopoulou and Economopoulos [11].

As a rough guide, the normalized capital investment and operating costs for the road transportation of the wastes to a central IWMF is in the order of 30 €/t/y and 33 €/t respectively for regions with a surface area of about 12,000 km^2 and 50 €/t/y and 72 €/t respectively for regions with a surface area of 60,000 km^2, Economopoulos [7].

3.2.3 Revenue From the Sale of Products

The revenue from the sale of recyclable materials recovered and/or electricity produced can be estimated on the basis of the material and energy balances of the treatment processes used (such as these in Figs. 5, 6, 7 and 8 and Table 1) and the market prices of these products.

An example estimation of the anticipated income from the operation of an Aerobic MBT plant is presented in Table 2.

The estimation is based on the material recovery factors listed in the diagram of Fig. 5 and on the recyclable material prices in the Greek market during August 2008 and March 2009. In this example, the significant drop in the recyclable material prices due to the economic crisis, reduced the anticipated plant revenue from 32.1 to 11.8 €/t of MSW.

3.2.4 Income From Incentives

Financial incentives are offered by some countries for promoting specific management objectives. These may affect the net cost of treatment and the comparative economic evaluation of alternative treatment methods and need to be taken into consideration.

In Greece for example, the incentives offered for the recovery of packaging materials from the MSW could generate a revenue to aerobic or anaerobic MBT plants as high as 22 €/t of MSW treated, Economopoulos [7]. It is interesting to note that the combined revenues from the sale of products (see Sect. 3.2.3) and the incentives mentioned above, could cover most of the treatment cost in aerobic MBT plants in Greece (Fig. 13).

Table 2 Normalized income from the sale of recyclable materials recovered from Aerobic MBT plants [7]

	Price, €/t of recyclable material		Recovery	Income, €/t of MSW	
	August-08	March-09	kg/t of MSW	August-08	August-08
Paper	55.8	17.6	164.7	9.19	2.90
Plastics	114	34.6	157.3	17.93	5.44
Metals	201	141	24.6	4.94	3.47
Glass	0	0	24.8	0.00	0.00
Overall	86.34	31.80	371.4	32.07	11.81

4 Scope and Purpose of Management Plan Optimization

The objective of this section is to address the questions of what is to be optimized, why optimization is required and how optimal plans can be developed.

4.1 Alternative Management Options

The diagram of Fig. 14 illustrates a number of management alternatives based on the treatment technologies presented in Sect. 2.2. More specifically:

- Recyclable materials, such as paper/cardboard, plastic, metal and/or glass, separated at the source, can be reused or recycled with minimal processing.
- The commingled MSW that remain can be treated in aerobic or anaerobic MBT plants so as to obtain recyclable materials and/or RDF and biostabilized organics. The inert residues can be landfilled. If RDF is produced, it can be used by waste incineration plants and (in limited quantities) by the clinker kilns of the cement industry.
- The aerobic MBT plants can be designed so as to accept source-separated recyclable material streams into their material separation units and source-separated kitchen and garden wastes into their biological stabilization units. This way, plants built for treating commingled MSW can also accommodate source-separated waste streams as their quantities increase with time.
- Alternatively, the comingled MSW can be processed, along with SRF or RDF, in waste incineration plants, possibly after the recovery of some recyclable material at material separation units. The exported energy can be in the form of electricity and/or heat for space heating or industrial use. The residue (grade ash, boiler slag, and fly ash or filter cake from the flue gas treatment system) contains toxic substances and need to be disposed of at appropriate facilities.
- The comingled MSW can also be pretreated in biological drying plants, possibly after the recovery of some recyclable material at material separation units. The SRF produced can be incinerated and/or used (in limited quantities) in the cement industry. The inert residues are landfilled.

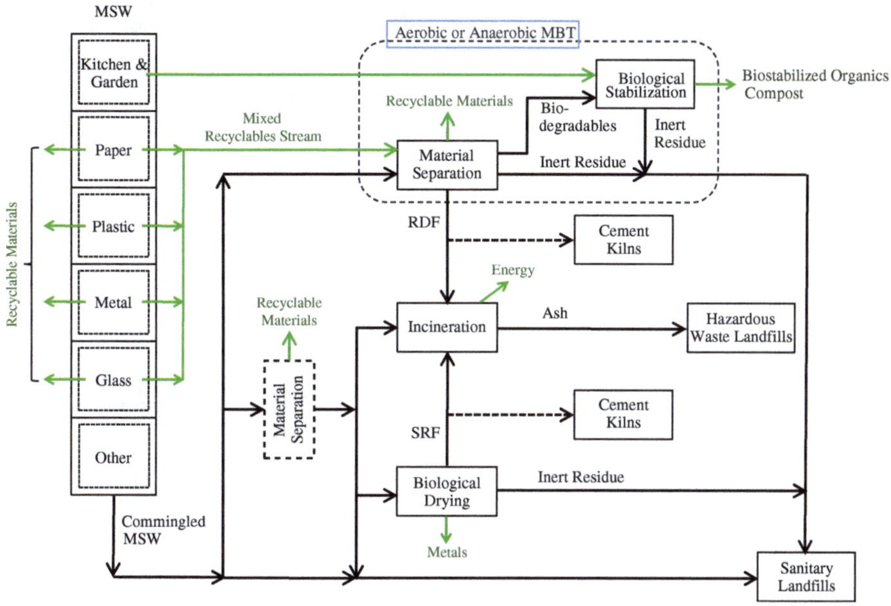

Fig. 14 Waste flows in alternative management plans based on the treatment technologies presented in Sect. 2

The options illustrated in the diagram of Fig. 14 and outlined above are not exhaustive as additional treatment methods exist (e.g. gasification, pyrolysis). Moreover, for most treatment methods, a number of alternative technologies exist, each with its own product types and qualities, product yields and economics.

Each management plan developed by planners can be represented by a waste flow diagram, similar to that of Fig. 14, but simpler in form. In this diagram, planners can perform material and energy balances for each type of installation, or each individual installation (see for example Sect. 5.5), which are essential for the evaluation of the plan under consideration (see Sect. 5.6).

4.2 Scope of Plan Optimization

The number of the alternative waste management options and their combinations are numerous, as illustrated by the waste flow diagram of Fig. 14. Among them, planners have to select the ones that minimize the sum of the annualized capital investment and annual operating costs of all transportation, treatment and disposal operations, taking into consideration the income from the sale of products and the possible financial incentives offered. Moreover, planners have to define the optimal location, type, size and operation of each transfer station, pre-treatment, treatment and final disposal installation, as well as the flow of the wastes, waste products and residues among them.

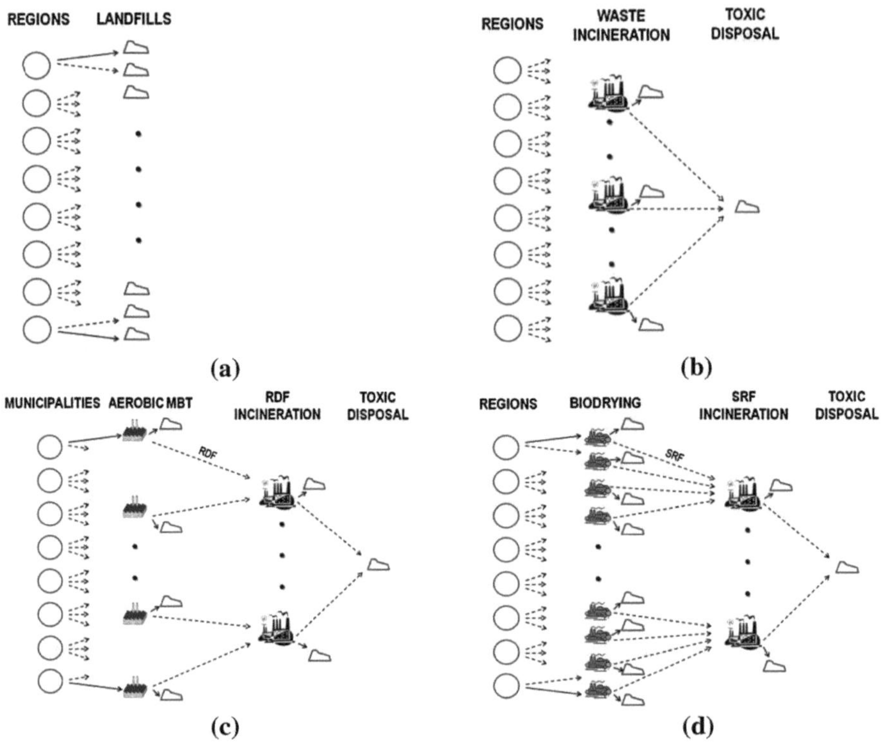

Fig. 15 Schematic representation of typical management plans involving **a** direct landfilling of the MSW, **b** direct incineration of the MSW, **c** aerobic MBT of the MSW and incineration of the RDF produced, and **d** biological drying of MSW and incineration of the SRF produced

The number, size and location of the installations reflect the optimal balance between the economy of scale offered by large central installations and the increased cost of waste transportation required for collecting the wastes to these installations. In order to illustrate this principle, let us consider four study areas with the management options illustrated in the diagrams (a)–(d) of Fig. 15.

Installations with strong economies of scale can optimally serve larger areas as they can compensate for higher transportation costs. Based on this:

- If landfilling is the sole objective of the MSW management, as in the diagram of Fig. 15a, a fair number of landfill sites are required as landfilling is a low cost operation offering limited economy of scale.
- If the wastes are to be incinerated, as in the diagram of Fig. 15b, a small number of plants is required, as they offer a strong economy of scale, and possibly fewer toxic waste disposal sites as the toxic wastes quantities generated are limited.
- If the wastes are to undergo aerobic MBT with RDF production and incineration, as in the diagram of Fig. 15c, a limited number of aerobic MBT plants are required, as they offer a fairly strong economy of scale. These have to be

combined with fewer RDF incineration plants and still fewer toxic waste disposal sites, as the RDF incineration plants offer strong economy of scale and the quantities of toxic wastes are limited.
- If the wastes are to be biologically dried and the SRF produced is to be incinerated, as in the diagram of Fig. 15d, numerous biological drying plants are required as they offer no economy of scale (they comprise parallel modules, each of which is capable of processing about 50,000 t/y of MSW) and as explained in the previous case, few SRF incineration plants and still fewer toxic waste disposal sites.

The above discussion serves to demonstrate that in all but the simplest cases, the development of optimal plans is a demanding task. As a result, planners can be significantly assisted by proper methodologies and software tools.

4.3 Purpose of Plan Optimization

Decision makers may wonder whether the effort required for the development of optimal plans is justified. In order to address this question, the costs and benefits of a relatively simple and a relatively sophisticated case study were reviewed and the conclusion was that the cost of proper planning represents only a small fraction of the annual cost savings it offers. This tends to be the rule and applies, not only when planners deal with large study areas and sophisticated management objectives, but also in cases of relatively small study areas and simple management objectives.

4.4 A Two-Stage Planning Approach

From the discussion in Sects. 4.1 and 4.2 it would appear that the development of optimal plans is, in most cases, a demanding operation. This is true, even in cases where appropriate software tools are available as the setup effort can be excessive. In view of this, a two-step approach has been developed, which comprises the strategic and the detailed optimal planning phases.

The strategic planning is based on the planning tools discussed in Sect. 3 and aims at the screening of alternative technologies, the definition of the number and approximate location of sites for treatment and/or disposal installations, the formulation of alternative management plans and the selection of the most prominent on the basis of their compatibility with the management objectives, cost, environmental problems and social acceptance characteristics (see Sect. 5). This exercise, which can be carried out reasonably fast and without the need for specialized software tools, provides valuable input to any relevant follow-up study.

The detailed optimal planning, which normally follows, focuses on the prominent management schemes identified in the previous phase and develops the plan that meets all legal and other requirements with the least cost. In addition, it performs sensitivity analysis involving the development of alternative optimal plans, each of which meets a set of constraints. The latter reflects the preferences and/or objections of the communities concerned (e.g. the exclusion or imposition of sites and/or technologies, the application of capacity limits etc.). The local authorities can then select among the alternative plans the one, which balances best their preferences and objections against the associated costs.

5 Development of Strategic Management Plans

The steps involved in the formulation of strategic management plans are illustrated in the diagram of Fig. 16 and outlined. A more detailed description, along with an application, is presented elsewhere, Economopoulos [7, 9].

5.1 Collection of Data and Information

For the formulation and analysis of strategic management plants, the collection of the following data and information is required:

- Data about the permanent population and the monthly visitors of the municipalities and communities in the study area over the design period. Future trend estimates are usually based on the analysis of relevant historical population and economic data and can be adjusted on the basis of development plans, Economopoulos et al. [10]. The main use of this information is for assessing the waste quantities and composition, as explained below.
- Data about the annual quantity of the wastes generated in each municipality and community and the typical composition of the wastes in each region over the design period. Estimates can be based on the analysis and extrapolation of past quantity and composition data, if they exist, Economopoulos et al. [10]. Alternatively, planners can use population data, along with typical normalized waste load and composition data from other similar regions of their country or relevant data from the literature, e.g. Eurostat [15], OECD [17], Riber et al. [18]. It is also possible to apply suitable statistical models, which provide waste load and composition estimates based on data and indices reflecting the present and future development of each region, e.g. Daskalopoulos et al. [4], Beigl et al. [1–3]. The required data and indices are usually available from the statistical services.
- Information about sites that could be particularly suitable for IWMFs, e.g. large lignite or bauxite mines. This information can be obtained from geological and other relevant services and is used in the IWMF sitting procedure discussed in Sect. 5.4.
- Definition of the existing management plan. This includes data on the quantities of recyclables recovered, on the location, design, capacity and function of all

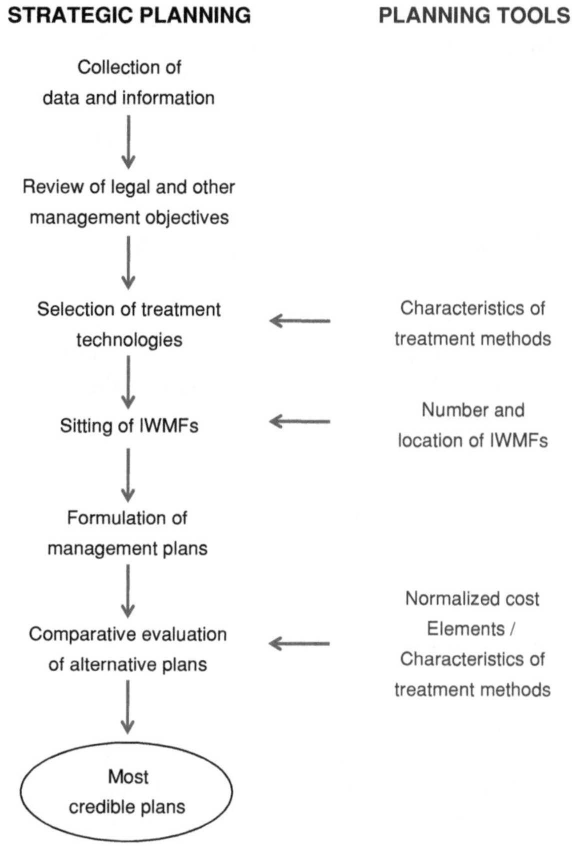

Fig. 16 Procedure for formulating strategic management plans

transfer, treatment and final disposal installations and the flow of the wastes among them.

5.2 Review of Legal and Other Management Objectives

In most countries there is legislation that defines the MSW management objectives. For example, in the European Union, the management of the MSW is controlled by a number of directives, such as Directive 1999/31, which stipulates the progressive reduction of biodegradable materials to be landfilled, Directive 2004/12, which stipulates the progressive recovery and use of the packaging wastes and the Directive 2008/98, which sets the hierarchy in waste prevention, management legislation and policy.

The applicable legislation in the study area must be reviewed and its stipulations must be quantified and expressed in a form compatible with the plan formulation procedure. If the legislative requirements are incomplete, outdated and relatively loose, planners can set additional requirements so as to improve the quality of management. Landfilling of the MSW is the minimum requirement for the protection of health and the environment. Prior to this, the recovery of recyclable materials at the source or from commingled wastes at material separation facilities represents a good, and often profitable, practice.

5.3 Selection of Treatment Technologies

The alternative treatment methods need to be screened so as to select these, and/or their combinations, that are capable of fulfilling all legal and other management objectives. The selection can be assisted by the description of alternative technologies and the presentation of typical material and energy balances in Sect. 2.2 and by the citation of alternative management schemes and their combinations in Sect. 4.1.

The selected methods and their combinations can be pre-screened, using for this purpose the cost data given in Sect. 3.2, so as to select the most cost-effective alternatives.

5.4 Sitting of IWMFs

The graphical methodology presented in Sect. 3.1 can be followed for determining the number and the approximate location (within the selected overlay areas) of the IWMF sites required for each type of technology considered. This methodology is directly applicable in cases where the management schemes involve landfilling, as in Fig. 15a, or simple treatment with landfilling of inert residues. The same methodology can be also used in more complex management schemes, if properly applied. For example:

- In management schemes with incineration plants, which generate toxic residues requiring costly disposal, as in Fig. 15b, the search for the IWMF sites can be based on the waste incineration plants. The toxic waste disposal facilities, which are very few due to their strong economy of scale and the small quantities of residues involved, can be built in the sites with the largest incineration plants.
- In management schemes involving aerobic or anaerobic MBT plants producing RDF, along with RDF incineration plants generating toxic residues, as in Fig. 15c, the search for the IWMF sites can be based on the aerobic MBT plants. The SRF incineration plants, which are far fewer due to their strong economy of scale and the limited quantities of RDF involved, can be built in the sites of the largest aerobic MBT plants. The still fewer toxic waste disposal facilities can be built in the sites with the largest incineration plants.
- In management schemes involving biological drying plants producing SRF, along with SRF incineration plants generating toxic residues, as in Fig. 15d, the search for the IWMF sites cannot be based on the biological drying plants as these do not

offer any significant economy of scale (see Sect. 3.2.1). Clearly, the optimal location of these plants is as close to the waste generation sources as possible. In view of this, the search for IWMF sites can be conducted for the SRF incineration plants and the analysis can be based on the transportation of the SRF, which is about 55% of the quantity of the bio-dried MSW (see diagram of Fig. 7).

The definition of the approximate, but not the exact, location of IWMF sites from the above procedure is sufficient for the purposes of the present strategic plan formulation and comparative evaluation phase, as no transportation system design is involved and only approximate transportation cost estimates are required (see Sect. 5.6.2). For the same as above reasons, there is no need to define the transfer stations sites at this stage.

5.5 Formulation of Strategic Management Plans

The selection of alternative treatment technologies in Sect. 5.3 and the definition of the relevant sites for treatment and disposal installations in Sect. 5.4, specify the shape of the alternative management schemes to be considered. For each such scheme, planners can proceed to define the flow of wastes, products and residues around each type of installation (e.g. primary treatment, secondary treatment and/or final disposal), taking into consideration the following:

- The annual waste loads generated in the study area.
- The quantities of materials recovered at the source.
- The normalized material and energy balances for each type of installation, such as these given in Figs. 5, 6, 7 and 8 and in Table 1, but adapted to the waste composition in the study area and to the product yields emanating from the legal and other objectives that have to be fulfilled.

If a different management plan already exists, it can be defined in a similar way so as to be evaluated, along with the newly formulated ones.

5.6 Comparative Evaluation of Alternative Plans

The alternative plans, formulated according to Sect. 5.5 above, need to be evaluated in relation to their compatibility with the applicable legal and other management objectives, implementation cost, environmental friendliness and public acceptance characteristics.

5.6.1 Compatibility with Legal and Other Management Objectives

Each alternative management plan should be reviewed, with its material and energy balances checked, to ensure that all legal and other management requirements defined in Sect. 5.2 are fulfilled. Possible uncertainties, assumptions and deviations must be noted and further elaborated and plans with unresolved compatibility problems should be rejected.

5.6.2 Economic Performance

The economic performance of each alternative plan can now be analyzed. The objective of the analysis is the estimation of the capital investment, the annual operating costs, the annual revenues from products and incentives and, based on the above, the net annual cost of transportation, treatment and final disposal. This can be based on the data from the material and energy balances (see Sect. 5.5) and on the cost functions and information given in Sect. 3.2, which may have to be adapted to local conditions.

The estimates from the use graphs in Figs. 12 and 13 can be based on representative (e.g. average) plant sizes. For assessing the latter, the number of installation defined in Sect. 5.4 need to be considered.

5.6.3 Environmental Friendliness

On the local scale, the potential air, water and land pollution problems can be reviewed on the basis of information provided in Sect. 2.2 and in the literature. The existence of local infrastructure and enforcement mechanisms should be evaluated so as to ensure that proper management practices can be implemented and maintained, especially when toxic emissions, effluents and solid residues are involved.

On a global scale, the EU Directive 2008/98 has made mandatory the following hierarchy in waste prevention, management legislation and policy: (a) prevention; (b) preparing for re-use; (c) recycling; (d) other recovery; and (e) disposal. The above management priorities lead into the formulation of sustainable policies, with due consideration to environmental pollution and resource conservation issues. Each alternative plan can be classified in a hierarchy level according to the conditions specified by the Directive, and this provides a sound measure of its relative environmental friendliness on a global scale.

5.6.4 Social Acceptance

The administrations often blame people for reacting unfavorably to plans for building waste treatment and disposal installation in their neighborhood, but rarely admit that the strongest objections are often justified and can be effectively addressed though proper planning. In view of the above, alternative management plans need to be evaluated in terms of their social acceptance characteristics, taking into consideration parameters, such as the following:

- The suitability of sites selected for treatment and final disposal installations. As only the approximate location of sites is known at this stage, their suitability cannot be assessed. This however does not apply in the case of existing management plans and, in the case of new plans, for sites that may be known, right from the beginning, as particularly suitable (see Sect. 3.1).
- The environmental friendliness of the management scheme, as discussed in Sect. 5.6.3.
- The beneficial use of products. For example, the use of biostabilized organics for the restoration of disturbed lands.
- The minimization of road traffic problems through the use of transfer stations and especially rain transportation.
- The direct and indirect economic and other development possibilities offered to the local communities hosting IWMFs. Of major importance are the new jobs created by the waste management installations and services involved.

6 Development of Detailed Optimal Plans

The present section deals with the management of the MSW that remain after the application of material recovery at the source programs and considers the waste transportation (from the point where the packer vehicles complete their collection programs), treatment and final disposal.

The objective here is to describe a procedure for developing optimal plans that meet all legal and other management objectives with minimal cost (i.e. the sum of the annualized capital investment and annual operating cost of all transportation, treatment and disposal operations, taking into consideration the possible income from the sale of products and/or financial incentives).

The development of optimal plants is assisted by software systems, which are designed to fulfill some or all of the interacting planning requirements highlighted in Sect. 4.2. Ideally, a relevant software system should be able to:

- Consider all alternative treatment technologies of interest, with their specific products and yields, and define the best combination of technologies and the waste flow through them. The solution yields the optimal location and size of each pre-treatment, treatment and final disposal installation, along with their input sources and output receivers.
- Define in a similar way the optimal transportation system for the MSW, intermediate products and rejects. The system often comprises a network of transfer stations and may combine road, railroad and marine transport media.
- Consider all existing infrastructure and define its optimal use within the overall management plan.
- Develop optimal solutions subject to constraints, such as landfill holdup capacities, site-specific capacity limitations etc.

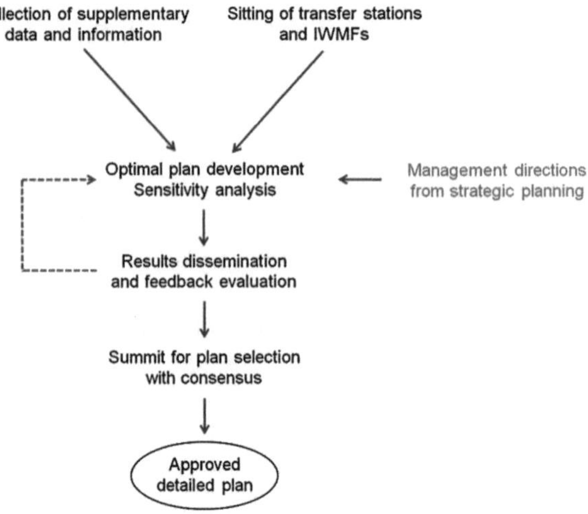

Fig. 17 Procedure for formulating detailed optimal plans

- Develop optimal dynamic plans in cases where some of the basic design parameters, such as the waste quantities and/or the management objectives, change significantly with time.
- Provide a comprehensive cost analysis for each optimal plan generated.

When an optimal solution is produced, the decision makers often wish to know the technical and economic impact of specific limitations reflecting the desires and objections of the local authorities (e.g. exclusion or imposition of a technology, exclusion or imposition of some sites, imposition of capacity limits etc.). To cope with this requirements, a series of optimal management plans can be developed, each of which meets a set of constraints. The comparison of these optimal plans at the municipal, prefectural and regional level constitutes the sensitivity analysis. The latter increases the planning efforts, but provides invaluable information to the decision makers, helping them to balance social and other preferences against costs and thus to select the most appropriate solution.

A rational procedure for the development and approval of optimal plants is illustrated in the diagram of Fig. 17.

According to this, planners need to define the exact location of the candidate sites, collect the necessary data and information and, on the basis of the management directions defined in the strategic planning phase, develop their detailed optimal plan and perform sensitivity analysis. The alternative plans, along with the associated cost data, are presented to the local authorities and the public and reviewed by relevant services (archaeological, forestry etc.). Depending on the feedback received, new plans may have to be developed so as to address more effectively public preferences and concerns and/or to comply with additional (e.g. land use) restrictions. A summit may then be convened with the objective of selecting one of the alternative plans presented, taking into consideration the relevant technical and economic data. The existence of alternative plans allows the participants to select the plan, not necessarily the least expensive one, which balances, in the best way, their preferences and objections against the associated costs. During this meeting, financial and other incentives can be negotiated until a decision is reached, hopefully with the concession of all parties involved.

The steps involved in the formulation of detailed optimal plans are described in more detail in the sections that follow.

6.1 Sitting of Installations

6.1.1 Sites for Transfer Stations

Sites for transfer stations are relatively easy to be located, even within congested urban areas, as the relevant nuisance problems are minor. In most cases, these sites are bordering highways, train terminal stations and/or ports, depending on the transport medium considered. The candidate sites should have a good geographic distribution and be easily accessible by the municipalities likely to be served. If possible, a somewhat excessive number of candidate sites is preferable, so as to allow the optimization program to select the most appropriate among them.

6.1.2 Sites for Treatment and Disposal Installations

The location of sites suitable for treatment and/or for disposal installations presents difficulties, especially for the latter, and requires considerable attention. Proper sites must meet a multitude of suitability criteria, depending on the kind of installation to be accommodated and must not create significant nuisance problems, Economopoulos [6]. More specifically:

The first step in the search of candidate sites is the definition of a carefully balanced set of suitability criteria. These must be strict enough for effective environmental protection, but not unduly strict so as to prevent the elimination of potentially suitable sites. Following this, the entire study area can be searched with GIS so as to locate the sites which meet all criteria. The GIS performs successive spatial operations using for this purpose spatial information relevant to the suitability criteria. The quality of the search results depends on the availability, completeness and accuracy of the relevant spatial information.

The second step involves field visits to the selected sites by a team of experts, so as to examine the site for apparent geological or other problems. Ideally, this step can be omitted if the available spatial information is complete, accurate and up-to-date, but this is rarely the case.

The third step involves the evaluation of the nuisance problems generated by each site. A key criterion is the visibility of the site from surrounding settlements, primary and secondary roads and archaeological sites. This task is aided by GIS, which can depict, in local maps, all areas with visibility to at least some part of the site under consideration. Additional criteria are the number, the size and the distance of the nearby settlements and the accessibility of the site through roads passing outside of the settlements. The morphology or the area needs also to be considered so as to detect potential problems, as for example conditions that affect adversely the dispersion of odors and emissions. GIS can provide significant input to all these evaluation tests.

Through the above procedure, the sites are screened so as to select the most suitable ones within or around the areas defined in Sect. 5.4. Unavailability of proper sites within or near these areas can be covered by sites in other locations that provide a reasonable overall spatial distribution. If additional sites turn out to be suitable, they can also be considered so as to allow the optimization program to select the most appropriate among them.

Generally, the total management cost of the resultant optimal plans is not very much dependent on the particular location of sites, as long as a sufficient number of candidate sites with good geographic coverage are available.

6.2 Collection of Data and Information

The kind and form of the input data required of optimal planning depend on the available computer system, its design and capabilities. For a computer system designed to perform the functions outlined in Sects. 4.2 and 6, input data, such as the following are required, Fig. 18:

- Quantities and composition of wastes. This information is normally collected at the strategic planning phase (see Sect. 5.1), but possible deficiencies in resolution, trends and general quality will have to be completed.
- Temporal distribution of the waste loads collected by the packer vehicles in each municipality. This is required for the design of the waste receiving and temporary storage units of transfer and treatment installations and can be expressed as maximum weekly loads, daily distribution of the weekly loads and hourly distribution of the daily loads.
- Location of existing installations, along with key data about their equipment, and performance characteristics. This information can be collected through specialized questionnaires for each type of installation.
- Location of candidate sites for the construction of transfer, treatment and final disposal installations, along with key characteristics (e.g. holding capacity of landfills) (see Sect. 6.1).
- Definition of alternative optimal routes that connect municipalities to sites and sites to sites, along with data about distances, mean velocities and mean transportation times. This information is normally produced through the use of GIS.

In addition to the above, information is required about the desirable sets of constraints related to the objections and/or preferences of the local communities so as to be used for the sensitivity analysis (see Sect. 6).

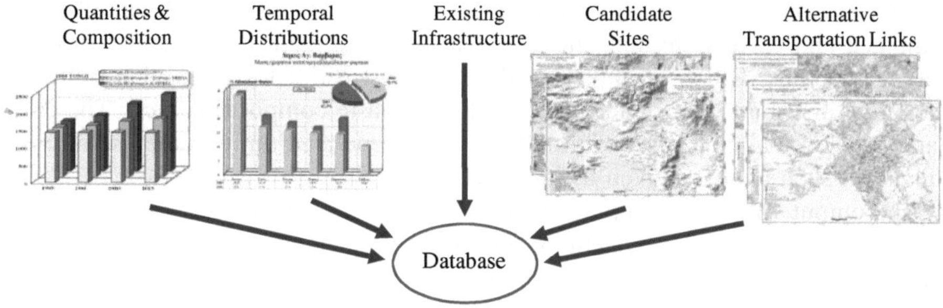

Fig. 18 Input data requirements by the plan optimization program

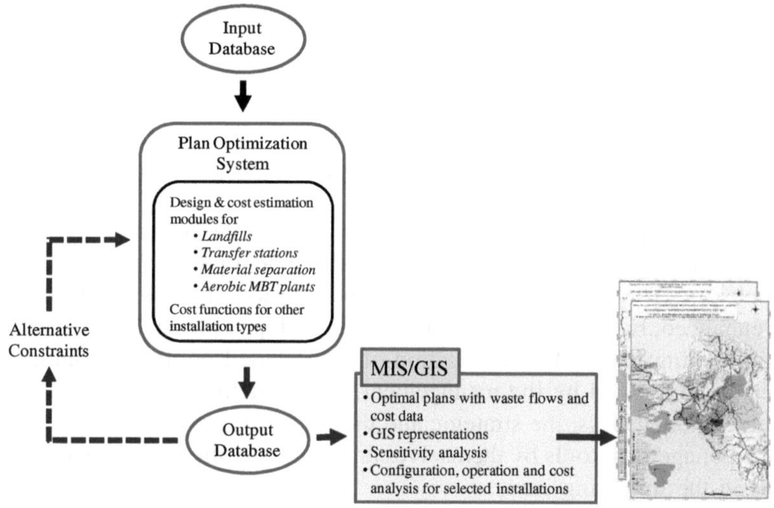

Fig. 19 Schematic flow chart of the plan optimization system illustrating selected eatures and functions

6.3 Development of Optimal Management Schemes

Depending on the sophistication of the available software some or all of the optimal planning aspects mentioned in Sects. 4.2 and 6 can be fulfilled. Most software tools can deal with the optimization of the waste transportation system, if everything else is specified.

The diagram of Fig. 19 illustrates some selected features and functions of a computer system designed to perform the tasks described in Sects. 4.2 and 6 and, in addition, to provide the optimal configuration and operation of selected types of installations, Economopoulos [5] and Economopoulou et al. [13].

This system comprises modules that estimate the capital investment and annual operating cost of each installation involved in the plan. The required cost data for transfer stations, material separation units, aerobic MBT plants and landfills are estimated analytically through the use of appropriate plant design and cost estimating modules, which are able to adapt the cost estimates to local conditions (temporal distribution of collected waste loads, waste composition etc.). The cost data for the remaining types of installations are generated through the use of cost functions, such as these given as in Sect. 3.1.

The output from the computer system is stored in a database, which feeds a Management Information System (MIS). The latter generates a number of reports, including reports with material flows and economic data that define the optimal solutions, reports that allow convenient comparison of the alternative plans at the Municipality, Regional and National level (sensitivity analysis), and reports with the indicative configuration, operation and cost analysis of selected installations (transfer stations, material separation units, aerobic MBT plants and landfills). Finally, GIS is used for the graphical representation of the optimal solutions, Economopoulos [5].

7 Discussion

The consecutive application of the two planning phases described in Sects. 5 and 6, enables planners to deal virtually with any management problem, irrespectively of its complexity. The strategic planning can be assisted, and the credibility of its results enhanced, by the use of software tools developed primarily for the detailed optimal planning phase. The application of the detailed optimal planning can be significantly simplified by the management options defined in the strategic planning phase. Nonetheless, the strategic and the detailed optimal planning constitute valuable management tools by themselves and can be applied independently, as for example in the following cases:

- If the resources are limited, planners could perform at least the strategic planning and, on the basis of its results, develop their detailed plan in the traditional way. The definition of the near optimal number and locations of sites, the screening of alternative treatment schemes and the formulation and comparative evaluation of alternative management plans of the strategic planning, contribute significantly towards the rationalization of the eventual plan.
- In not too complex studies, in which the alternative management options are not excessive, planners can proceed directly in the formulation of the detailed optimal plan, skipping the strategic planning phase.

The detailed optimal planning procedure is normally used for the formulation of integrated management plans, but it can be also used for optimizing some aspects of an existing plan, e.g. for optimizing the waste transportation system.

References

1. Beigl P, Wassermann G, Schneivder F, Salhofer S (2003) Draft waste generation prognostic model/the use of life cycle assessment tool for the development of integrated waste management strategies for cities and regions with rapid growing economies. Contract number: EVK4-CT-2002-00087 EU
2. Beigl P, Wassermann G, Schneivder, Salhofer S (2004) Forecasting municipal solid waste generation in major European cities. International environmental modeling and software society (IEMSS). In: Conference proceedings, 14–17 June 2004. University of Osnabruck, Germany
3. Beigl P, Lebersorger S, Salhofer S (2008) Modeling municipal solid waste generation: a review. Waste Manage 28:200–214
4. Daskalopoulos E, Badr O, Probert SD (1998) Municipal solid waste: a prediction methodology for the generation rate and composition in the European union countries and the united states of America. Resour Conserv Recycl 24(2):155–166
5. Economopoulos AP (2004) Formulation of optimal solid wastes management plans/ methodology and application in Attica. Book in Greek. Technical chamber of Greece
6. Economopoulos AP (2009) Sitting of integrated waste management facilities/a new methodology with an application in crete. In: 11th international conference on environmental science and technology, 3–5 Sept 2009, Chania, Greece
7. Economopoulos AP (2009) Formulation and comparative evaluation of alternative plans for the management of municipal solid wastes in Greece. Book in Greek. Technical university of Crete. http://library.tee.gr/digital/books_notee/book_60264/book_60264.pdf
8. Economopoulos AP (2010) Technoeconomic aspects of alternative municipal solid wastes treatment methods. Waste Manage 30(2010):707–715
9. Economopoulos AP (2010) A methodology for developing strategic municipal solid wastes management plans with an application in Greece. Waste Manage Res 28(11):1021–1033
10. Economopoulos AP, Naxakis G, Gouskos Z (2008) Quantities and composition of the MSW in Greece and management requirements in accordance with EU directives. In: 1st international conference on hazardous wastes management, 1–3 Oct 2008, Chania, Greece
11. Economopoulou AA, Economopoulos AP (2005) Transport distances versus economies of scale for municipal solid wastes treatment and disposal installations. In: 9th international conference on environmental science and technology, 3–6 Sept 2005, Rodos island, Greece
12. Economopoulou AA, Economopoulos AP (2011) A graphical method for defining the near optimal number and the approximate location of waste treatment and/or disposal installations. In: Proceedings CEMEPE/SECOTOX 2011 conference, 19–24 June 2011, Skiathos island, Greece, 467–473
13. Economopoulou MA, Economopoulou AA, Economopoulos PP, Economopoulos AP (2005) Optimal solid wastes management/part i: methodology and software infrastructure. In: 5th international exhibition and conference on environment "HELECO'05", Paper O-C31. Technical chamber of Greece, 3–6 Feb 2005, Athens, Greece
14. European Commission (2006) Integrated pollution prevention and control/reference document on the available techniques for waste incineration. http://eippcb.jrc.es/reference/
15. Eurostat http://ec.europa.eu/eurostat/waste
16. Juniper Consultancy Services Ltd. (2005) MBT: a guide for decision makers processes, policies and markets. http://www.juniper.co.uk/Publications/mbt_report.html
17. OECD (2007) OECD Environmental data, compendium 2006/2007, waste. OECD
18. Riber C, Pedersen C, Christensen TH (2009) Chemical composition of material fractions in danish household waste. Waste Manage 29:1251–1257

A Methodological Framework for Integrating Waste Biomass into a Portfolio of Thermal Energy Production Systems

Eleftherios Iakovou, Dimitrios Vlachos and Agorasti Toka

Abstract The integration of Renewable Energy Sources (RES) within the contextual framework of existing thermal energy production systems has emerged as a promising and sustainable policy towards addressing the growing global energy demand. Especially for developing countries, as they are characterized by decentralized energy systems, locally available RES are a viable option for generating thermal energy. In this chapter, we provide a methodological framework for integrating waste biomass into a portfolio of supply chains for thermal energy production, by presenting the relevant drivers for waste biomass usage making especially the case for developing countries, the associated systems and the supply chain operations. A generic strategic optimization model is proposed for determining the optimal mixture of energy sources for a specific region. This model could be employed by a system's regulator to conduct various 'what-if' analyses, in order to develop comprehensive effective policies that also integrate waste biomass into the existing energy system. Finally, a real-world case study is presented, and interesting managerial insights are discussed.

1 Introduction

Global energy consumption has been increasing steadily over the past few decades in response to population growth, economic development and improvement of life standards throughout the world. At the same time, the usage of fossil fuels causes numerous well-known environmental problems, while their reserves are

E. Iakovou (✉) · D. Vlachos · A. Toka
Aristotle University of Thessaloniki, P.O. Box 461 54124
Thessaloniki, Greece
e-mail: eiakovou@auth.gr

A. Karagiannidis (ed.), *Waste to Energy*, Green Energy and Technology,
DOI: 10.1007/978-1-4471-2306-4_3,

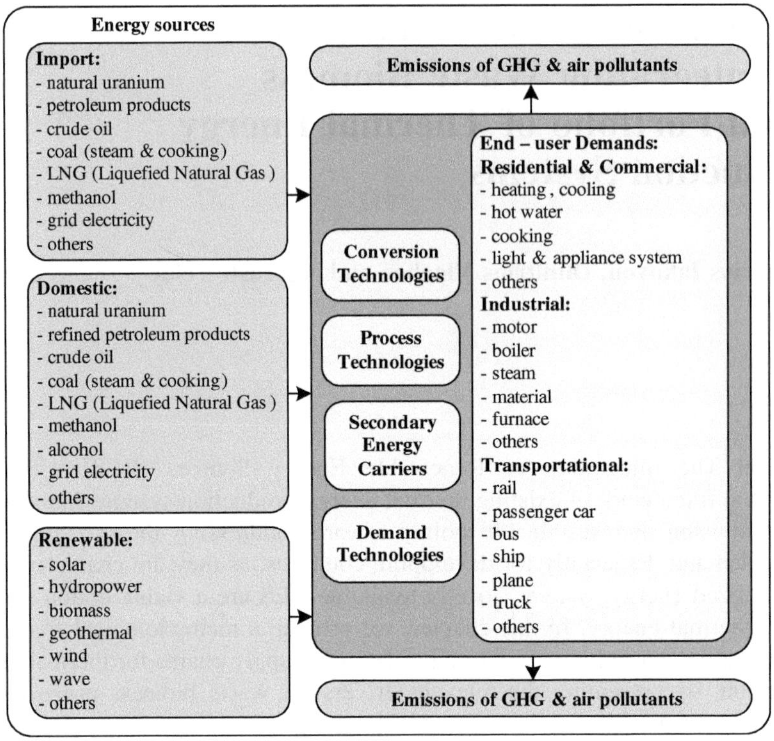

Fig. 1 Graphical representation of an integrated energy system (based on [4])

continuously shrinking and their price is increasingly volatile. This poses the challenge for decision-makers and regulators to allocate multiple energy resources within the existing energy management systems. A diversified energy mix can ensure a competitive market of fuels, and thus reasonable energy prices, while securing energy supply and managing the environmental and climate impacts [13]. Therefore, the integration of Renewable Energy Sources (RES) within existing energy systems appears as a promising policy. In this context, expanding access to clean and affordable energy into fast growing economies, as well as in many developing countries, has emerged as a critical issue. The complexity of contemporary energy systems with its range of potential energy sources for fulfilling multiple demand sectors is depicted in Fig. 1.

Especially, demand for heating accounts for a significant portion of the worldwide total energy demand. Locally available RES emerge as a viable alternative for thermal energy production even more so for transitional economies, as these are often characterized by a lack of centrally regulated energy systems. In this chapter, we address the integration of waste biomass in existing energy production systems for the production of thermal energy. Biomass utilization is hindered severely by its associated cost and the complexity of its logistics operations, as biomass is bulky and to develop the necessary capacity dispersed geographical generation points have to be coordinated. Moreover, weather dependent biomass quality, and the seasonality of agricultural crops pose unique challenges for that biomass supply chain management.

In this context, a comprehensive methodological framework is proposed for incorporating waste biomass into existing thermal energy production systems, taking into account all major managerial and technological aspects. At first, the main drivers along with the impact of such policies on developing countries are discussed (Sect. 2). The technologies of waste biomass thermal energy systems are presented in Sect. 3. More specifically, the alternative conversion routes for thermal energy production are presented, along with up-to-date applications with a focus on heating buildings. Following that, the several waste biomass supply chain operations are presented in conjunction with their key variables, unique characteristics, and relevant sustainability challenges. In Sect. 4, we present a first effort strategic optimization model that could be employed by a regulator to identify the optimal mixture of heating energy sources for various types of end-users on a regional level that minimizes total cost, while satisfying renewable and conventional energy sources' availability, demand, and CO_2 emissions constraints. Such a model could be used for the conduct of various "what-if" analysis by the system's regulator in order to develop effective interventionary policies (e.g. by offering tax-breaks, subsidies etc.), and thus promote the integration of waste biomass into existing thermal energy systems. For the case of Greece, one of the relevant regional sites that such a sensitivity analysis makes sense is that of the Prefecture of Pella. To that end, we implement the proposed model on the real-world case of this specific region, while results and managerial insights are discussed (Sect. 5). Finally, we wrap-up by summarizing in the last section.

2 Motivation

Biomass heat has a long tradition, often in domestic stoves or furnaces fed with round-wood and is estimated to account today for around 95% of the renewable heat produced [25]. In this section, the main drivers for incorporating waste biomass into thermal energy supply chains are presented. The European Union's (EU) relevant regulatory track is also presented to illustrate the effect of the current regulatory environment on the implementation of bio-heat applications in the Member States. Finally, the impact of waste to energy policies on developing countries is also discussed.

2.1 Waste Biomass for Thermal Energy Production

Below, we present the case for waste biomass utilization and its advantages compared to other alternative options.

Waste Biomass vs. Energy Crops Despite the attention that the production of biofuels from energy crops (also referred as first-generation biofuels) has attracted, a number of issues have emerged questioning the feasibility of this policy. According to the OECD (Organisation for Economic Co-operation and Development) and the United Nations' Food and Agriculture Organization [43], the increased demand for biofuels is causing fundamental changes to agricultural markets that drive up world prices for many farm products. Furthermore, concerns about the environmental degradation and economic sustainability of this production are also affecting the bio-ethanol business growth and its social perception [35].

On the other side, second-generation biofuels obtained by waste biomass are not plagued by these problems, while at the same time they can support effectively waste management policies. Second-generation biofuels are obtained from feedstocks not traditionally used for human consumption. As a result, there is much less concern about their use leading to famine in developing countries, or adversely affecting consumer prices in the developed nations. Aside from reducing the threat of food supplies being diverted to fuel production, second-generation biofuels are argued to be more environmentally friendly than first-generation biofuels [9]. In addition, the choice of feedstock is wide, including non-food parts of current crops, such as stems, leaves and husks that are left behind once the food crop has been extracted, as well as other crops that are not used for food purposes, such as switchgrass or cereals that bear little grain, and industry waste including among others wood chips, skins and pulp from fruit pressing [31]. Large quantities of agricultural residues are produced annually worldwide and are vastly underutilized. These residues are often burnt, left to decompose, or grazed by cattle, while they could be collected and utilized effectively for energy production [48]. Today, a new generation of biofuels is currently being investigated by modern companies to overcome the freezing of the biofuels industry. Instead of ethanol, companies plan to make hydrocarbons, which are molecules chemically much more similar to those that already power planes, trains and automobiles [51].

Waste Biomass vs. Other RES State-of-the-art scenario studies on energy supply and mitigation of climate change agree that all sustainable energy options are needed to meet the future world's energy needs and simultaneously reduce significantly greenhouse gas emissions. Sources such as wind and solar energy are promising, but their utilization is also constrained by their integration into electricity grids. In addition, electricity production from solar energy is still expensive. Hydropower has a limited potential and the utilization of geothermal and ocean energy has proved to be rather complex. Biomass on the other side can play a vital role in the production of carbon-neutral transport fuels of high quality as well as in providing feedstocks for various industries, thus being a prime alternative to the use of mineral oil. Therefore, biomass utilization can contribute significantly in improving the security of energy supply by substituting efficiently a large portion of oil demand.

Utilizing biomass for heat generation is also associated with high environmental benefits. When efficient combustion systems are employed, the emissions of pollutants with impact on the air quality can be significantly reduced compared to coal and heating oil. This is particularly important for densely populated urban areas, where central heat generation systems are located in close proximity to the consumers, in order to avoid large transmission losses. Furthermore, biomass can provide a constant energy supply, as it can be easily stored over long periods of time, unlike wind and solar, and price stability, unlike oil and natural gas. It is therefore expected that it will remain one of the most important renewable energy carriers for many decades to come [24].

Waste Biomass for Heating vs. Other Applications Certain core advantages of the use of waste biomass for heat generation can be identified, compared to the alternative applications of biomass and the utilization of other renewable and fossil fuels for heat generation on a regional level. With the recent progress in technologies, heat generation from biomass is roughly two times more energy efficient than the electricity generation from biomass. In addition, for technological reasons efficient power generation typically needs fuels with higher quality than that of the average quality of biomass fuels, and larger-scale systems than heat generation. With regards to the production of transport biofuels, the energy efficiency superiority of bio-heating seems to be even more evident [33]. Therefore, the integration of biomass in thermal energy supply chains can expand the share of the renewable energies in the energy mix, thus strongly improving the security and diversity of a region's energy supply.

The EU Paradigm According to the Intelligent Energy Executive Agency of the European Commission (IEEA) [29], the heating/cooling sector consumes almost half of the final energy in the European Union (EU), and almost as much as transport and electricity combined. Most of this thermal energy is produced from fossil fuels (oil, gas, coal). According to IEEA, in 2004, renewable sources of energy such as biomass, solar and geothermal energy provided 49 Mtoe (1 toe = 11.63 MWh), corresponding to 8.4% of total heat consumption. RES heating and cooling in the EU is dominated by biomass (>50% of which is household heat). Regarding solar energy, in the period up to the end of 2005 there have been over 16 million m^2 of collectors installed in the EU, but over 70% of this capacity has been installed just in three European countries, namely Germany, Greece and Austria. Moreover, in 2005, the share of geothermal energy sources was below 0.5% of the overall consumption of thermal energy in the EU-25 providing around 1.8 Mtoe.

In the last two decades considerable changes have taken place in the energy patterns of buildings due to higher standards of living, widespread utilization of domestic appliances and higher space heating demands. In addition, the increasing number of households following demographic and life-style changes towards smaller household sizes is expected to contribute to a rise of the energy demand in the near future. Thus, there is a promising potential for reducing carbon dioxide emissions and for saving energy by means of an integrating approach with traditional energy savings for buildings combined with environmentally friendly energy generation. The adopted solutions can allow less energy needs (and subsequently less CO_2 emissions), and clear economic benefits for the end-user. In fact, there is a close link between energy efficiency and the use of RES in buildings: few successful projects have demonstrated that by combining energy efficiency and renewable energy systems, it is possible to reduce drastically the demand for conventional energy sources in commercial and residential buildings [30].

The European policy framework clearly promotes renewable energy sources although unlike in the area of RES electricity or liquid biofuels for transport, the sector of renewable heating and cooling is not subject to dedicated European legislation. This is partly due to the major differences in heating and cooling demands, the existing infrastructure and availability of RES among various EU Member States. In this context, the Directive 2009/29/EC aims to improve and extend the greenhouse gas emission allowance trading scheme of the Community by further strengthening the incentives for the increased use of alternative energy sources [14]. The EU existing legislation that already promotes renewable energy heating includes the Directive on the energy performance of buildings (EPBD), which promotes the improvement of the energy performance of buildings, within the Community [12]. Among other directions, it is documented that for new buildings with a total useful floor area over 1000 m^2, Member States shall ensure that the technical, environmental and economic feasibility of alternative systems such as decentralized energy supply systems based on renewable energy, Combined Heat and Power (CHP), district or block heating or cooling, heat pumps, is considered and is taken into account before construction starts. The recently published Directive 2010/31/EU clearly states that new buildings should have a very high energy performance; the nearly zero or very low amount of energy required should to a very significant level be covered to a large extent by renewable sources, including energy produced on-site or near-by [15].

The interest for upgraded wood fuels, and consequently its European market, has improved and it is still growing as a consequence of the improved awareness for sustainable development, the behavior of the market of traditional fossil fuels, the availability of advanced boilers and burners, the development of new wood pellets production plants and combined heat and power plants, and the availability of advanced governmental subsidies [10]. Countries with a high percentage of biomass heat in the energy mix tend to have good resource availability and a strong district heating infrastructure such as in Sweden, Austria and Denmark [26]. The utilization of pellets (i.e. solid fuels with high energy content produced from biomass) in Europe is currently focused on a small number of member states including Sweden, Denmark, the Netherlands, Belgium, Germany, Austria and Italy. Indicatively, the market share of pellet heating systems in Austria grew within the last decade to over 12% of all new sold boilers for residential heating.

2.2 The Case for Developing Countries

The increased access to sustainable and affordable energy into fast growing and other developing economies has emerged as an issue of critical importance during the last decade. According to the International Energy Agency [28], non-OECD countries account for 93% of the projected increase in world energy demand, reflecting faster rates of growth of economic activity, industrial production, population and urbanization. China is a representative example, with its energy demand expected to contribute 36% of the projected growth in global energy use, rising by 75% between 2008 and 2035. India's energy consumption is also projected to more than double during the same time period. Thus, the growing need of fast developing countries for importing fossil fuels in order to meet their rising domestic demand is expected to have an increasingly large impact on the international market.

On the other hand, it is estimated that over 1.4 billion people currently have no access to electricity, the large majority of which are living in rural areas of developing countries, and depending mainly on traditional fuels to cover their energy needs. Access to electricity and other forms of energy is correlated to the achievement of the Millennium Development Goals (MDGs), a set of global targets promoted by the United Nations towards the reduction of poverty and improvement of living conditions [38]. Conventional rural electrification programs based on the extension of the electricity grid and/or decentralized schemes with diesel generators appear economically infeasible [20].

Although there is a wide-spread among developing economies in terms of socio-economic conditions, certain common (and rather unique) energy system characteristics can be found in most of them. These include amongst others: reliance on traditional energy sources, urban–rural divide with prevalence of inequity and poverty, structural changes of the economy and accompanying transition from traditional to modern lifestyles, inefficient energy sector characterized by supply shortages and poor performance of energy utilities, existence of multiple social and economic barriers to capital flow, and slow technology diffusion [3, 44, 52]. Rural remote areas in developing countries are especially characterized by low population densities, highly dispersed location of populated centres, low energy consumption levels per capita and thus, lack of access to centralized energy systems. Furthermore, in rural areas where waste biomass is widely available, decentralized thermal energy supply using local resources is a common practice, but implemented in a rather unorganised and non-systemic way. To that end, integrating waste to thermal energy production policies can have a wide impact on developing economies.

Developing regions could further benefit by developing waste biomass supply chains for heat production through substantial secondary advantages and synergies with other sectors. Biomass production can lead to the creation of direct and secondary employment, and business activity in rural areas, thus contributing to regional development and social cohesion. Due to the low energy density of biomass, the transportation, handling and storage costs of biomass are crucial for its competitiveness on the energy market. The closer to the biomass production sites the biomass consumption takes place, the lower the total bioenergy costs are. Because of the larger-scale units needed for efficient electricity generation and the lower distribution losses compared to those of heat, employing a larger part of the locally available biomass resource for heat generation (but not for electricity generation), appears to be more effective within the overall energy system taking into account energy saving, emissions and cost reduction points of view [33]. Finally, the production of second-generation biofuels obtained by waste biomass could support effectively regional waste management policies.

3 Thermal Energy from Waste Biomass

In this section, we first discuss generically thermal energy production from RES, and then follow with up-to-date applications in heating buildings. Then, the relevant supply chain operations are described in conjunction with their key variables and unique characteristics. Sustainability issues that emerge throughout thermal energy supply chains are also discussed.

3.1 Thermal Energy Production

RES utilization for thermal energy production is considered as an emerging opportunity of renewable energy potential on a global scale. Mature RES technologies using solar, biomass and geothermal resources are currently available as cost-effective means of reducing both carbon dioxide emissions and fossil fuel dependency under many circumstances. Demand for heating accounts for a significant portion of world total energy demand. The building sector itself consumes 35.3% of final energy demand, of which 75% is for space and domestic water heating.

According to the International Energy Agency, world solar thermal heat use is currently around 200–210 PJ/yr (4.8–5 Mtoe), geothermal heat is 260–280 PJ/yr (6.2–6.7 Mtoe), while heat from modern bioenergy probably almost 10 times the total of solar thermal and geothermal together (~4,000 PJ/yr; 80–100 Mtoe). Solar water heating, biomass for industrial and domestic heating, and geothermal heating systems are amongst the lowest cost options for reducing both CO_2 emissions and fossil fuel dependency [25]. RES converted to energy carriers that are then used to provide useful heating services are wide ranging (Fig. 2).

Generating energy from biomass (bio-energy) heat can involve complex pre-treatment, upgrading and conversion processes that can follow many alternative conversion routes from raw feedstock material to energy carriers (Fig. 3).

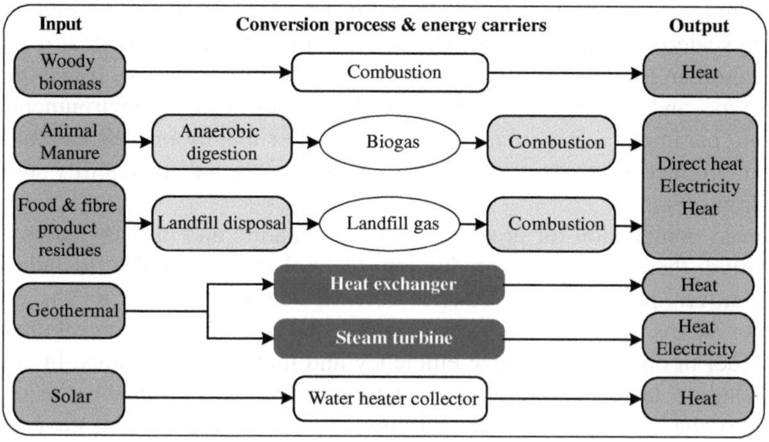

Fig. 2 Examples of RES for direct heat and combined heat and power (CHP) generation [25]

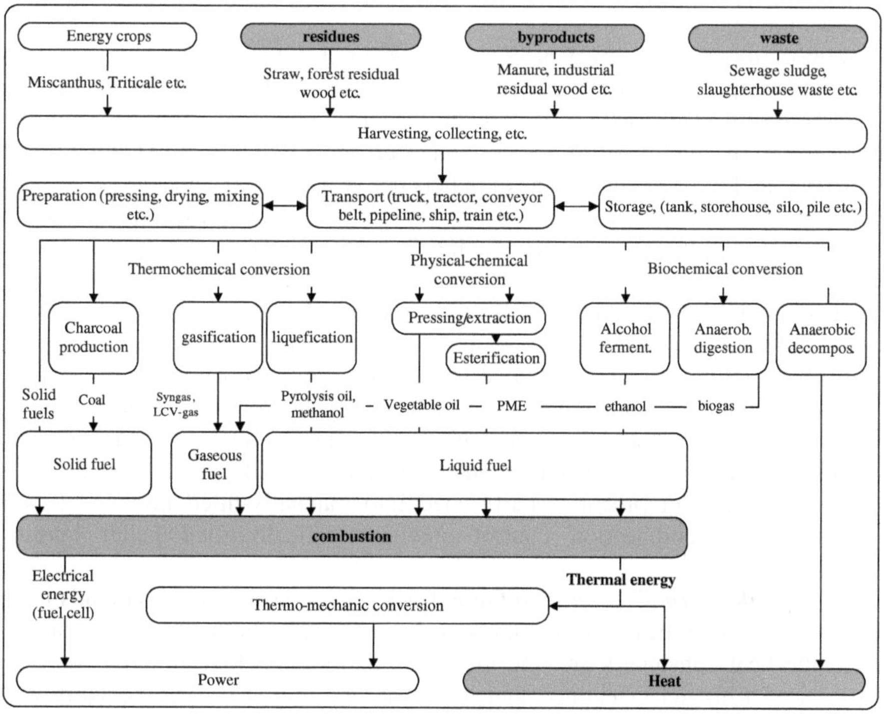

Fig. 3 Overview of renewable energy production from biomass (based on [32])

The choice of biomass-to-thermal energy production process depends mainly upon the scale of the production (whether it is intended for individual residential or/and commercial buildings, or for public or industrial systems). It also depends on the type and quantity of biomass feedstock, as well as environmental and economic parameters. Raw material products differ mainly in their suitability for different production processes, but also in their regional availability and their conversion costs.

Biomass combustion for producing heat is a mature and in many cases a quite competitive thermal energy production technology. Examples include wood burning stoves, pellet boilers and anaerobic digestion to produce biogas. Modern biomass systems such as wood burning stoves and pellet boilers control the mix of air and fuel in order to maximize efficiency and minimize emissions. In addition, they include a heat distribution system to transport heat from the site of combustion to the demand point. It is important that well-designed stoves and/or boilers are selected such that the combustion process is better controlled, thereby reducing carbon monoxide, pollution, hydrocarbons and particulate matter that may be associated with the burning of traditional biomass [25]. Another application of bio-energy is tri-generation for producing electricity, heating and cooling simultaneously, thus maximizing the overall conversion efficiency per unit of biomass [46].

Small-scale thermal energy production The simplest biomass burning system is the standard open fireplace. However, this primitive combustion concept has been almost abandoned, because of its very low energy efficiency, high dust and polluting emissions. Small-scale consumption refers mainly to residential pellets users with a demand of less than 10 tons of pellets per year. Pellets, which are solid fuels with high energy content produced from biomass, are used extensively for the heating of individual houses using pellet stoves and/or pellet boilers for warm water heating systems. A small-to-medium scale consumption refers to users with a demand between 10 and 1,000 tons per year (typical users are companies, hotels, larger residential units). Currently, a great variety of small-scale combustion facilities (with typical heat output of 6–25 kW, up to 50 kW for multi-family houses) is available. The smaller-scale facilities are less efficient than the larger-scale systems. The automatic systems using pellets and in fewer cases using straw, are the efficiency leaders in the small-scale sector. Finally, modern small-scale biomass burning equipment is partitioned into manually-filled firewood stoves with electronic combustion control and automatically-filled pellet burning facilities.

Large-scale thermal generation On a district level, heat generation tends to be less centralized, as heat distribution losses are typically high (the more concentrated the heat consumers are, the lower the distribution losses). Therefore, the capacity of a district's heating plant is determined by the nearby availability of heat consumers and on the nearby availability of feedstock, unlike e.g. the heat generation from fossil fuels. The heating plants, employing biomass, should always be designed based on a careful and thorough assessment of the feasible feedstock availability within a small radius, of around 80–150 km [33].

3.2 Waste Biomass Supply Chain Operations

Waste biomass-to-thermal energy supply chains possess several distinctive characteristics that differentiate them from conventional supply chains. Firstly, agricultural biomass types are usually characterized by seasonal availability, thus dictating the need of storing large amounts of biomass for lengthy time periods; this in turn leads to high inventory holding costs during the year-round operation of a power plant. Moreover, weather related variability and competing uses of waste biomass in a dynamically changing market have to be considered when determining the flows of the material supply network. The complexity of biomass supply chains is even higher for perishable products, as perishability affects profoundly both the acceptable transportation lead times and the length of storage time. Furthermore, most forms of biomass tend to have a relatively low energy density per unit of mass compared with fossil fuels. This often makes the handling, storage and transportation more costly per unit of energy carried. Waste biomass supply chains need to be robust and flexible enough to adapt to unpredictable changes in market conditions, as the demand of the produced energy depends on the type of the conversion facility and/or the price of competitive fuel substitutes. Iakovou et al. [22] present the natural hierarchy of this decision-making process and provide a relevant taxonomy of all research efforts as these are mapped on the relevant strategic, tactical and operational levels of the decision-making process. Iakovou et al. [23] present a review of advanced modeling techniques for biomass supply chains and a taxonomy of the existing modeling efforts.

The different operations that take place in the context of waste biomass-to-thermal energy supply chains span in general the following: biomass harvest-ing/collection (from single or several locations), pre-treatment (first-stage processing and/or pellets production), storage (in one or more intermediate locations), transport (using a single or multiple echelons), and small-scale or large-scale thermal energy generation processes (Fig. 4). Following, the system's operations are described and their special characteristics are discussed.

Harvesting and Pre-treatment The collection of biomass residues represents one of the most significant cost factors in waste biomass-to-thermal energy supply chains. The harvesting process is energy-intensive, primarily due to transport fuel costs. The moisture content of the biomass varies with the time of harvest and for some crops it can introduce additional processing costs, due to the need to pre-dry, before further processing [36]. The processing of biomass improves its handling efficiency and the quantity that can be transported, through increasing the bulk density of the biomass (e.g. processing forest fuel or coppice stems into wood chips) or unitizing the biomass (e.g. processing straw into bales). First-stage processing can occur at any stage in the supply chain but often precedes road transport and is generally cheaper when integrated with the harvesting [2]. The harvesting approach fundamentally affects the storage, handling and transport requirements in the biomass supply chain, as there is a very high level of interconnection among the different operations that take place. Several authors have

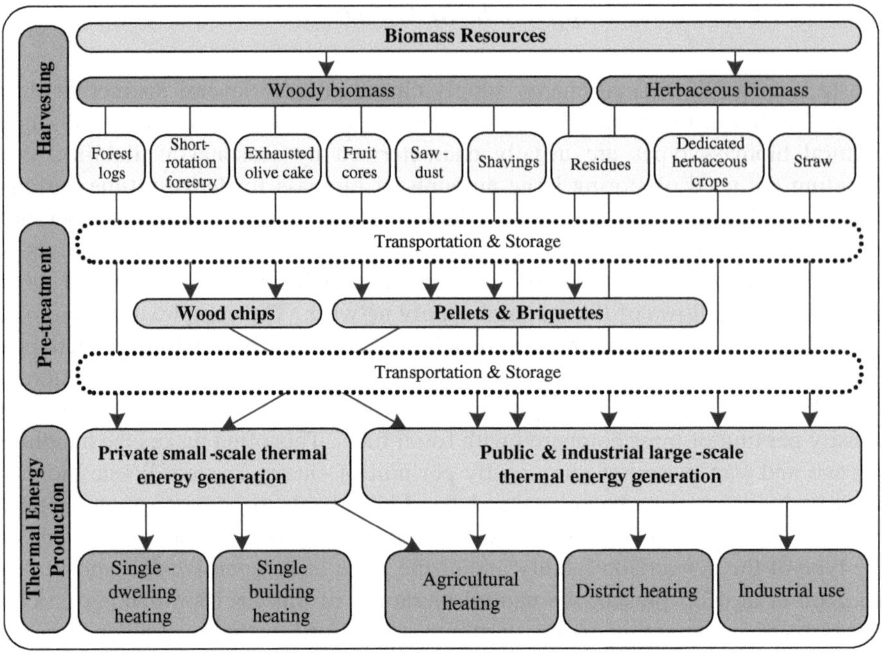

Fig. 4 Graphical representation of feasible waste biomass supply chains for thermal energy production

discussed harvesting with focus on methods, collecting machines, relative costs etc., along with storage or transportation issues, for specific biomass raw materials, such as switchgrass [7], forest fuel [11], cotton plant residues [16–18], herbaceous biomass in general [6], logging residue [42] and corn stover [47]. The additional processing of biomass on subsequent stages after harvesting is related to the conversion technology employed into the energy production facility or the pellet production plant. Unprocessed biofuels are essentially used as harvested for residential use to cover cooking, space heating, or electricity production needs. Processed biofuels can be used for a wide range of applications, including transport and high-temperature industrial processes [32].

Pelletizing Techniques for upgrading solid biomass fuels, which may be bulky with high moisture content, include natural drying, pelletising and briquetting [26]. Pellets are the outcome of a second-stage processing that takes place either in dedicated pellets production plants or in central district heating plants. Pellets are a solid fuel produced from biomass, at present mainly from wood residues (Fig. 2). However, as a result of the limited supply of wood residues for pellet production, attention is turning to using a variety of agricultural products as raw material. Such raw materials include cultivated energy crops and agricultural wastes and byproducts. Pellets are produced by a simple and fairly cheap process of milling, drying and compacting which requires small amounts of energy. The main advantages of pelleted biofuels, in comparison to unprocessed biofuels, are the higher energy density, with lower transport and storage costs, the more even quality for a constant moisture content, the higher mass fluidity, which allows for automatic feeding equipment being used even in small-scale boilers, and the smaller fuel particles, resulting in more even boiler feeding, lower emissions and longer boiler utilisation times [41]. As technologies for pellet production and pellets use are fully developed, the usage of pellets appears poised to grow dramatically [1].

Storage In the case of biomass that is harvested over a relatively short period of the year, such as straw and short rotation coppice, large quantities need to be stored in order that the supply of fuel is spread evenly on a year-round basis. This requires storage facilities that can be located on the farm/forest, at the pellet production plant, at the conversion facility or the final consumer, and/or at an intermediate site.

According to Rentizelas et al. [46], in most cases low cost storage solutions are chosen, such as on-field biomass storage [2, 21, 49]. Both ambient and covered on-field storage has also been examined [8]. On-field storage has the advantage of low cost. However, biomass material loss is significant and biomass moisture cannot be controlled and reduced to a desired level, thus leading to potential problems in the power plant equipment. Various studies consider the use of intermediate storage locations between the fields and the final conversion plant [2, 40, 50]. According to Allen et al. [2], using an intermediate storage stage may add 10–20% to the delivered costs, as a result of the additional transportation and handling costs incurred. However, in some cases intermediate storage may be inevitable as there are many biomass collection areas that cannot be easily accessed by road transport vehicles during wet winter periods, while on-farm storage does not constitute a viable solution.

Finally, the option of settling the storage facility next to the biomass plant has also been examined by several authors [45, 50]. Using storage facilities attached to the power plant is the only viable option of accelerating the drying process of the biomass, as dumped heat may be used without the need for extra energy consumption. However, most power stations or other energy production facilities to which biomass is supplied have limited on-site storage facilities, mainly due to the space required to stock large quantities of seasonal products that bears the physical and financial costs of holding stock [2]. In this case, inventory management should be effective enough in order to ensure that a few days of supply are available on-site with low risk of stock-out.

Rentizelas et al. [46] compared the above mentioned three biomass storage solutions found in the literature, in terms of total system cost. The authors suggest the development of a multi-biomass system, i.e. the joint utilization of various types of biomass or/and from different sources, aiming at reducing the storage space requirements. Finally, for small-scale biomass-to-heat production systems, the storage of pellets is considered as a barrier when densely populated areas are involved. Although in a compressed form, the bulky nature of the fuel requires large storage spaces, which are often not available in urban residential communities.

Transport The transport element of the biomass supply chain links together all the activities that have to take place between the point of production through to the point of use at the energy conversion facility or the individual buildings, and the locations at which they occur. Specifically, after the harvesting and processing of the material on first-stage, in-field/forest transport takes place to move the biomass to a point where road transport vehicles can be used. Once the biomass has been moved to the roadside, it will either have to be stored for some time or be directly transferred and loaded to road transport vehicles for transferring to the demand point or warehouse, where it has to be unloaded.

The transportation costs of supplying biomass to energy production facilities are mainly a function of the distance over which the material has to be moved, the type of transportation means selected to be used (trucks, ship or train), the type of biomass and the form in which it is transported (e.g. chopped or coppiced timber, compared with baled cereal straw), as well as the time spent for loading and unloading vehicles. Moreover, the size of the storage facility either on an intermediate location or at the final point of use affects the transport arrangements. The catchment area for the biomass collection points and hence the transport distance over which biomass has to be moved depends on a number of factors, including the size of the facility and the conversion technology used and thus, the quantity of biomass fuel required, the crop yield that is achieved, and the availability of the material for biomass resource.

Considering the typical locations of biomass fuel sources (i.e. in farms or forests) the transport infrastructure is usually such that road transport will be the sole potential mode for collection and transportation of the fuel. Additional factors that reinforce the use of road transport include the relatively short distances over which the fuel is transported and the greater flexibility that road transport can offer in comparison with other modes [46].

3.3 Sustainability

Energy supplied from biomass cannot be considered truly carbon-neutral even though the direct carbon emissions from combustion are offset by carbon fixation during feedstock photosynthesis. Indirect carbon emissions are also generated along the supply chain—especially by processing, transportation and burning which release emissions. The spatially distributed typical sources of biomass (farms, forests, etc.) along with their relatively low energy density, require the development of extensive logistical infrastructure and significant transport capacities for the efficient design of biomass supply chain networks. For regional biomass supply chains road transport is the usual mode for collection and transportation, affecting the level of sustainability of the biomass-based energy [34]. The environmental impact of biomass fuel supply is of pivotal importance as it relates to the main driver for using biomass fuel, namely being eco-friendly as opposed to traditional fossil fuels.

According to Allen et al. [2], public perception is often a significant factor in the acceptability and future development of an industrial or commercial activity and can influence location choices, land-use and transport planning decisions. Transport activities should therefore be planned as efficiently as possible in order to minimize their environmental impacts. Moreover, interdependencies among biomass logistics operations affect the decisions regarding the environmental impact of a biomass supply chain. For example, even if an alternative approach to storage and transport is known to be less harmful to the environment, the chosen harvesting system can prohibit its selection.

While around 2.5 billion people living in developing countries rely on biomass for household uses, an estimated 1.3 million die each year from the resulting indoor air pollution (from the emitted carbon monoxide, hydrocarbons and particulate matter) [27]. These fatalities could be reduced by the usage of stoves that can better control the combustion process and the filtering of exhaust gases. Emissions from modern wood stoves as used for domestic heating can also produce emissions which can cause local air pollution, especially when firewood with higher moisture contents is used. When well-designed, enclosed, domestic stoves are operated correctly, these emissions can be minimized. Finally, an additional advantage of pellet utilization is its low environmental impact which transcends its production process, its transport and its usage [1].

4 A Strategic Optimization Model

The integration of RES within the contextual framework of existing thermal energy production systems appears as a meaningful option for addressing the numerous problems that arise from the utilization of fossil fuels. In this section, we investigate the strategic problem of determining the optimal mixture of a system's energy sources for various types of end-users on a regional level, while exploring policies for increasing the contribution of RES in the energy sources mix.

Specifically, we present a first effort strategic optimization model that could be employed by a regulator to identify the optimal blend of heating energy sources for various types of end-users on a regional level. The objective function of the model is cost-based, while capacity, demand and environmental constraints are satisfied. The model could be used for the conduct of various "what-if" analyses by a system's regulator in order to develop effective regulatory policies (e.g. by offering tax-breaks, subsidies etc.), and promoting the integration of waste biomass into existing thermal energy systems. Furthermore, such analyses could also provide managerial insights regarding the merit of improving a region's access to alternative fuel sources (e.g. developing appropriate infrastructures and/or supporting investments for pellet production plants that utilize waste biomass). Finally, the proposed model could be used by a regulator to study the impact of setting CO_2 emissions quotas on the region's optimal mix of the energy sources.

4.1 System Under Study

We consider a geographical region with alternative energy sources i, where $i \in I$ including renewable and conventional forms and different types of local end-users j, where $j \in J$, which includes types of buildings further characterized by their level of annual heat consumption and energy efficiency. We assume that the energy consumption/demand for the different types of end-users is deterministic and constant, while the availability of the several types of fuels is limited. The latter makes sense for cases that either the energy source is too expensive to be transported over long distances, or there is no logistical network (e.g. there is no pipeline network in the region that could be used for the distribution of natural gas), or regulation prohibits the utilization of a specific type of fuel due to environmental concerns. Finally, due to cost limitations we assume that each type of end-users is supplied by a single type of fuel. In Table 1 we display the rest of all the employed nomenclature. In this context, it would be of interest to a decision-maker/regulator of the regional system, to identify the optimal mix of energy sources for utilization by the different types of end-users over a strategic time horizon. This decision-making tool could assist in developing policies supporting the increased usage of renewable energy sources into the existing portfolio. To that effect, public subsidy is a rather common practice for promoting the utilization of renewable energy sources both at national and regional levels. Public subsidies could be provided to the end-users either directly, including e.g. cash payments, or loans at below market interest rates or loan guarantees or indirectly, through the reduction of fees or taxes.

The decision variables are defined as follows:

x_{ij} :Quantity of fuel type i, consumed by end-users of type j (MJ)

$$y_{ij} = \begin{cases} 1, & \text{if fuel type } i \text{ is selected by end - users of type } j, \text{ and} \\ 0, & \text{otherwise.} \end{cases}$$

The objective of the optimization model is to minimize total fixed and operational costs of thermal energy production for all types of end-users utilizing a mix of locally available energy sources. The constraints of the model include demand, capacity, CO_2 emissions, modeling and logical constraints as well. The total fixed cost and operational cost parameters are defined by:

$$\mathrm{C}^{\mathrm{ij}}_{\mathrm{fixed}} = \mathrm{C}^{\mathrm{ij}}_{\mathrm{inv}} - \mathrm{C}^{\mathrm{i}}_{\mathrm{sub}} \tag{1}$$

$$\mathrm{C}^{\mathrm{ij}}_{\mathrm{oper}} = \mathrm{C}^{\mathrm{i}}_{\mathrm{fuel}} + \mathrm{C}^{\mathrm{ij}}_{\mathrm{maint}} \tag{2}$$

Following, we present our optimization model, (**P**):

$$(\mathrm{P}) \quad \min \sum_i \sum_j \left(C^{ij}_{fixed} \times y_{ij} \right) + \sum_i \sum_j \left(C^{ij}_{oper} \times x_{ij} \right) \tag{3}$$

s.t.:

$$\sum_i \left(a_{ij} \times x_{ij} \right) \geq D_j, \quad \forall j \tag{4}$$

Table 1 Nomenclature

i	Type of fuel, $i \in I$
j	Type of end-user, $j \in J$
C^{ij}_{fixed}	Total fixed cost of end-users of type j that consume fuel type i
C^{ij}_{inv}	Total investment cost (€) of the equipment of the heating system of fuel type i installed by end-users of type j
C^{i}_{sub}	Public subsidy (€) for installing heating system of fuel type i
C^{i}_{fuel}	Supply cost (€/MJ) of fuel type i
C^{ij}_{maint}	Net present value of the maintenance cost (€/MJ) of the equipment of the heating system of fuel type i installed by the end-users of type j
D_j	Thermal energy demand (MJ) of end-users of type j
Z_i	Capacity (MJ) of fuel type i
e^i_f	CO_2 emissions (tons/MJ) produced by consumption of fuel type i
Q_{co_2}	CO_2 emissions (tons/MJ) allowed to be produced by all end-users
a_{ij}	Efficiency factor of the heating system of fuel type i installed by end-users of type j

$$\sum_j x_{ij} \leq Z_i, \quad \forall\, i \tag{5}$$

$$\sum_i \sum_j \left(e^i_f \times x_{ij} \right) \leq Q_{co_2} \tag{6}$$

$$\sum_i y_{ij} = 1, \quad \forall\, j \tag{7}$$

$$x_{ij} \leq M y_{ij}, \quad \forall i, j \tag{8}$$

$$x_{ij} \geq 0, \quad y_{ij} = \{0, 1\} \tag{9}$$

The objective function (3) captures the total fixed and operational costs of the system. Constraints (4) ensure that the total fuel quantity consumed by each type of end-users satisfies the required thermal energy demand, while constraints (5) enforce that the total fuel quantity consumed by all types of end-users does not exceed the total capacity of each type of fuel. Constraint (6) maps the total CO_2 emissions constraint, as set by the regulator. Constraints (7) model the assumption that each type of end-users is supplied by a single type of fuel. Finally, constraints (8) and (9) are typical modeling constraints that link the x_{ij}'s with the binary variables y_{ij}'s, and logical constraints, respectively (with M being a satisfactorily large number).

(**P**) is then Mixed Integer Linear Programming (MILP) problem with a rather small number of binary variables $|I| \times |J|$, as the potential types of fuels and end-users are limited in general (<10 and <20, respectively), and thus can be solved easily by readily available software, such as LINGO®.

5 A Real-World Case Study

In this section, we demonstrate the applicability of the developed model with a real-world case study. Throughout the study, we employ real data and quite accurate quantitative estimates. More specifically, the optimal energy sources mix for heating non-industrial users is identified for the case of all 51 Prefectures in Greece, with the strategic horizon of 15 years. The energy sources considered are waste biomass (specifically pellets produced from locally available agricultural and forest residues), oil, natural gas and electricity. Herein, only due to space limitations we present the application of one typical Prefecture, namely that of Pella (Northern Greece).

Data relevant to the availability of various types of biomass were obtained from an open-access online database (National Information System for Energy) [39], developed by the Centre for Renewable Energy Sources and Saving (CRES) [5] of the Ministry of Environment, Energy and Climate Change [37] (www.ypeka.gr). Moreover, the National Information System for Energy provides updated information regarding the thermal energy demand of specific type of end-users per Prefecture. These data in conjunction with national statistical data, as recorded by the Hellenic Statistical Authority (EL.STAT.) [19] led to the identification of 14 types of end-users, as defined by the level of their annual heat consumption, the type of the building, the number and size of households per building, the year of construction and thus the energy efficiency per type of building. Additionally, the applied model cost parameters have been estimated based on the current state of fixed and proportional fuel prices of the market in Greece.

Moreover, certain regional regulatory and geographical constraints are considered. We assume that the energy consumption for the different types of end-users is deterministic but increases annually at a constant rate of $\sim 2\%$, based on historical data of energy demand (Ministry of Environment, Energy and Climate Change). The identification of the appropriate heating systems per type of building has been conducted under extensive local market research, while the relevant costs reflect current market prices in Greece. At first, the CO_2 emissions constraint is not taken into account. Specifically, for the application of the model to the case study of Prefecture of Pella, the optimal solution selects a single energy source, that of oil, for all the types of end-users (13 types appear in this Prefecture). However, the Prefecture is a region with significant agricultural production, and thus it is characterized by high levels of waste biomass supply. Therefore, it would be of great interest to assess the impact of potential interventionary policies that would support the usage of pellets (produced by the local biomass feedstock).

To that end, we proceeded by taking into account the national target of 20% reduction of CO_2 emissions, as dictated by the relevant EU Directive [14]. We consider as baseline quantity of the produced CO_2 emissions, the one that corresponds to the scenario of oil consumption by all types of energy consumers. The application of the developed model provides a solution that indicates the participation of waste biomass in the fuel mix at the percentage of 21.58%, while the rest of the energy demand is covered by oil. In this case, only half of the total

Table 2 Optimal solution (20% CO_2 emissions reduction considered)

Type of end-user		Fuel	% of total energy consumption
1	Detached: Local heating, <224 and 225–650 m^2 (ins[1]), 0–15 KW	Oil	49.65
2	Duplex: Local heating, <274 and 275–800 m^2 (ins)/Central heating, <149 and <440 m^2 (ins) 0–15 KW	Oil	21.44
3	Block: Local heating, <274 and 275–800 m^2 (ins)/Central heating, <99 m2 (ins) 0–15 KW	*Pellets*	20.07
4	Detached: Local heating, 225–390 m^2, 15–22 KW	Oil	0.09
5	Duplex: Local heating, 275–440 m^2/ Central heating, 150–199 m^2, 15–22 KW	Oil	0.18
6	Block: Local heating, 275–440 m^2 (ins)/Central heating, <49 and 100–149 m^2 (ins), 15–22 KW	*Pellets*	1.50
7	Duplex: Central heating, 200–350 m^2, 22–34 KW	Oil	0.05
8	Block: Central heating, 50–74 m^2 and 150–249 m^2 (ins), 22–34 KW	Oil	1.70
9	Block: Central heating, 75–99 m^2 and 250–350 m^2 (ins), 35–48 KW	Oil	3.26
10	Block: Central heating, 100–149 m^2, 48–64 KW	Oil	1.96
11	Block: Central heating, 150–199 m^2, 64–85 KW	Oil	0.08
12	Block: Central heating, 200–249 m^2, 85–100 KW	Oil	0.01
13	Block: Central heating, 250–355 m^2, 100–140 KW	Oil	0.01

[1] insulated

capacity of local waste biomass is utilized. In Table 2 the optimal solution is presented: the type of fuel and fuel consumption (GJ) per type of end-users. The optimal solution indicates that the 21.57% (20.07 + 1.50%) of waste biomass is allocated to the end-users of type 3 and 6 [building blocks of low (<15 kW/y) and medium (15–22 kW/y) consumption, respectively].

Various 'what-if' analyses related to the CO_2 emissions constraint to the levels of public subsidy on the fixed cost of equipment for utilization of pellets, and oil and pellets costs were conducted, to identify their impact on the optimal mixture of energy sources. Figure 5 exhibits the effect of the reduction of CO_2 emissions cap on the increased participation of biomass in the optimal energy mix.

As illustrated in Fig. 5, for a reduction of CO_2 emissions above 42%, no change is observed in the optimal energy mix, as at this point the capacity of waste biomass is fully utilized. Electricity is not included in the mix for any percentage of CO_2 emissions' reduction; this is intuitively sound, as the electricity's emissions per MJ are higher compared to those of oil (due to the high utilization of carbon-based raw materials for its production in Greece).

Conducting additional sensitivity analysis for the effect of public subsidy reveals that there is no effect on the optimal energy mix for any subsidy at levels of up to 40% of the total investment costs, for the current related values of the related equipment and operational costs. Thus, for this economic environment, public subsidy is proven inefficient for promoting waste biomass in the energy sources mix.

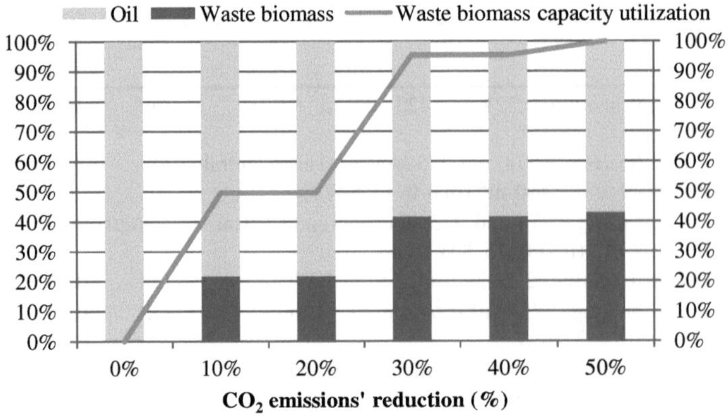

Fig. 5 Impact of reduction of CO_2 emissions on the optimal energy mix

Fig. 6 The impact of increase of oil cost on the optimal energy mix

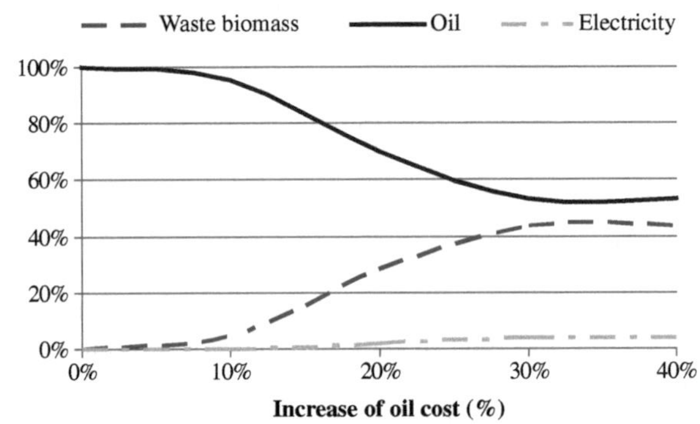

Fig. 7 Impact of decrease of pellet price on the optimal energy mix

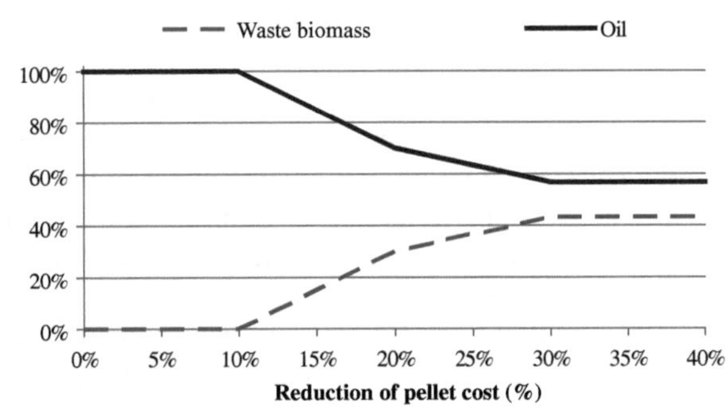

Figure 6 exhibits the optimal energy mix for various levels of oil cost. We observe that the portion of waste biomass increases as oil cost increases. When the increase in oil price is more than 20%, electricity emerges into the optimal energy mix. The optimal mix attains an equilibrium for oil price increases over 30% due to the usage of all available waste biomass.

Similar results are obtained when we examine the reduction in pellet cost. Waste biomass appears in the optimal energy mix at a 10% reduction of pellet cost and it continues increasing for pellet cost reduction of up to 21% (Fig. 7). While the equilibrium again is attained when all available biomass is used, in this case electricity never appears in the optimal solution.

6 Summary

The integration of Renewable Energy Sources (RES) within the existing thermal energy production systems has emerged as a promising and sustainable policy towards addressing the growing global energy demand. Locally available RES emerge as a viable alternative for thermal energy production even more so for developing countries, as these are often characterized by a lack of centrally regulated energy systems. In this chapter, we addressed the integration of waste biomass in existing energy production systems for the production of thermal energy. At first, we discussed the main drivers along with the impact of such policies on developing countries. Then, we presented the alternative conversion routes for thermal energy production, along with up-to-date small-scale and large-scale applications with a focus on heating buildings. Following that, the several waste biomass supply chain operations were discussed in conjunction with their key variables, unique characteristics, and relevant sustainability challenges.

Recognizing the emerging challenge for energy systems' decision-makers to integrate RES in the existing portfolio of energy sources, we investigated the strategic problem for determining the optimal mixture of a system's energy sources for various types of end-users on a regional level. Employing this, a system's regulator could obtain valuable managerial insights about the impact of potential regulatory interventions on waste biomass for thermal energy production systems. We proposed a first effort MILP optimization model that could be used for the conduct of various "what-if" analyses and the evaluation of the merit of various interventionary policies.

We demonstrated the applicability of the proposed model on a real-world case of the Prefecture of Pella, Greece, and documented that given the current fuel costs of the Greek market, oil consumption by all end-users is the optimal solution. We further conducted extensive sensitivity analyses on the merit of public subsidies, oil prices and pellet cost obtaining valuable managerial insights in order for a policy maker to effectively promote the inclusion of waste biomass into an existing energy portfolio mix.

Acknowledgments The authors would like to acknowledge the contribution of Mrs Anastasia-Loukia Grigoriadou in the collection and analysis of all relevant data, as part of her undergraduate thesis at the Department of Mechanical Engineering of the Aristotle University of Thessaloniki.

References

1. AEBIOM (European Biomass Association) (2008) A pellet roadmap for Europe. Available at: http://www.aebiom.org/IMG/pdf/Pellet_Roadmap_final.pdf. Accessed 24 Dec 2010
2. Allen J, Browne M, Hunter A, Boyd J, Palmer H (1998) Logistics management and costs of biomass fuel supply. Int J Phys Distrib Logistics Manage 28:463–477

3. Bhattacharyya SC, Timilsina GR (2010) Modelling energy demand of developing countries: Are the specific features adequately captured? Energy Policy 38:1979–1990
4. Cai YP, Huang GH, Lin QG, Nie XH, Tan Q (2009) An optimization-model-based interactive decision support system for regional energy management systems planning under uncertainty. Expert Syst Appl 36:3470–3482
5. Centre for Renewable Energy Sources and Saving (CRES). Available at: http://www.cres.gr/kape/index_eng.htm. Accessed 23 Dec 2010
6. Cundiff JS, Grisso RD (2008) Containerized handling to minimize hauling cost of herbaceous biomass. Biomass Bioenergy 32(4):308–313
7. Cundiff JS, Marsh LS (1996) Harvest and storage costs for bales of switchgrass in the southeastern United States. Bioresour Technol 56(1):95–101
8. Cundiff JS, Dias N, Sherali HD (1997) A linear programming approach for designing a herbaceous biomass delivery system. Bioresour Technol 59(1):47–55
9. Deurwaarder EP (2005) Overview and analysis of national reports of the EU biofuel directive: prospects and barriers for 2005. ECN (Energy research centre of the Netherlands). Available at: http://www.ecn.nl/docs/library/report/2005/c05042.pdf. Accessed 23 Dec 2010
10. Di Giakomo G, Taglieri L (2009) Renewable energy benefits with conversion of woody residues to pellets. Energy 34:724–731
11. Eriksson LO, Björheden R (1989) Optimal storing, transport and processing for a forest-fuel supplier. Eur J Oper Res 43(1):26–33
12. EU (2002) Directive 2002/91/EC of the European parliament and of the council of 16 December 2002 on the energy performance of buildings
13. EU (2006) Opinion of the European economic and social committee on the energy supply of the EU: a strategy for an optimal energy mix (2006/C 318/31)
14. EU (2009) Directive 2009/29/EC of the European parliament and of the council of 23 April 2009 amending Directive 2003/87/EC so as to improve and extend the greenhouse gas emission allowance trading scheme of the community
15. EU (2010) Directive 2010/31/EU of the European parliament and of the council of 19 May 2010 on the energy performance of buildings
16. Fischer T, Gaderer M (2000) Processes for storage and energetic use of cotton plant residues. In: Proceedings of the 1st world conference and exhibition, biomass for energy and industry. Seville, Munchen, Germany
17. Gemtos T, Tsiricoglou T (1992) Cotton residue harvesting and storage in Greece for energy production. In: Proceedings of the 2nd world renewable energy congress, vol 3. Pergamon Press, Oxford, pp 1409–1413
18. Gemtos TA, Tsiricoglou T (1999) Harvesting of cotton residue for energy production. Biomass Bioenergy 16(1):51–59
19. Hellenic Statistical Authority (EL.STAT.). http://www.statistics.gr/portal/page/portal/ESYE
20. Herran DS, Nakata T (2009) Optimization of decentralized energy systems using biomass resources for rural electrification in developing countries. In: Proceedings of the 32nd IAAE international conference, San Francisco. Available at: http://www.usaee.org/usaee2009/submissions/OnlineProceedings/Silva-Nakata-Proceeding.pdf. Accessed 23 Dec 2010
21. Huisman W, Venturi P, Molenaar J (1997) Costs of supply chains of miscanthus giganteus. Ind Crops Prod 6(3–4):353–366
22. Iakovou E, Karagiannidis A, Vlachos D, Toka A, Malamakis A (2010) Waste biomass-to-energy supply chain management: a critical synthesis. Waste Manage 30(10):1860–1870
23. Iakovou E, Vlachos D, Toka A (2010) Waste biomass supply chains for energy production: a hierarchical decision-making framework. In: Haghi AK (ed) Waste management: research advances to convert waste to wealth. Waste and waste management, Nova Science, New York, pp 81–120
24. IEA Bioenergy (2007) Potential contribution of bioenergy to the world's future energy demand. IEA bioenergy, France. Available at: http://www.ieabioenergy.com/MediaItem.aspx?id=5586. Accessed 23 Dec 2010
25. IEA (2007) Renewables for heating and cooling–untapped potential. OECD/IEA, Paris

26. IEA (2007) Member country database, international energy agency. OECD/IEA, Paris. Available at: http://www.iea.org/country/index.asp. Accessed 23 Dec 2010
27. IEA (2008) World energy outlook 2008. OECD/IEA, Paris
28. IEA (2010) World energy outlook 2010. OECD/IEA, Paris
29. IEEA (2006) Renewable energy heating and cooling: 21 innovative projects supported by the IEE programme. Available at: http://ec.europa.eu/energy/intelligent/library/doc/ka_reports/renew_heat.pdf. Accessed 23 Dec 2010
30. IEEA (2006) Small scale renewable applications: 15 innovative projects supported by the IEE programme. Available at: http://ec.europa.eu/energy/intelligent/library/doc/ka_reports/small_scale.pdf. Accessed 23 Dec 2010
31. Inderwildi OR, King DA (2009) Quo vadis biofuels? Energy Environ Sci 2:343–346
32. Kaltschmitt M, Thrän D, Smith K (2004) Renewable energy from biomass. Encyclopedia of physical science and technology, pp 203–228
33. Kavalov B, Peteves SD (2004) Bioheat applications in the European union: an analysis and perspective for 2010. European communities. Available at: http://ie.jrc.ec.europa.eu/publications/scientific_publications/2004/EUR%2021401%20EN.pdf. Accessed 23 Dec 2010
34. Lam HL, Varbanov PS, Klemes JJ (2010) Optimisation of regional energy supply chains utilising renewables: P-graph approach. Comput Chem Eng 34(5):782–792
35. Londo M, Lensink S, Wakker A, Fischer G, Prieler S, van Velthuizen H, de Wit M, Faaij A, Junginger M, Berndes G, Hansson J, Egeskog A, Duer H, Lundbaek J, Wisniewski G, Kupczyk A, Könighofer K (2010) The REFUEL EU road map for biofuels in transport: application of the project's tools to some short-term policy issues. Biomass Bioenergy 34: 244–250
36. McKendry P (2002) Energy production from biomass (part 1): overview of biomass. Bioresour Technol 83(1):37–46
37. Ministry of Environment, Energy and Climate Change. Available at: http://www.ypeka.gr/Default.aspx?tabid=37&locale=en-US&language=el-GR. Accessed 23 Dec 2010
38. Modi VS, McDade S, Lallement D, Saghir J (2005) Energy and the millennium development goals. Energy sector management assistance programme, United Nations, Development programme, UN millenium project, World Bank, New York
39. National Information System for Energy, Ministry of Environment, Energy and Climate Change, Greece. Available at: http://195.251.42.2/cgi-bin/nisehist.sh. Accessed 23 Dec 2010
40. Nilsson D, Hansson PA (2001) Influence of various machinery combinations, fuel proportions and storage capacities on costs for co-handling of straw and reed canary grass to district heating plants. Biomass Bioenergy 20(4):247–260
41. Nilsson D, Bernesson S, Hansson P (2010) Pellet production from agricultural raw materials—a systems study. Biomass Bioenergy. doi:10.1016/j.biombioe.2010.10.016
42. Nurmi J (1999) The storage of logging residue for fuel. Biomass Bioenergy 17(1):41–47
43. OECD/FAO Agricultural Outlook 2007–2016. Available at: http://www.oecd.org/dataoecd/6/10/38893266.pdf. Accessed 23 Dec 2010
44. Pandey R (2002) Energy policy modeling: agenda for developing countries. Energy Policy 30:97–106
45. Papadopoulos DP, Katsigiannis PA (2002) Biomass energy surveying and techno-economic assessment of suitable CHP system installations. Biomass Bioenergy 22(2):105–124
46. Rentizelas AA, Tolis AJ, Tatsiopoulos IP (2009) Logistics issues of biomass: the storage problem and the multi-biomass supply chain. Renewable Sustainable Energy Rev 13(4): 887–894
47. Shinners KJ, Binversie BN, Muck RE, Weimer PJ (2007) Comparison of wet and dry corn stover harvest and storage. Biomass Bioenergy 31(4):211–221
48. Sims R (2002) The brilliance of bioenergy: in business and practice. James & James (Science Publishers) Ltd, London
49. Sokhansanj S, Kumar A, Turhollow AF (2006) Development and implementation of integrated biomass supply analysis and logistics model (IBSAL). Biomass Bioenergy 30(10): 838–847

50. Tatsiopoulos IP, Tolis AJ (2003) Economic aspects of the cotton-stalk biomass logistics and comparison of supply chain methods. Biomass Bioenergy 24(3):199–214
51. The Economist (2010) The future of biofuels. The post-alcohol world. Biofuels are back. This time they might even work. 28 Oct 2010
52. Urban F, Benders RMJ, Moll HC (2007) Modelling energy systems for developing countries. Energy Policy 35:3473–3482

Modeling Waste Characteristics and WtE Plants as a Tool for Optimum Operation Conditions

Jasmin Kornau and Henning Albers

Abstract Material and energy flow networks are an applicable optimization tool for waste treatment and recycling management. Once the models have been complied and fed with waste and process data, the actual operation conditions and states can be calculated and visualized. Furthermore possible treatment results can be simulated, in case operation conditions have been modified. When modeling a plant as a material flow network using UMBERTO® all process steps can be described in detail, their treatment purposes be shown and potentials for the optimization be indicated. Specific optimization strategies for several approaches can be established, e.g. adjusting plant configurations, determining the optimal waste mixture or deducing material flow management strategies in order to reach optimal treatment results. This chapter will give an insight how the modeling for WtE plants has been performed and optimization potentials have been generated on basis of these models.

1 Introduction

Waste numbers are increasing worldwide. Most of the generated waste is dumped under insufficient conditions, often without any utilization of the material and energetic potential [4]. But even though environmental and health risks come along land filling is still the most common way of disposal. Especially in

J. Kornau (✉) · H. Albers
Hochschule Bremen–University of Applied Sciences,
Neustadtswall 30, 28199 Bremen, Germany
e-mail: jasmin.kornau@hs-bremen.de

H. Albers
e-mail: henning.albers@hs-bremen.de

A. Karagiannidis (ed.), *Waste to Energy*, Green Energy and Technology,
DOI: 10.1007/978-1-4471-2306-4_4, © Springer-Verlag London Limited 2012

developing countries these risks are distinctive due to a lack of minimum landfill standards. Treatment costs and low technical standards are the major causes of a poor waste disposal and treatment. However, there is an uprising market for waste-to-energy plants worldwide. Within the last decade the global treatment capacities have gone up from 180 to 350 Million tons. Further increase up to 420 million tons is expected within the next 5 years [2]. That development is basically driven by the force to find waste disposal solutions and sources of substitute fuels. By the application of those plants, land filling can be reduced and the energetic and material potential of waste will be enhanced. Environmental and health hazards are being minimized [1].

Optimum operation conditions are crucial for the operation of a WtE plant to run the process as effective as possible. The material and energy efficiency of a plant or waste treatment pathway is coming more in focus. The *material efficiency* indicates the level how intensive the material potential of wastes has been used and how economically the plant uses auxiliary supplies. The *energy efficiency* on the other hand expresses the relation between energy input (including energy from waste) and a plant's energy output. The evaluation of the material and energy efficiency is a suitable way to rate the plant or treatment pathway.

Using material flow networks modelled in UMBERTO®[1] operation conditions, actual operation states and all relevant material and energy flows can be displayed. In focus on minimizing environmental and health risks but rising the material and energy efficiency the presented flow networks of this chapter are a supportive management tool:

1. Material flows can be managed–it allows the control of waste streams, the determination, description and evaluation of the impact of waste input streams on the operating state and operating conditions of a plant.
2. Plants and complex treatment pathways can be evaluated–by rating a plant's material and energy efficiency decisions about the optimal treatment pathway for a specific waste fraction or mixture can be made.
3. The operation of a plant can be optimized—e.g. reaching the maximum energy decoupling, maximum availability or a higher efficiency. As each treatment module within a plant can be modeled in detail, optimization potentials of a single process step within the plant can be detected.
4. Simulations of future scenarios can be carried out—that may include the predictions about process outcomes when changing waste inputs or technical settings. These outcomes may also have an effect on the construction of a plant. The early assessment of optimization strategies can be involved into the construction plans.

Because the utilized waste composition has a major influence on the treatment effect it is important to lodge as much waste parameters into the models as possible.

[1] Umberto is a software tool for modeling, calculation, visualization, and evaluation of material and energy flows [3].

Table 1 Waste groups and fractions feeding WtE plants

Waste groups	Waste fractions[a]
Plastics-sorting residues	Plastics
Industrial and construction waste residues	Composite materials
High calorific fraction (HCF)	Composite packaging
Sewage sludge	Paper, cardboard, carton
Hospital wastes and hazardous wastes	Disposable nappies
Production wastes and specific industrial wastes	Textiles
Shredder light fraction	Furniture and wood
Municipal solid waste	Glass
Household-type commercial waste	Ferromagnetic metal
Bulky waste	Non-ferrous metal
Substitute fuel (RDF)	Inert
	Organic waste
	Fine particle fraction <40 mm

[a] As the waste composition is varying during different periods within a year (e.g. summer and winter) it may be necessary to repeat those analyses within the respective balance scope regularly and update all data in the model

Depending on the culture, local conditions, legal matters and consumption behaviors the waste compositions and characteristics can vary widely. Hence a local waste definition is necessary to create and run the models as a management tool.

As a long-term result the waste's energy potential will be used as efficient as possible to decouple the most achievable energy, save resources and minimize environmental and health hazards. By the successful use of material flow networks those effects can be predicted in advance and early action may be taken to improve operation conditions of WtE plants.

2 Waste Characteristics

An extensive waste characterization is necessary to manage wastes in a strategic way and to reach optimum operation conditions of a waste treatment plant. Just if conversions of different waste fractions within the system are known, strategies how to feed, control and operate a plant can be developed.

Waste is very heterogenic all over the world. Its qualities may vary from dry to wet, from organic to inert, from high calorific to low calorific. Therefore, the treatment result may be different even though the same technique is applied.

Firstly the waste needs to be distinguished by waste groups depending on its origin and nature. Those wastes need further be divided into its individual fractions. That will basically be realized by doing comprehensive waste sorting analyses. For the feeding of a WtE plant the following waste groups and fraction have been assumed (Table 1).

Table 2 Chemical and physical parameters for the waste characterization

Chemical parameters	Physical parameter
Arsenic	Particle size distribution
Lead	Aggregate state
Cadmium	Density
Chrome	Ash content
Chlorine	Calorific value
Copper	Moisture content
Nickel	Magnetic
Mercury	Non-magnetic
Antimony	Optical appearance
Sulfur	
Zinc	

All waste fractions and relevant materials can be set up as a structured material list within a project in UMBERTO®. That list can be copied and used for the modeling of other WtE plants. Subsequent changes are possible.

Afterwards chemical and physical parameters for each fraction need to be analyzed. They include indicators such as calorific value, ash content, heavy metals and moisture content as listed in Table 2.

The data will be entered into the model. Every process step gets assigned its relevant parameters. On request they can be integrated into the calculations carried out by the model.

For example–the process step *bunker* will be fed with several fractions of wastes having different calorific values. On basis of the mass and calorific value of each fraction, the formulas hidden in that process step will calculate the medial calorific value of the whole waste mixture in the bunker. In practice such scenario could be realized mixing the delivered waste by crane.

That mixture is transferred into the next process step e.g. combustion and influences basically the outcome of this step. The procedure runs throughput the model step by step. If the waste mixture changes, by the delivery of new fraction, different result in the combustions and consequently the whole remaining process can be expected.

3 Material and Energy Flow Networks

Material and energy flow networks are an applicable way to manage, balance and visualize waste flows within a specific system. All material and energy flows within the system boundary can be implemented into the model which can finally be used as a management tool.

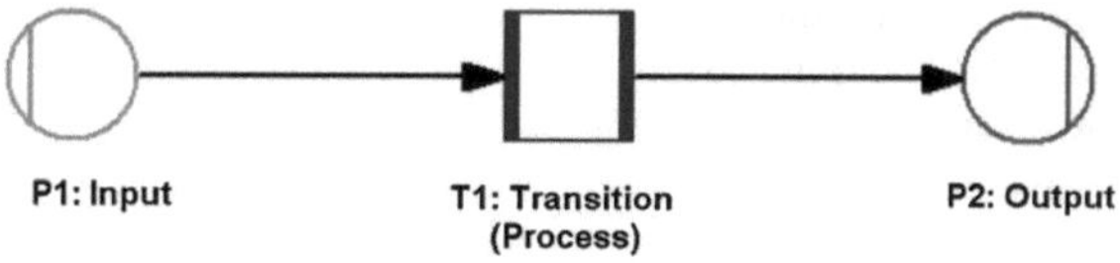

Fig. 1 Elements of a material flow network

The models are formed by input and output places, transitions and connection arrows as Fig. 1 shows.

Input places contain all input flows such as different waste fractions, additive fuels, electric energy, additional materials and operation supplies like air and water. The *output places* contain all process outputs like residues e.g. slag, fly ash, flue gas and waste water, as well as electricity, thermal heat for central district heating, process steam and energy losses. In the *transitions* all physical and chemical conversions of a process step are specified linear or nonlinear. Actual conversions from the plant will be assigned as functions to the transition. The connection between places and transitions can be done by *arrows*. They can either transmit data released by the transitions or work as manual flows where a specific amount can be entered. That allows limiting a throughput through an indicated flow line e.g. supply of a specific waste fraction or a maximum flow of combustion air.

As a plant model can get quite complex subnets may be created within a main network:

Figure 2 shows a subnet containing a complex furnace system of a WtE plant which is hidden behind the main layer (for clarity just one of four furnaces is displayed). With each layer a system can become more detailed. Via double-click those subnets can become visible. Their inputs and outputs are linked to the main model on the upper layer.

Subnets might even embody a whole plant. By the application of several plants working for the overall treatment aim those plant subnets can also be linked to complex process chains (Fig. 3). Waste fractions from a mechanical pre-treatment can be transferred into the WtE plant and residues can automatically be put on a landfill. After several plants have been linked together it will be possible to balance the process data of the complete system or just of a single plant within the system.

Figure 3 shows a complex process chain network containing several treatment plants which are linked to each other. Waste is treated in a mechanical–biological waste treatment (MBT) plant. The biological treated fraction goes directly to a landfill. The light weight fraction is transferred into the RDF plant and the remaining heavy weight fraction will be treated in a waste incineration plant.

After a model has been accomplished all lodged material and energy flows can be calculated and released as a balance. Balances can be visualized by UMBERTO® or exported into programs like Microsoft Excel®. Specific material and energy flows can also be displayed using *Sankey Diagrams*. Figure 4 illustrates the distribution of energy within the system. The waste's energy content is divided onto the furnaces. Furnace 1 (left) and 4 are fed with the same amount of

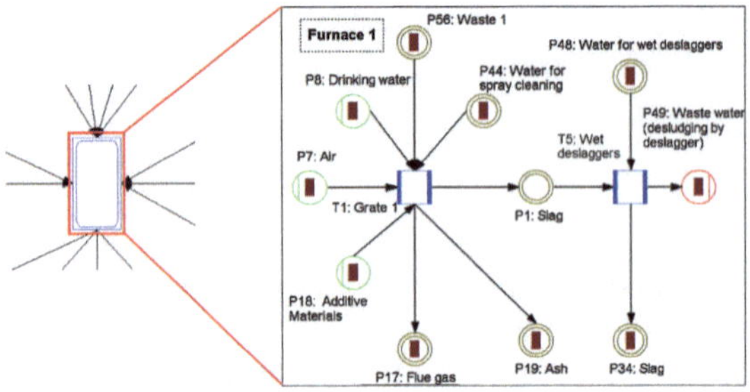

Fig. 2 Implied subnet on the main layer with its hidden subsystem

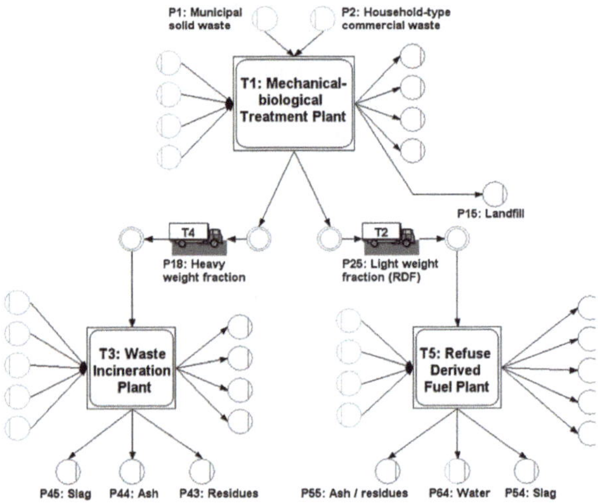

Fig. 3 Process chain containing several treatment plants

waste's energy. Furnace 2 receives less input while furnace 3 is not in operation. Energy containing steam is generated. Some energy gets lost through ash and slag.

3.1 Modeling a Network

Generally there are some basic steps which need to be followed by the development of a material and energy flow network:

1. Model structuring—building up the structure of the plant of interest.
2. Data pooling—generating, collecting and implementing waste and process data.
3. Balancing and displaying—all material and energy flows can be balanced and displayed upon request.

The plant of interest can be created by building up a new *model structure*. The modeling will follow the pattern as lined out in the section above using places,

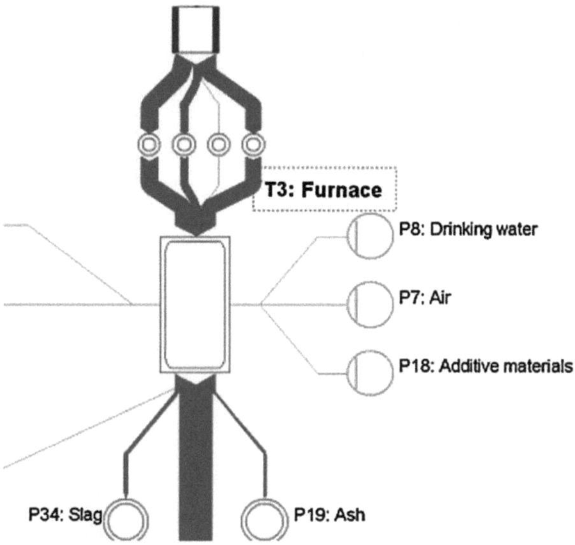

Fig. 4 Displaying energy flows as a Sankey diagram

transitions and connection arrows. Alternatively prepared modules for a WtE plant can be used and configured to specific demands and conditions. After a structure has been built up specific parameters, generated by an extensive *data pooling*, need to be implemented into the model. That includes:

- Waste parameters—e.g. composition, physical and chemical characteristics.
- Process parameters—e.g. days of operation, operating hours, air demand, electricity usage and maximum feeding rate as well as detailed information about the conversions within each process steps.
- Economical parameters—e.g. costs for additive materials, disposal and external obtained energy.

Waste parameters need to be generated by a waste characterization as already indicated in this chapter. Process data need to be generated by an extensive technical data pooling at the plant. Some information might also be taken from the UMBERTO® database. Economical data can be ascertained by actual resource costs and accounting records.

Based on the functions and parameters entered into the model all material and energy flows in the system will be calculated on request. As a result the input places will show how much material is needed and what amount of waste fraction is fed into the plant. On basis of the used material and combusted waste the model is calculating the amount of residues and the emissions. Afterwards *balances* of the inputs and outputs can be shown by UMBERTO® or exported into a Microsoft Excel® worksheet. Results may be *displayed* visually. As a further option optimization strategies can be developed on basis of the available results.

Step 1 and 2 may vary as some elements of the system structure might only become visible while data generation.

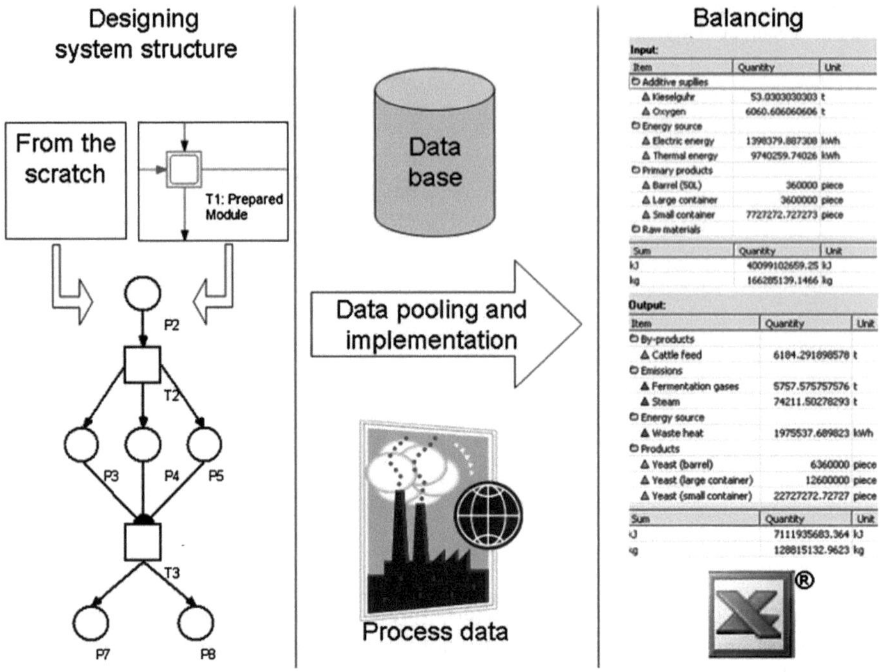

Fig. 5 Building up networks by stages

Figure 5 gives a schematic overview about the procedure. But not just existing plants can be modelled. Moreover, plants in construction or planning can be modelled by using prepared modules or cognate plants as pattern. It allows simulating future scenarios for the operation of a plant or the material flow management, especially waste streams. Optimization potentials can already be implemented in the construction of a plant, dimensioning of a whole treatment pathway or a strategic waste management.

3.2 Waste-to-Energy Plant Models

A plant will be divided into various sections within the model. Each section is formed as a module. The basic modules are:

1. Waste delivery and storage—waste fractions are delivered by different providers and stored in a bunker or remedy depot.
2. Combustion (including wet deslagger)—waste is combusted by the furnace system producing flue gas, slag and ash.
3. Steam generation—flue gas is heating up steam which is used for the operation of the turbines and internal requirements.

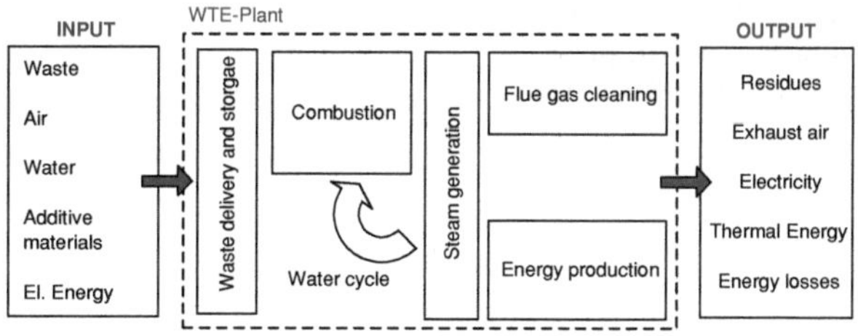

Fig. 6 Sections of a WtE plant

4. Energy production—electricity is produced by the operation of turbines; thermal process energy or energy for district and internal heating is recovered from waste steam.
5. Water cycle—condensates and communal water is purified for steam generation.
6. Flue gas cleaning—flue gas is cleaned by different techniques such as spray absorber and different filters (electric filter or baghouse filters).

Figure 6 shows schematic basic modules of a WtE plant.

Depending on the plant these modules can be configured to the individual plant constructions, structures and operators' needs. The actual feeding conditions as well as the usage of remaining material and energy flows need to be implemented into the new developed model. Independently, models can also be modeled individually without using any prepared modules. Because of its flexibility that tool could be used by plant operators all over the world.

In the following passage two models of different waste-to-energy plants will be illustrated. It includes a conventional waste incineration plant as well as a refused derived fuel (RDF) plant. Both plants are located in Germany; however, the approach will give an insight how WtE plants could be modeled and used to optimize operation conditions. Furthermore it presents typical WtE plant modules which could be used to create new models for different kind of plants at other sites.

3.2.1 Waste Incineration Plant

The presented waste incineration plant is quite old and was basically built to reduce waste masses. The production of electric and thermal energy used to be a positive side effect. Due to legal regulations and the demand to increase the production of energy using substitute fuels, the generation of power has come more into focus. Therefore its operation has been improved over the last decade. In the future further optimization strategies shall be developed using the model. That might not only include the technical operation but also the overall waste management. The influence of co-incineration of sewage sludge could for example be simulated with the model.

Technically the plant works with a stocker-fired furnace and is fed with municipal solid waste. Its steam parameters reach about 22 bar and 215°C.

Fig. 7 Waste delivery and storage of a WtE plant modeled in UMBERTO®

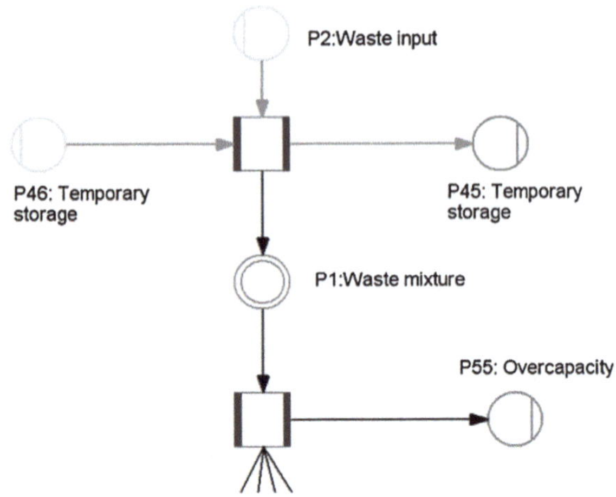

Fig. 8 Furnace system and wet deslagger of a WtE plant modeled in UMBERTO®

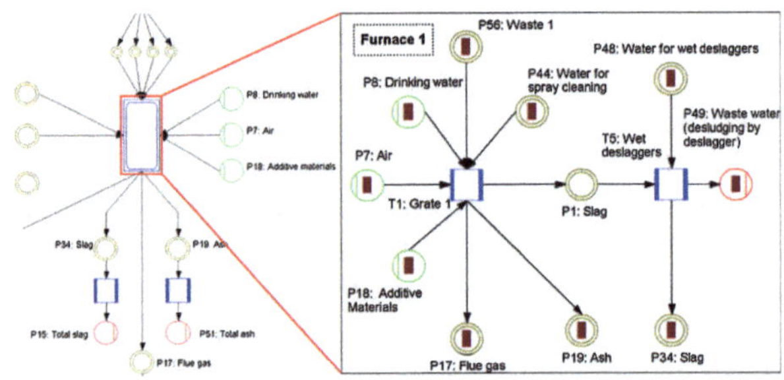

The Model

A detailed WtE plant model is rather complex. For this reason the following passage will show each single section modelled in UMBERTO®.

The *waste delivery and storage* is the first section (Fig. 7). Different wastes may be delivered by various suppliers. Depending on the maximum feeding rate some fractions might be taken from or put into a temporary storage. After the waste has been put into the system the waste gets mixed up and is transferred as a mixture into the plant. As physical and chemical characteristics are stored in the model's matrix, the mixture contains a calculated emission and pollution load which will influence further conversion in the following process steps.

Section two embodies the *combustion*. As that considered plant has implemented four furnaces, the waste mixture is distributed into four combustion lines. The subnet covers four furnaces as well as the wet deslagger (one furnace system is exemplary displayed in Fig. 8). On the input side waste, water, air and operation materials are fed into the plant. On the output side flue gas, fly ash and solid residues will be emitted. The flue gas will later be transferred into the flue gas

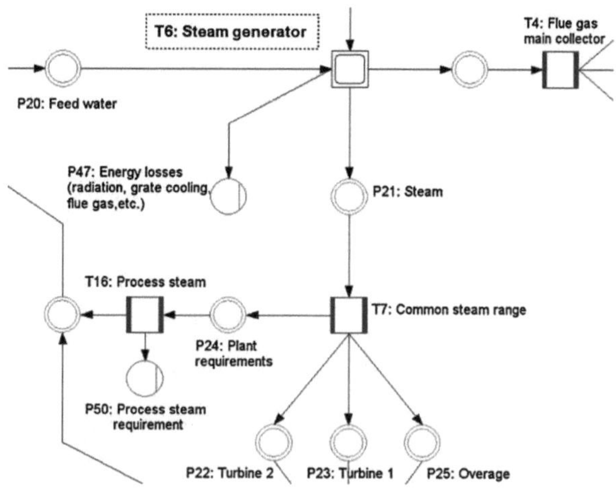

Fig. 9 Steam generation of a WtE plant modeled in UMBERTO®

cleaning. Solid residues are transferred into the wet deslagger which is operated with water. Its final products are waste water and slag.

Section three displays the *steam generation* (Fig. 9). In the steam generator transmitted flue gas is heating up feed water which is fed by the feed water tank. The steam is partly used internally but its major part will be utilized to run two turbines. The distribution of the steam is set by the transition.

Section four outlines the *energy production* (Fig. 10). Electric energy is produced by two turbines. Electricity is either used internally or fed into the main grid. A heat condenser and high pressure heat exchanger convert exhaust steam into heating water. That is used internally or pumped into the pipe system for district heating. An additional oil driven heating station secures the delivery of central heating during operation standstills.

The *water cycle* (Fig. 11) forms the fifth section of the plant. It draws through many parts of the plant and includes not only the liquid water but also high pressure and waste steam. Firstly ground water is conditioned for the use as feed water which is stored in a feed water tank. In the steam generator, feed water is heated up by the flue gas to produce high pressure process steam. That is partly used internally and for the operation of the turbines. Waste steam which could not be used as thermal energy for internal and district heating is cooled down by air capacitors. The condensate as well as the condensate from internal process steam usage is returned to the feed water tank. Waste water will be discharged from the system. Well water in the cycle is conditioned for the use in the spray absorber, spray cleaning and wet deslagger.

The *flue gas cleaning* as the last section of the presented plant model is performed by a spray absorber and a baghouse filter. The spray absorber is fed with water and operation materials like lignite coke and white fine lime. Some by the baghouse absorbed coke and lime will be fed back into the spray absorber. Residues of the flue gas cleaning are discharged of the system. One of three flue gas cleaning lines is exemplary shown in the following Fig. 12.

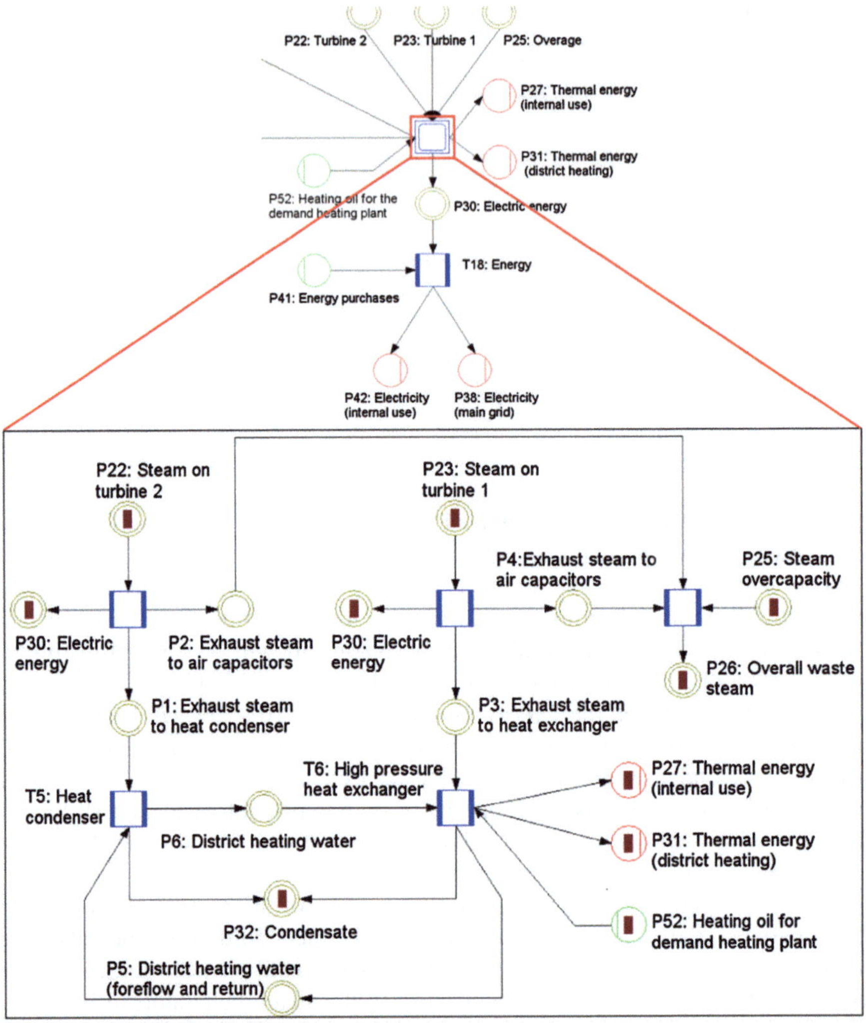

Fig. 10 Energy production of a WtE plant modeled in UMBERTO®

3.2.2 Refused Derived Fuel Plant

The presented refused derived fuel plant is quite new and was built to supply an external customer with process steam. By the combustion of high calorific refused derived fuels, high pressure process steam is generated and via heat exchanger supplied to the customer. Even though the thermal energy takes the major part of the overall energy production one turbine is driven to produce electric energy which is used internally, delivered to the external customer or fed into the main grid.

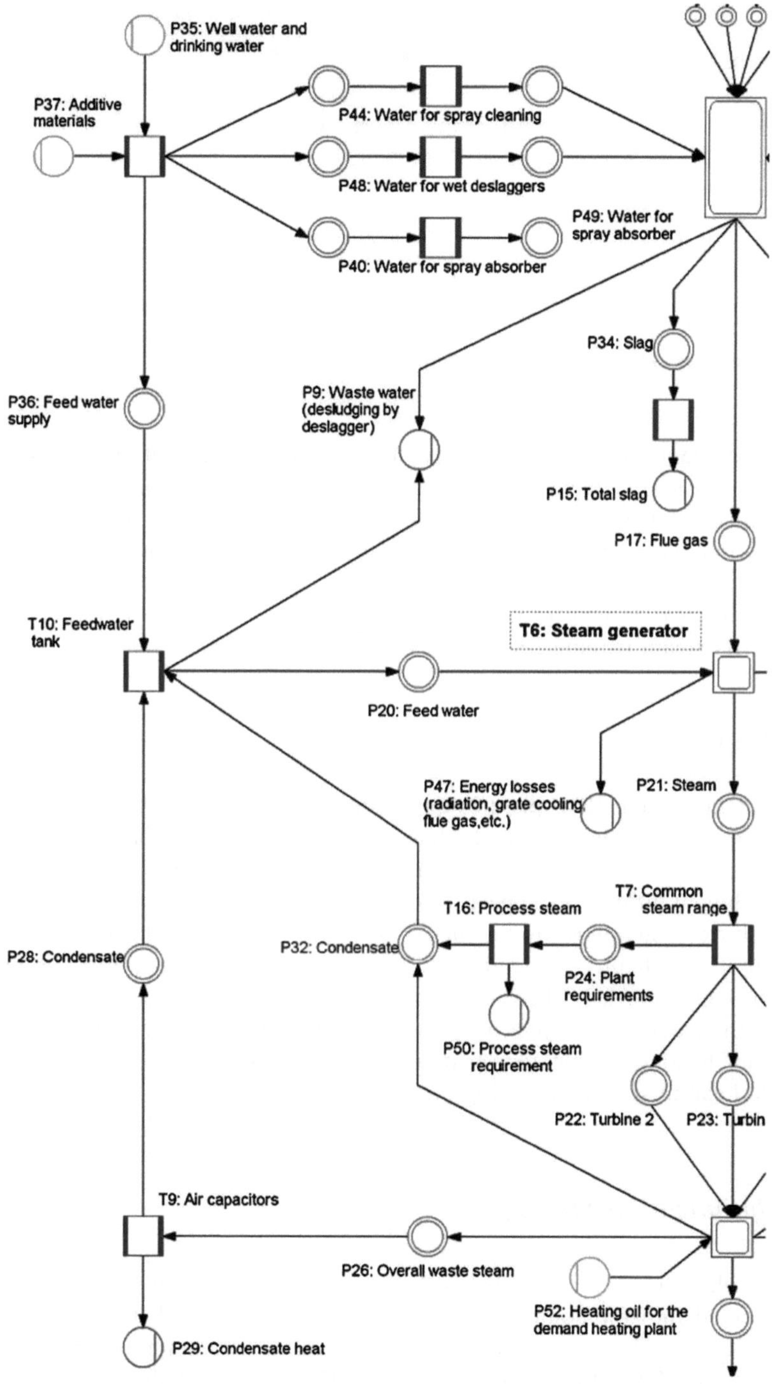

Fig. 11 Water cycle of a WtE plant modeled in UMBERTO®

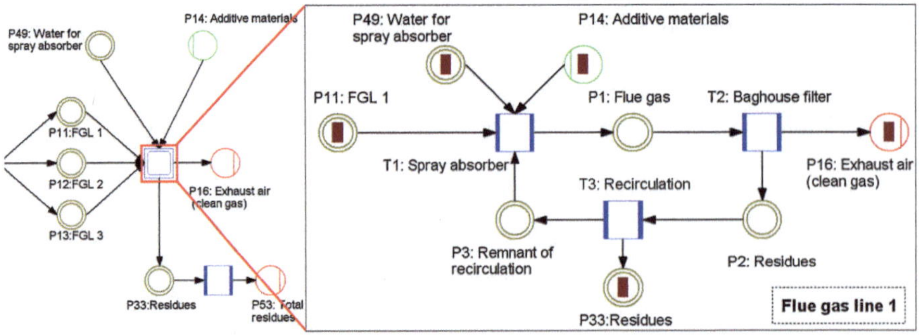

Fig. 12 Flue gas cleaning of a WtE plant modeled in UMBERTO®

As that plant has a short running time its operation conditions are not completely stable so far. Especially during the performance test the operation conditions varied a lot. In the future, operation conditions could be optimized using a material and energy flow networks and results implemented into the real existing plant.

Like the waste incineration plant it technically works with a stocker-fired furnace. Its steam parameters reach about 42 bar and 400°C.

The Model

The structure of the RDF plant is quite similar to the conventional waste incineration plant. It includes modules like waste delivery and storage, combustion, steam generation, energy production, flue gas cleaning and water cycle. However, some technical differences appear. The RDF plant has just one furnace. Hence no subnets have been applied to the model and the wet deslagger becomes visible on the main layer. Steam is divided into a bypass and a turbine line. Just one flue gas cleaning line is needed. There is no district heating because all thermal energy is used by the external customer or cooled down by air capacitors. The plant has no oil driven heating station for operation standstills as the customer has its own separate heating station supplying process steam Fig. 13.

3.3 Optimization Strategies

When a plant has been modeled, new operation and feeding conditions can be implemented into the computer-model. These conditions will be defined by the local situation and operators demands. Afterwards a simulation of any ecological and economical results becomes possible. That may include all material and energy flows, as well as emissions and costs. Several scenarios containing different

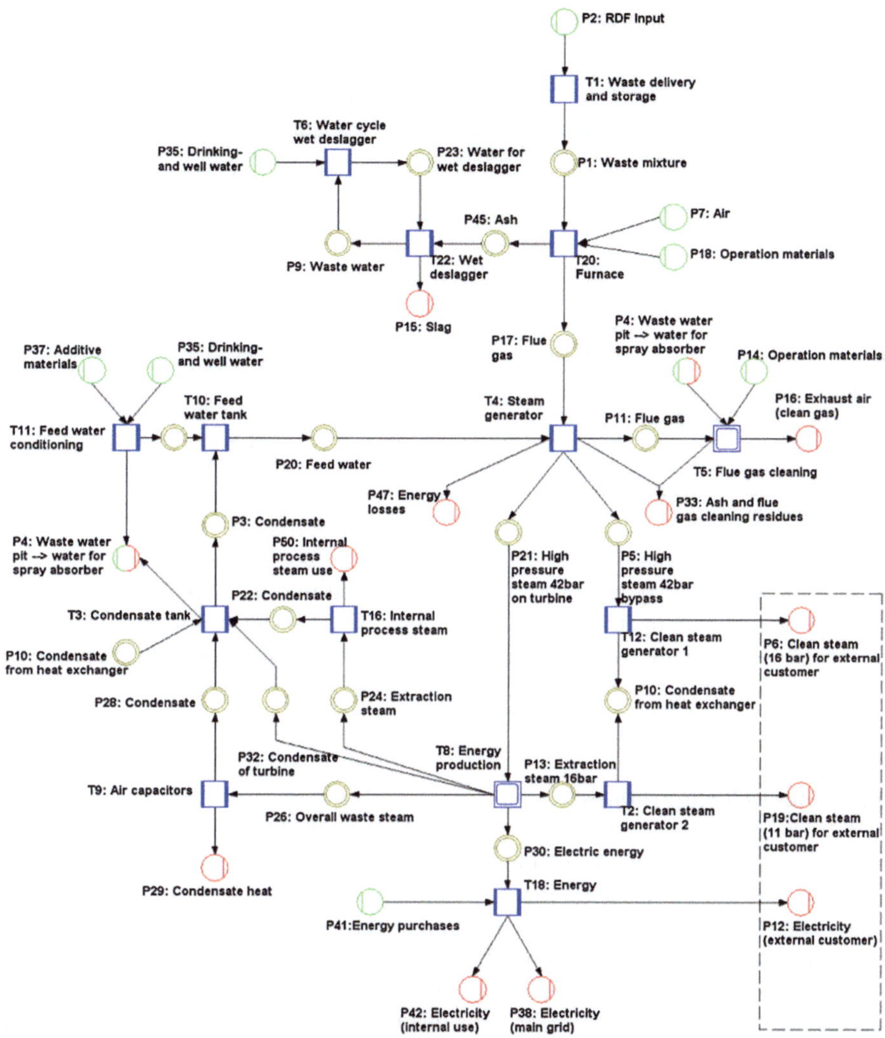

Fig. 13 RDF plant modeled in UMBERTO®

approaches can be established. Specific optimization strategies for each approach can be developed. For example:

- Plant configurations—Modification of a plant or treatment pathway to optimize treatment results.
- Waste mixture—Calculation of the appropriate waste mixture to achieve optimum treatment results.
- Material flow management—Strategic management of waste flows to achieve their optimum treatment.

Table 3 Treatment results with the application of different plant configurations

	Unit	Scenario 1		Scenario 2	
		Household and commercial waste		Household and commercial waste	
Waste input–MBT—plant	Mg	500,000		500,000	
Calorific value	MJ/kg	10.2		10.2	
Mechanical treatment		*No* RDF preparation		RDF preparation	
		HWF[a]	RDF	HWF[a]	RDF
Mass output mechanical treatment	Mg/a	341,900	x	156,600	185,300
Calorific value	MJ/kg	11	x	11	14
Slag	kg/Mg	198	x	198	154
Ash	kg/Mg	14	x	14	57
Flue gas cleaning residues	kg/Mg	39.6	x	43	
Electric energy	MWh	68,800	x	31,500	75,800
Thermal energy	MWh	147,500	x	67,600	232,200

[a] Heavy weight fraction

- These strategies may afterwards be implemented into the actual operation of a plant or the associated waste management system. The following subsequent chapters will demonstrate optimization strategies for different approaches mentioned above.

3.3.1 Plant Configurations

Modifying plant configurations mostly cause changes in the treatment outcomes. The models can be used to identify configurations which will improve these outcomes. That may be the mode of action of the flue gas cleaning improving the exhaust gas qualities or implementing a super heater to create higher steam parameters resulting in a higher energy production. But not just in the WtE plant but also in the early pre-treatment steps technical alterations have an influence on the overall process outcome. Depending on the mechanical pre-treatment, fractions of different qualities for diverse utilization purposes may be generated.

According to the example listed in Table 3, it could be proven that different mechanical pre-treatment-constructions will affect the results of all further treatment plants. In scenario 1 calorific waste (heavy weight fraction) is generated by the mechanical treatment and utilized in a conventional waste incineration plant. For scenario 2 an additional air separator has been integrated to extract more high calorific fractions such as foils. The light weight fraction is utilized in a RDF plant whereas the middle calorific fraction is still treated in a conventional waste incineration plant. Treating a mixture of 500,000 Mg of household and commercial waste using different plant configurations will result in different process

Table 4 Treatment results incinerating different waste mixtures

	Unit	Household waste	Commercial waste	Household waste	Commercial waste
Waste input	Mg/a	500,000	500,000	260,000	240,000
Calorific value	MJ/kg	8	13	8	13
Calorific value waste mixture	MJ/kg	8	13	10.4	
Slag	kg/Mg	–	221	208	
Ash	kg/Mg	–	15.5	14.5	
Flue gas cleaning residues	kg/Mg	–	218	24.3	
Electric energy	MWh/a	–	100,600	91,100	
Thermal energy	MWh/a	–	215,700	135,500	

outcomes. The organic fraction which is discharged off the balance scope stays the same but the mass distribution of the calorific fractions varies.

In scenario 1 341,900 Mg heavy weight fraction are utilized in a waste incineration plant producing 68,800 MW h electric and 147,500 MW h thermal energy. In sum the RDF and waste incineration plant of scenario 2 are producing 107,300 MW h of electric energy; far more than scenario 1. Even more thermal energy could be produced. With a total decoupling of 299,800 MW h nearly twice as much as energy in scenario 1 is produced.

As a result of this specific example it is recommendable from energy perspectives to integrate an air separator into the mechanical treatment and treat the RDF fraction in a refused derived fuel plant if it is available in the surrounding area.

3.3.2 Waste Mixture

The optimization of a waste treatment process may not only be realized by modifying specific plant configurations. Combusting the optimal waste mixture has also a significant influence in improving the treatment results. That is exemplary shown by a simulation feeding a WtE plant with different fractions of waste:

Step 1: *Scenarios*: A conventional waste incineration plant is either fed with household waste, commercial waste or a mixture of both (Table 4). Before entering those wastes into the models they have been characterized by an extensive waste definition to allow comparability.

The combustion of household waste would not be possible in the plant and model considered. With an average calorific value of 8 MJ/kg the minimum combustion heat performance will not be reached. As a result the model does not calculate any data because the entered minimum combustion heat performance in a transition is not fulfilled. The combustion-simulation of commercial waste is possible without any problems. Commercial waste contains more dry fractions than organic compounds and is usually much more inflammable. By the input of 500,000 Mg with a medial calorific value of 13 MJ/kg 100,600 MW h of electric energy could be produced and 215,700 MW h thermal energy could be decoupled. If household waste gets mixed up with commercial waste the combustion heat performance will be fulfilled but the energy production is pretty much inhibited. About 9, 4% less electric and 37, 2% thermal energy is produced by the admixture of household waste. As a result finding the right mixture is a major key for a successful treatment.

Of course household waste needs to be treated in a way but it might be useful to verify if another treatment option might be chosen, e.g. using a mechanical–biological waste treatment plant, treating and disposing the organic fraction and just burn the extracted high calorific fraction. Otherwise commercial waste needs to be used as an additive for the combustion of household waste creating a combustible mixture.

Step 2: *Optimization Algorithms*: The optimal waste mixture might also be generated by the application of optimization algorithms. They are calculating the optimal waste mixture for a specific plant instead of entering different waste mixture into the models and testing possible results manually. Under the determination of decision variables and inclusion of restrictions, the optimal waste mixture is calculated in order to raise the plant's economic benefit and environmental impacts.

The restrictions are basically set by the amount of available waste fractions. Decision variables could include:

- Maximum energy extraction.
- High throughput.
- High availability of the furnaces.
- Compliance with the emission limit values.
- High overall efficiency of a plant.

A case study Optimal Input Mix for a Waste-to-Energy Plant has been done by Lambrecht and Schmidt [5] within the research project KOMSA: to find the optimal operating conditions, it is necessary to solve a parameter optimization problem with the waste input mix as decision vector and the company's benefit as objective. The model-based analysis revealed that the mix of treated waste fractions is crucial for improving the operating efficiency of the plant. The most important characteristics of the waste fractions regarding economic benefit and environmental impacts are the specific incineration fee charged by the company, their calorific value and ash content. Environmental impact and treatment costs (e.g. flue gas cleaning and final disposal of sludge and ashes) depend further on pollutant contents, such as chlorine, sulphur, heavy metals, etc.

The waste incineration plant model as presented in this chapter has been taken as prime example. Due to the model's complexity just the waste input mix, i.e., nine out of several hundred material flows have been set as independent decision variables.

The numerical optimization carried out by Lambrecht and Schmidt [5] has lead to an increase of a specific objective function by about 420%, which is remarkable improvement for this field of application.

3.3.3 Material Flow Management

By the application of material flow networks entire waste systems like cities, counties, islands or even whole states can be displayed. That allows developing location-based material flow management strategies. Depending on the wastes composition and characterization different handling pathways might be necessary to reach optimal treatment results. In order to prove the applicability of existing or planed plants, the models can be used to simulate possible treatment outcomes of local available wastes. By that assessment the models act as a decision tool. Assertions which plant to select and how to direct waste flows can be made in order to allow an environmental friendly treatment and to create optimal operation conditions; especially to reach the highest possible energy and material efficiency.

Therefore characterized waste of a selected region is put into several models of those plants available at site. All results are getting calculated and balanced by the models. Finally the data can be compared. With local limitations and demands in mind a decision for the plant to chose can be made.

The following example as shown in Fig. 14 will demonstrate two scenarios where municipal solid waste can either be treated in a combination of mechanical pre-treatment, RDF plant and waste incineration plant or just in a conventional waste incineration plant. The balance scope will just include the plants considered. The transport for example has been neglected.

By the application of the RDF plant, the mechanical pre-treatment needs to be taken into account as the refused derived fuel plant has a higher requirement for the waste quality. In this particular case a mechanical–biological waste treatment plant is used to pre-treat the waste mechanically. Firstly heavy substances and impurities will be removed manually. The remaining waste gets shredded. Afterwards metal will be discharged. Before the waste gets air separated household waste is additionally sieved. The sieve underflow which contains basically low calorific wastes like organics is treated in a rotting-system and later be put on a landfill meeting legal requirements. Sieve residues will be separated by air flow into the light weight fraction which forms the RDF and heavy weight fraction which is combusted in a conventional waste incineration plant tolerating lower fuel qualities. The untreated waste mixture on the other hand can directly be combusted in the conventional waste incineration plant.

Each scenario is fed with the same amount and composition of characterized waste. By the simulation the following outcomes as shown in Table 5 can be expected:

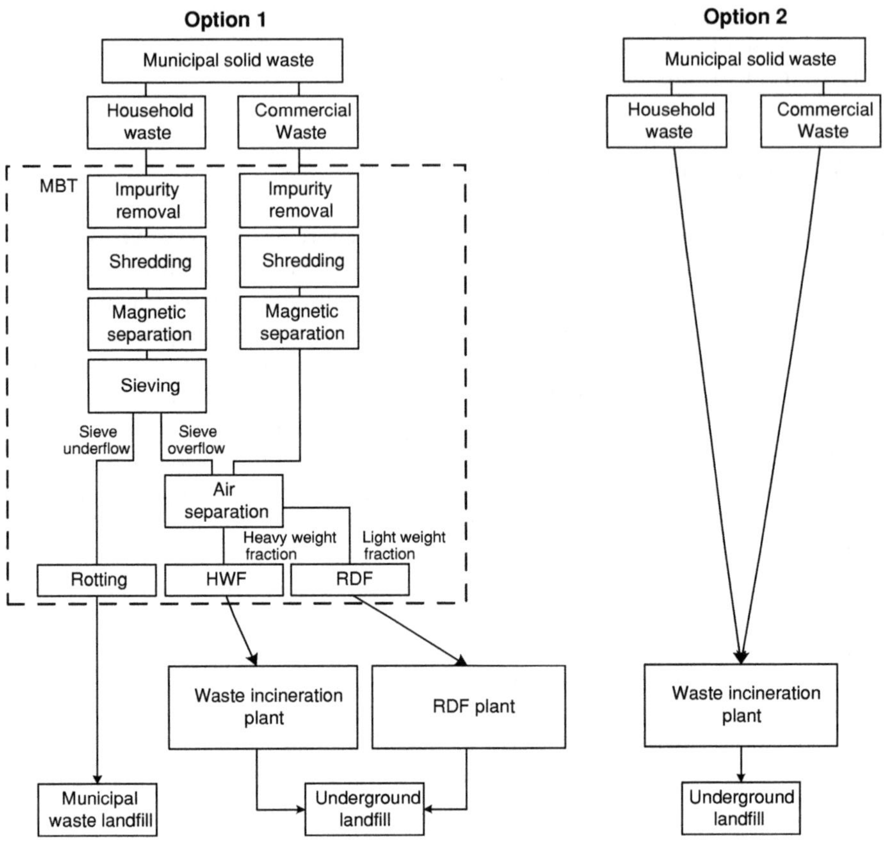

Fig. 14 Different treatment options for municipal solid waste in comparison

For the treatment of 500,000 Mg municipal solid waste option 2 would need 37% more water compared to option 1. On the other hand 22% less air is used by the conventional waste incineration plant. Some additive materials are more used in option 1, others more in option 2. A significant difference can be observed regarding the landfill material. As option 1 produces–due to its preceded mechanical–biological waste treatment plant–more landfill material the mass loss is about 23% higher using the waste incineration plant. The amount of waste water is leveled; the RDF option is producing less exhaust air.

The rating for the optimal treatment option depends on local requirements and operators demands. That's why the decision is defined by the individual significance of each factor. If, for example, a region (e.g. an island) is short of landfills option 2 would be the best. If rather more recyclables shall be extracted option 1 needs to be chosen. Those decisions have to be done by an individual assessment each time applied for each plant and waste.

Table 5 Material balances of various treatment options of municipal solid waste

Material	Unit	Option 1 MBT plant, RDF plant and waste incineration plant	Option 2 Waste incineration plant	Difference in percent
Overall waste input	Mg	500,000	500,000	
Operation supplies				
Water	m^3/Mg	0.31	0.5	+37
Air	m^3/Mg	5,700	4,400	−22
Additive materials				
For feed water conditioning	kg/Mg	0.52	1.20	+57
For exhaust air cleaning	m^3/Mg	0.25	–	−100
	kg/Mg	21.1	19.8	−6
Other additives	m^3/Mg	0.0004	0.0005	+33
Recyclables	kg/Mg	51.9	20.8	−60
Residues				
Landfill residues	kg/Mg	320	247	−23
Waste water	m^3/Mg	0.07	0.07	0
Exhaust air	m^3/Mg	4,400	5,100	+13

Observing the energy flows the following results have been simulated Table 6:

Even though both scenarios are fed with the same waste, meaning same calorific value the RDF plant will produce more thermal and electric energy. Moreover the RDF and MBT plant need less energy for operation but slightly more energy from additional fuel.

To obtain the overall efficiency, regarding to the VDI guideline 3460[2] [6], the sum of the thermal and electrical target energy (benefit) refers to the total supplied input energy (demand) of the particular balance space. This concept does not distinguish between the characteristics of the different forms of energy. Putting that in a simple formula (1), the energy efficiency of the demonstrated plants (2, 3) is becoming comparable:

$$\text{Energy efficiency} = \frac{\text{produced energy} - \text{plant's requirments}}{\text{energy}_{(\text{waste})} + \text{energy}_{(\text{additional fuel})}} \tag{1}$$

$$\text{Energy efficiency option 1} = \frac{537,300 - 128,100}{1,450,000 + 4,500} = \eta = 0.28 \tag{2}$$

$$\text{Energy efficiency option 2} = \frac{393,000 - 166,900}{1,450,000 + 3,700} = \eta = 0.16 \tag{3}$$

[2] VDI Guideline 3460 'Energy Conversion in thermal solid Waste Treatment', Part 2, Chapter 4.6 'Total efficiency'.

Table 6 Energy balances of various treatment options of municipal solid waste

Energy	Unit	MBT plant, RDF plant and waste incineration plant	Waste incineration plant
Energy from waste	MWh	1,450,000	1,450,000
Energy from additional fuel	MWh	4,500	3,700
Produced energy (th, el)	MWh	537,300	393,000
Plant's energy requirements	MWh	128,100	166,900

Option 1 is reaching a higher efficiency than option 2. As a result it would be advantageous to pre-treat municipal solid waste and utilize the light weight fraction in a RDF plant. High calorific fractions will be generated through mechanical pre-treatment allowing a much better combustion than burning the whole waste mixture in a conventional waste incineration plant. However, a biological process is needed for treatment of organic fractions to complete the process chain. But even though the mechanical–biological treatment plant is using energy as well the combination of MBT and RDF is still more efficient than combusting the untreated waste mixture.

In consequence for that specific example it would be recommendable to use a RDF plant in correlation with a mechanical pre-treatment.

That conclusion is based on those illustrated waste-to-energy plants. Because of the fact that all plants are different and the waste mixtures vary a lot, the presented results are only applicable for those specific plants and not be generalized to all WtE plants. The example will just give an insight how management strategies can be developed by using material flow networks.

4 Conclusion and Outlook

Material and energy flow networks present an applicable optimization tool for waste treatment and recycling management. The models can be used to present the actual operation conditions and states as well as simulating possible treatment results in case operation conditions have been modified. That might be after technical alterations or varying the waste input e.g. the co-incineration of sewage sludge or the change of operation settings. Furthermore plants in planning or construction phases may be modelled in advance and process outcomes may be predicted. As a result strategies for the optimization of operation conditions can be developed and adapted in the operation, planning or construction of a plant. Moreover specific material flow strategies can be developed. Waste can be managed with focus on the most efficient and environmental friendly treatment. It allows locating suitable plants for specific types and fractions of waste.

Several optimization strategies which have been developed and indicated in this chapter could prove that those strategies may lead to an improvement of treatment results. One opportunity to optimize operation conditions of a WtE plant is to modify the plant's configurations. Treatment results could be improved by installing an air separator into the system. Through the additional preparation of calorific fractions into a higher quality fraction, that waste can be utilized in an RDF plant reaching higher energy decoupling. Furthermore, it could be proven that a defined waste mixture results in optimized treatment results than combusting waste in an undefined mixture. Characterized waste fractions may be put together in order to reach an optimal conversion of the mixture in the process. Additionally the overall energy efficiency of a treatment pathway can be influenced by the application of a specific material flow management.

Each plant can be modelled individually or linked to complex process chains. The waste input into the model/models can be variable (specific waste analyses necessary), input parameters and transfer factors can be configured anytime e.g. after technical alteration.

Especially in developing countries material flow networks are a supportive tool for the optimization of the waste and recycling industry. It helps to structure and manage waste streams in selected areas and to improve operation conditions of WtE plants. As a consequence treatment results will be improved, environmental and health risk minimized and the plant's economy enhanced.

Having global warming and strategies for the reduction of CO_2 emissions in mind, CO_2 calculation could be realized by the extension of the models taking more parameters and chemical conversions into account. Additionally UMBERTO® for Carbon Footprint could be used to develop comprehensive CO_2-models for WtE plants.

Acknowledgments The authors gratefully acknowledge the financial support by the Federal State of Bremen, Germany from the Ecological and Applied Environmental Research Fund and the European Regional Development Fund. Furthermore the authors like to thank all colleagues working onto the research project RessourceMan (www.ressourceman.de), especially Anke Schmidt, Tobias Brinkmann (eco!ogix, Germany) and Sebastian Wolff (Institute for Energy, Recycling and Environmental Protection at Bremen University of Applied Sciences, Germany). The author likes also to thank Hendrik Lambrecht and Mario Schmidt from Pforzheim University, Germany for the cooperation between the research projects RessourceMan and KOMSA.

References

1. CEWEP (2011) Waste-to-energy plants. http://www.cewep.eu/. Cited 18 Jan 2011
2. Ecoprog & Fraunhofer UMSICHT (2010) The worldwide market for waste incineration plants. In: Study—waste to energy; The worldwide market for waste incineration plants 2010/2011. Cologne, Germany
3. Ifu Hamburg GmbH (2011) Umberto—the software for process optimization. In: Umberto Product Flyer. Available via http://www.umberto.de/en. Cited 19 Jan 2011
4. Jaron (2009) Global waste–borders, evils, perpectives. Müll und Abfall 9:436–443
5. Lambrecht Schmidt (2010) Material flow networks as a means of optimizing production systems. Chem Eng Technol 33(4):610–617
6. Verein Deutscher Ingenieure (2007) Total efficiency. In: VDI guideline 3,460 energy conversion in thermal solid waste treatment, part 2, chapter 4, vol 6. VDI-Verlag, Düsseldorf, p 21

Anaerobic Digestion of Waste

Martin Kranert, Sigrid Kusch, Jingjing Huang and Klaus Fischer

Abstract All sustainable development closely links to the context of energy and to appropriate solutions to cope with challenges arising from trends of increasing urbanisation, by at the same time allowing for development of rural areas. Biogas production through anaerobic digestion of biomass, including the organic fraction of waste materials and residues, is a particularly promising choice and experiences increasing interest worldwide. It does not only supply a clean and versatile energy carrier, but is well suited to contribute towards appropriate waste management schemes in urban areas and in agriculture. Biogas production has high potential worldwide, and in this chapter special focus is given to its implementation in countries with economies in development or transition. China and India are countries where biogas production is already well-known and often adapted, and more widespread implementation is to be expected. This book chapter also highlights the topic anaerobic digestion in countries in Latin America and Africa.

1 General Aspects of Anaerobic Digestion

Biomass is a renewable resource. Its use for energy generation releases no (or only few) additional carbon dioxide into the environment, which is favorable against the background of the climate debate.

M. Kranert (✉) · S. Kusch · J. Huang · K. Fischer
Institute for Sanitary Engineering,
Water Quality and Solid Waste Management (ISWA),
University of Stuttgart, Bandtaele 2, 70569 Stuttgart, Germany
e-mail: martin.kranert@iswa.uni-stuttgart.de

A. Karagiannidis (ed.), *Waste to Energy*, Green Energy and Technology,
DOI: 10.1007/978-1-4471-2306-4_5,

Especially for organic wastes from households, local industry or agriculture, which have a high water content, anaerobic digestion (AD) is a good way to tap the energy content. While generating energy for covering demands in heat, electricity or gas, a valuable fertilizer is generated simultaneously, given that the input materials have a low content of impurities and pollutants.

Anaerobic digestion was practiced 2,000 years ago in Mesopotamia. Today it is widely applied throughout the world. In Europe alone, currently more than 5 million tons of bio-waste per year are treated anaerobically. Globally, a sharp increase of biogas production from agricultural wastes is to be expected especially in India and China. In both countries there are efforts to increase the share of renewable energies. In China, a potential of 200 million biogas plants [17], and in India, a capacity of 15,000 MW within the next 15 years are expected [13].

In addition to production of renewable energy and valuable digestate, anaerobic digestion has further benefits. The amount of organic waste destined for landfills is thus reduced significantly. According to the EU Landfill Directive and some state laws this is necessary to avoid harmful emissions (gas, leachate).

In developing countries AD also has social and societal impact. Especially in rural areas, self-sufficient supply with fuel for cooking and heating purposes can be partially achieved. The often very time consuming and exhausting collection of firewood can be omitted, which would have positive effects on the environment, but above all, have a positive influence on housework and cooking. Considering the fact that in most cases women carry out these activities, AD may also contribute to improve work conditions of women.

2 Basics of the Anaerobic Process

2.1 Biochemistry and Microbiology

Anaerobic digestion (or anaerobic fermentation) is a biological process in which organic material in liquid or solid phase is biodegraded by several groups of microorganisms in absence of free oxygen. Outputs of this process are (1) biogas, which is a mixture of methane, carbon dioxide and trace elements and which has a high calorific value, and (2) a stabilized residue, which in most cases can be used as organic fertilizer. In some process types waste water can also be an effluent.

The general biochemical reaction follows the Eq. 1 [5]:

$$C_cH_hO_oN_nS_s + xH_2O \rightarrow yCH_4 + (c - y)CO_2 + nNH_3 + sH_2S \tag{1}$$

with

$$x = \frac{1}{4}(4c - h - 2o + 3n + 2s)$$
$$y = \frac{1}{8}(4c + h - 2o - 3n + 2s)$$

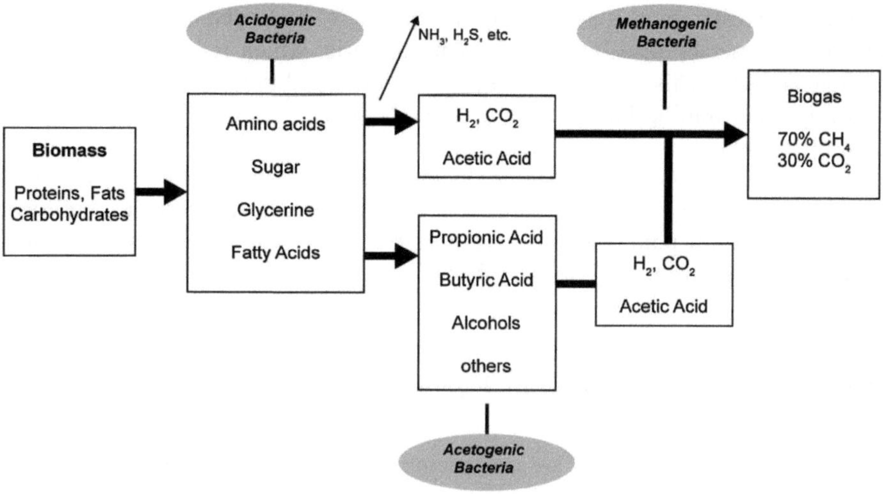

Fig. 1 Reaction chain of anaerobic digestion

As an example the anaerobic biodegradation of glucose (carbohydrate) [24] is:

$$C_6H_{12}O_6 \rightarrow 3CO_2 + 3CH_4$$
$$\Delta G_f' = -420\,kJ/mol$$

By combustion of biogas CO_2 and H_2O are produced:

$$3CH_4 + 6O_2 \rightarrow 3CO_2 + 6H_2O$$
$$\Delta G_f' = -2448\,kJ/mol$$

If glucose is fully oxidized into carbon dioxide and water, a free energy of 2,868 kJ/mol is released (which is equal to the energy contained in glucose after having been formed in the photosynthesis pathway). Burning of methane in the reaction mentioned above leads to a free energy of 2,448 kJ/mol glucose, which is equivalent to 85% of the energy content.

The anaerobic reaction chain includes several steps, each involving specific groups of microorganisms (Fig. 1) [6, 9, 27].

Hydrolysis: In this first stage the substrate, which consists of complex organic structures (carbohydrates, proteins, fats) is hydrolysed into monomeric components (monosaccharides, amino-acids, long chain fatty acids).

Acidification: In this second stage the monomers are transformed into alcohols and volatile fatty acids by acidification bacteria. Hydrogen and carbon dioxide are released as well.

Acetogenesis: In this stage the volatile fatty acids are transformed by acetogenic bacteria in acetic and formic acid by releasing also hydrogen and carbon dioxide.

Methanogenesis: In this forth stage acetic acids, hydrogen and partly carbon dioxide are metabolized by methanogenic bacteria, which releases methane and carbon dioxide. This requires strictly anaerobic conditions (no oxygen) and in general is the limiting kinetic step, as these organisms have a slower growth rate than other groups involved in the steps mentioned above.

In the reaction chain hydrogen producing and hydrogen consuming bacteria are working in synergy and are dependent of each other (interspecies hydrogen transfer). To allow both reactions to be possible, hydrogen partial pressure must be in a small range from 10^{-6} to 10^{-3} bar [28].

2.2 Process Parameters and Factors

Many parameters and factors are influencing the microbial metabolism process. In order to ensure process stability and to optimize the process, some of these parameters have to be controlled [4, 11, 25, 26]. One of the challenges is that optimum conditions for hydrolysis and acidification differ from requirements of methanogenesis. This, among other factors, has to be taken into account when designing and operating AD plants. Most important parameters are shown as follows.

2.2.1 Substrate

The process itself, optimal choice of process technology and plant operation is determined by the type of substrate. The substrate must provide all vital components and should not have limiting or inhibiting substances. Micronutrients as nickel, vanadium, iron, sodium, potassium etc. are necessary, normally given in the substrate [6, 9].

Inhibitory and toxic effects can occur during the digestion process, some examples are as follows: high concentration of ammonia in basic milieu ($pH > 8$) [6], presence of sulphur (competition between sulphur reducing and methanogenic bacteria, undissociated H_2S concentrations >50 mg/L, heavy metals in high concentrations (depending on pH), antibiotics and aromatic compounds (e.g. phenol) [9]. It should also be avoided, that intermediate metabolic products limit the process (for example accumulation of volatile fatty acids). The biodegradability determining the retention time is shown in Fig. 2.

C:N ratio (based on dry mass) should be in a range of 16:1–25:1, at too high C:N ratio nitrogen is deficient (metabolism is malfunctioning). The ratio N:P:S should be about 100:20:20 [6].

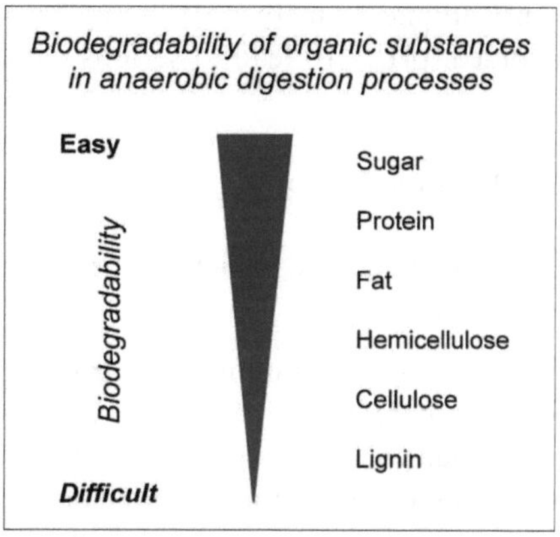

Fig. 2 Biodegradability of organic substances in anaerobic digestion processes

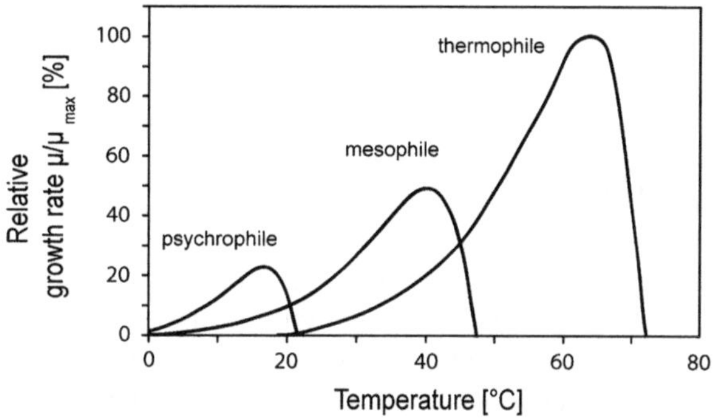

Fig. 3 Effect of temperature on the reaction rate of methanogenic bacteria [4, 31]

2.2.2 Temperature

Temperature range for microorganisms in anaerobic processes is between 0 and 65°C. Enzymatic and bacterial activity increase with temperature (*Arrhenius equation*). Depending on the bacteria three optima can be mentioned. In technical processes for acidogenic and acetogenic bacteria high microbiological activities can be found at 35 and 55°C [6]. Compared to methanogenic bacteria the sensitivity to temperature changes is lower because of the high growth rates. For methanogenic bacteria the optima are in a small range, the intervals are smaller (mesophilic 30–40°C, thermophilic 55–65°C) [36] (Fig. 3).

2.2.3 pH

pH is influencing the enzymatic systems and indirectly the process, because the dissociation equilibrium of the acids is shifted (volatile fatty acids in undissociated form penetrate cell membranes) [10]. Hydrolysing and acidifying bacteria prefer a pH around 6, methane forming bacteria a pH of 6.7–7.5. A pH < 6 has to be avoided, pH > 10 lead to serious damage of the microbial system.

2.2.4 Other Factors

Additional factors, which influence the microbial metabolism processes are as follows [13]:

- buffer capacity (ammonia, carbonate)
- redox-potential
- mechanical factors (stirring, mixing)
- concentration of microorganisms
- specific surface area of substrate
- disintegration
- light.

2.3 Biogas Characteristics

Biogas consists of methane and carbon dioxide; in addition it contains impurities which can affect its use, emissions, and operation and life time of the technical equipment. Gas composition and gas production rate are depending of the substrate (see also Sect. 2 in chapter "Planning Tools and Procedures for Rational-Municipal Solid Wastes Management"). Table 1 shows typical parameters of biogas.

2.4 Substrates for Biogas Production

Biomass of different types can be used as input material for anaerobic digestion. Considering the specific demands of the AD process (described above) a mixture of substrates is advantageous to allow optimum conditions for the microorganisms concerning biodegradable substances, micronutrients, water and avoiding harmful and inhibiting substances.

It has to be mentioned, that lignified biomass (e.g. in wood) and synthetic organic polymers (plastic) are not or very slowly biodegradable. In general theoretical biogas yield can be calculated based on organic compounds (Table 2).

Examples for substrate from waste and residues for anaerobic digestion are: organic waste from households and gastronomy, municipal solid waste, organic waste from industry/commercial waste, sludge and excreta, agricultural residues. Table 3 shows typical biogas yields and composition of selected substrates.

Table 1 Components and characteristics of biogas [6, 13, 15, 27]

Parameter	Data
Methane	50–75% by volume
Carbon dioxide	25–45% by volume
Water (vapor)	2–7% (saturated) by volume
Nitrogen	0–2% by volume
Hydrogen	0–1% by volume
Oxygen	0–2% by volume
Ammonia	0–0.05% by volume
Hydrogen sulfide	10–30,000 mg/m^3 [a]
Siloxanes	0–50 mg/m^3 [a]
Calorific value (60% CH_4)	21 MJ/m^3 [a]
Range	6.0–6.5 kWh/m^3 [a]
Density	1.2 kg/m^3 [a]
Explosion limits	6–12% (in air)

[a] Standard reference conditions

Table 2 Biogas yields of organic compounds [6]

Lignin	0 L/kg VS
Starch	830 L/kg VS
Protein	890 L/kg VS
Cellulose	960 L/kg VS
Fat	1,420 L/kg VS

3 Planning and Successful Operation of Biogas Plants

Successful operation and economic viability of biogas plants highly depend on adequate planning of the project, skills of the plant operator and knowledge about the biological process.

3.1 Biogas Plant Types

Many different biogas plant types have been developed and are to be found in full-scale for various applications and in different regions. The following overview is restricted on types typically implemented for digestion of solid waste materials, agricultural substrates and household wastes. It needs to be pointed out that a wide variety of additional plant types exists related to waste water treatment, especially for digestion of sewage sludge and industrial liquid wastes rich in COD, e.g. the upflow anaerobic sludge blanket reactor (UASB), the anaerobic baffled digester, the anaerobic contact digester, the fixed film digester, the upflow fixed film

Table 3 Biogas yields and composition of selected substrates [2, 12, 13, 15], (Kranert et al. 2010, Waste analyses of the laboratory of solid waste management. Institute of Sanitary Engineering, Water Quality and Solid Waste Management, "Unpublished")

Substrate	TS (% FM)	VS (% TS)	Biogas (m^3/kg VS)	Methane (% by Vol.)
Municipal solid waste				
Bio waste	30–65	45–70	0.15–0.6	58–65
Market waste	35–60	75–90	0.4–0.6	60–65
Leftovers	10–35	80–98	0.2–0.5	45–60
Green waste	35–70	45–90	0.5–0.65	55–65
Grass	9–13	80–90	0.2–0.7	50–56
Solid waste	65–75	55–65	0.15–0.5	55–65
Fat (fat separator)	2–70	75–90	0.6–0.7	60–70
Food and beverage industry				
Spent fruits	25–45	90–95	0.4–0.7	55–65
Apple mash	2–3	95	0.5	55–65
Potato mash	6–18	85–96	0.3–0.9	55–60
Oilseed				
Residuals	92	97	0.9–1.0	60–68
Rumen	11–19	80–90	0.2–0.4	58–62
Sludge and excreta				
Human excreta				
Without urine	25–30	72	0.2–0.4	45–65
With urine	5	63	0.2–0.4	55–75
Sewage sludge	2–10	50–80	0.3–0.6	40–75
Agricultural residues				
Liquid manure				
Cattle, cow	7,5–13	75–82	0.15–0.6	53–62
Pig	2,3–11	75–86	0.3–0.88	47–68
Poultry	10–29	67–77	0.3–0.8	55–63
Solid manure				
Cattle, cow	24–26	68–76	0.15–0.55	42–68
Pig	15–25	75–80	0.27–0.45	55–62
Poultry	35–86	60–80	0.25–0.45	51–60

TS total solids (dry mass), *FM* fresh mass, *VS* volatile solids (organic mass)

digester, the anaerobic filter digester. Some of the mentioned concepts are occasionally found as one element in hybrid systems when digesting wastes or agricultural materials, but they are not to be considered as a standard in anaerobic digestion outside the waste water treatment branch.

The following table provides an overview on different technology concepts applied for anaerobic digestion (Table 4). In batch systems digestion and methane production start anew with each filling of the reactor and biogas supply therefore is not continuous. For commercial operation it is in general necessary to have several reactors run off-set (alternative loading and unloading), at least three reactors should be operated. A continuous system in most cases is judged to be better suited

Table 4 Types of digesters

Operation of mode: batch/fed-batch or continuous	In *batch systems* substrate is digested over a pre-defined period. Once digestion is complete material is removed and the process is started with a fresh load of substrate In *fed-batch mode* material is added to the digester by and by until the space is used up. Then the digester is emptied to provide new reactor volume In a *continuous* system (or more precise semi-continuous) substrate is regularly fed into the reactor, there is no interruption of either loading the fresh material or unloading the effluent
Transport of material and homogenisation in reactor	The most common types of AD plants are based on the concept *continuously stirred tank reactor* (CSTR). Those plants are equipped with facilities for stirring the digester content (continuously, or in most cases semi-continuously), resulting in homogenization of reactor content but also in differing retention times for different particles, with part of the material leaving the reactor after very short digestion time *Plug flow digesters* are long narrow tanks (typically 5 times as long as the width) with no internal agitation. Inlet and outlet are at opposite ends, feeding is carried out semi-continuously and typically with a thick substrate of 11–15% total solids. In theory, reactor content in this type of digester does not mix longitudinally on its way through the reactor, advancing towards the outlet whenever new manure is added (but actually material does not remain as a plug and portions flow through the digester faster than others–but minimum retention time is assured far better than in CSTR concepts, thus allowing for better hygienisation)
Total solids content (TS)	So-called *wet digestion plants* are most common in agricultural biogas production, they are operated at TS $<$ 15%. When digesting higher amounts of solid materials in this process type, water content needs to be adjusted (addition of liquid substrates, water or recirculation of digester effluent) For digestion of organic materials available mainly in solid form, implementation of technical processes designed to be able to cope with higher TS contents was a logical step (solid municipal bio waste, solid agricultural substrates, etc.). So-called *dry digestion plants* are typically operated at TS $>$ 20%, water content often is not adjusted to a specific value but is a result of the digesting substrates
One-, or two-stage (multi-stage) AD systems	By far the most AD plants are *one-stage processes*, with one single reactor for the digestion process (in general followed by a storage tank). In *two-stage systems* (or multi-stage systems, which however are very rare) process conditions can be optimized for the different groups of microorganisms in order to improve overall efficiency of biogas production. While during the first phase conditions can be optimized in order to achieve a rapid liquefaction, the second phase converts soluble matter into biogas. Compared to single-stage systems the process is more rapid and more stable, but investment and maintenance costs are considerably higher

for large-scale operations [35], drastic changes of input composition should be avoided.

A batch digester is the least expensive to build and easiest to operate. Single-stage batch digesters (often operated in fed-batch mode) need little skills, little familiarity and are often preferred in rural areas due to simplicity of operation. Batch dry digestion is particularly attractive, this is the simplest process where operation involves merely charging solid waste in an air-tight digester, seeding with inoculum and in some cases adding alkali in order to maintain pH. The most successful biogas programme using the batch system is conducted in the Philippines [35]. When operating several reactors off-set, biogas production can be equalized, and batch dry AD plants have now also become a standard in industrialized countries such as Germany (e.g. system Bekon), operating with digestion times of 4–6 weeks (while in developing countries batch reactors are often loaded only a few times per year).

Different variants of continuous digesters mainly implemented in developing countries include the following systems [30, 35]:

1. *Floating dome digester.* This type is of Indian origin and was promoted back in the 1950s by Khadi and Village Industries Commission (KVIC). Digesters of this design are now being used extensively throughout the world. Biogas produced in the digester is trapped under a floating cover. The volume of the gas cover is around 50% of the total daily gas production. The system was originally made of mild steel until fiberglass reinforced plastic gas holders were introduced. Historically, cattle manure was fed for AD of night soil; typical feedstock is cattle dung, agricultural residues, night soil, aquatic plants. Cattle dung is diluted to 10% TS before feeding as inoculum. The system has undergone many efforts to optimize efficiency (heating, mixing, insulation, modifications in geometric configuration, inlets/outlets). In case of a high height:diameter ratio of the digester, a central baffle is included to prevent short circuiting in mixing.
2. *Janata model digester.* This digester was introduced by an Indian non-governmental organization to be 20–30% cheaper than the floating dome model. Most common capacities are 2–6 m^3 biogas/day. The system is well suited for domestic operation or for community size digesters e.g. used in rural/hilly areas.
3. *Fixed dome digester* (*Chinese model*). This type is by far the most commonly used in developing countries. The unit consists of a gas-tight chamber constructed of bricks, stone, or poured concrete. The inside is applied with many thin layers of mortar to make it gas-tight. Level of digestion material is at 95% of total reactor volume at ambient pressure, the digester is fed semi-continuously. Biogas is stored under the dome and displaces some of the contents in the effluent chamber.
4. *Bag design digester* (*Taiwanese model*). The reactor is a long cylinder (length:diameter 14:3), made of PVC or neoprene coated nylon fabric, and acting as plug flow digester (unmixed). Digester walls are thin, which facilitates heating, heating with solar energy might offer additional potential for increasing performance.

Dissemination of technology types has experienced different success in different countries. Even when applying the same technology, often different results were obtained in different countries, and even in different regions [1], which highlights that besides technology other factors have decisive influence on successful implementation of AD projects.

In addition to further potential for installation of AD plants at household levels, biogas facilities at large scale gain importance in countries with economies in development or transition. The need of sustainable waste management schemes, trends of massive urbanisation, industrialisation of agriculture branches (such as animal breeding) are some of the associated factors. In China for example, where pig manure is the traditional substrate for biogas plants, household swine production experiences significant decrease, while industrial pig farming is on the rise and has already taken over 30% of China's pork supply [22]. Such facilities experience problems to dispose the huge amounts of pig wastes, and environmental risks are generally high (waste leakages, foul water flowing into rivers), which makes a biogas plant particularly interesting.

In Brazil and the Philippines large-scale crop-based digesters using sugar-cane residues as feedstock are most common.

India is highly aware of the potential of biogas, and in search of solutions to deal with air pollution from transport (one of its major problems) the country is also planning to use compressed biogas as automotive fuel. AD has been assessed to be a suitable technology to treat organic household waste in urban and peri-urban areas. Based on research, development and implementation activities now much knowledge and experience in AD of kitchen/market waste and organic household waste is already available e.g. in South India, and a range of plants have been developed and implemented by Indian institutions specifically designed to treat organic solid waste rather than manure [37]. Large-scale facilities with capacities up to 100 tonnes per day exist. Main motivation of developers and operators is the need to find waste treatment solutions, generation of biogas is often perceived as an added value. Though not yet fully implemented, the existent innovative legal framework (city authorities are liable to promote waste segregation at source to avoid landfilling of biodegradable waste) can significantly contribute to further dissemination of AD in the country.

Technology is not directly transferable from household scale to large scale at municipal level, specialised expertise is required. The need for hygienisation, additional infrastructure including reception areas, required pre-treatment equipment, management of digestates, safety issues are some points to be considered.

3.2 Dimensioning of AD Plants

Main parameter for dimensioning a biogas plant is the retention time. The retention time can only be accurately defined in batch-type plants. For continuously operated facilities the mean retention time (hydraulic retention time HRT) is

Fig. 4 Hydraulic retention time and organic loading rate

$$HRT = \frac{V_R}{\dot{V}}$$

HRT hydraulic retention time [d]
V_R reactor volume [m³]
$\dot{V}$ daily influent rate [m³/d]

$$OLR = \frac{\dot{m} * c}{V_R}$$

OLR organic loading rate [kg VS/(m³*d)]
c concentration organic matter [kg VS/kg]
$\dot{m}$ daily influent substrate mass flow [kg/d]

VS: volatile solids = organic matter

approximated by dividing the digester volume by the daily influent rate, and this dependency can be applied when deciding about the necessary plant volume (Fig. 4). It needs to be considered that each additional m^3 of reactor volume results in higher investment costs, and therefore the cost-optimum in biogas production is in general below the biological optimum.

The organic loading rate (OLR) is limited by the biological conversion capacity of the AD system, and depends on the plant type, the mode of operation and the digested substrates. When feeding the system above its sustainable OLR gas yield will decrease due to accumulation of inhibitory substances such as fatty acids. OLR is one of the key control parameters in continuous systems, it is closely linked with HRT.

Typical organic loading rates are as follows [12]:

- Stirred reactor run on sludge, manure: 2.0–4.5 kg VS/(m^3*d)
- Stirred reactor with co-fermentation of bio-waste: 0.5–3.5 kg VS/(m^3*d)
- Plug-flow reactor operated with source-segregated biowaste: 7–9 kg VS/(m^3*d)
- AD plant at house-hold scale run on waste, excreta: 0.8–1.2 kg VS/(m^3*d)

For digestion of liquid manure in the mesophilic temperature range the following approximate values apply for choice of HRT [18]:

- Liquid cow manure: 20–30 days
- Liquid pig manure: 15–25 days
- Liquid chicken manure: 20–40 days
- Animal manure mixed with plant material: 50–80 days

Too short retention times can result in a situation where microorganisms are "washed out" faster than they can be reproduced, which means that biogas production comes to a standstill. While this rarely occurs in agricultural plants run on slurry/manure, it can be a problem when treating plant material or industrial wastes.

Retention times are highly dependent of process temperature and digested substrates. Typical HRT for agricultural plants run on slurry are 15–20 days for thermophilic operation (48–55°C), 30–40 days for mesophilic plants (30–42°C) and 70–80 days for psychrophilic digestion (<20°C) [11]. Those retention times refer to a degree of biodegradation of around 50%. Higher degrees are possible but not common due to economic reasons.

In biowaste plants typical retention time is between 20 and 30 days [12, 27], in waste water treatment plants around 20–30 days, in some cases high reactor volume is available, so that retention time is increased up to 60 days [23]. Intensive thermophilic degradation processes are also found with HRT of 10–15 days [6]. For rural biogas plants in developing countries retention time is normally around 150–200 days, sometimes more than a year.

Climatic conditions need to be taken into account especially when planning a facility without heating equipment. It is not suitable to fully rely on general literature data. Large areas in developing countries are highlands or have a continental climate characterised by warm summers but cold winters. Low ambient temperatures decrease microbiological activity in the digester and consequently the rate of biogas production. This can be overcome by increasing either digester volume or digestion temperature. As an example, in India the typical retention time of the feedstock in the tropical south is around 30 days, while in the north material is typically digested for 50–55 days, resulting in a digester volume which is around 1.8 times larger (= 55/30). This works well at an average temperature of 15°C, at lower temperatures heating is required [10].

3.3 Planning Phase

The planning phase is most important in each biogas project. Many failures in biogas production are caused by planning errors and unsuitable decisions. General points to be considered are [32]:

- Siting of the biogas plant and layout of the facility is as important as the construction itself. A well-planned AD plant will be a useless facility when installed at the wrong place.
- Substrates need to be appropriate for the installed AD technology. Filling a plant with unsuitable material will result in an unproductive unit (and may also cause technical problems).
- Careless planning of the site may unnecessarily require additional equipment or may cause further labour input.

Knowledge about technology and biological processes is essential for the future biogas plant operator. Literature, biogas seminars and training courses, biogas study tours, individual visits to existing AD plants and direct contact of plant operators are most suitable. Problems to overcome in developing countries are also related to availability of such opportunities (and quality of seminars, courses), costs implicated, high degrees of illiteracy, availability of knowledge and advice at local/regional level.

Among the most important factors when planning an AD plant is a careful and realistic assessment of available substrates and their potential biogas yield.

Labour requirement to run the digester and manage the material flows needs be taken into account. It can vary very much from one site to another, depending on design, general management, knowledge and skills of the operator, the characteristics of available substrates, weather conditions, biogas utilisation paths, etc. Some of the activities are as follows: inspect the digester and all pipes regularly, management of substrates, shredding and pre-composting of farm residues, control of water content and addition of manure/liquid, management of effluent/digestate, control of mixing, monitoring of the biological process, and management of problems.

A survey in southwest China revealed that 61% of the members of a rural community did not believe to have enough labour to run a household digester [22]. Main labourers leaving the village for wage jobs in the city, is one crucial factor. Emigrant labourers often are the young and well educated, who would have been the most likely to adopt biogas technology. It is not only labour requirement for the biogas unit which needs to be considered. Technically it would need more than four pigs to run a household biogas plant in rural China, so that it can meet the household's need for cooking and lighting. Even among skillful pig farmers many villagers are unable to raise more than two pigs [22].

A biogas plant is a long-term fixed investment. When looking at domestic biogas plants in developing countries, investment requirements of some hundred dollar ($) are a significant investment barrier. It is this investment barrier which needs to be decreased, and among possible solutions the following are to be mentioned [10]:

- Investment subsidies. Different programmes exist in different countries. In Cambodia for example, $150 are refunded from the total investment of a family size biogas plant (funds are provided by donors).
- Micro financing. This can be attractive if it allows for an affordable credit for the biodigester.
- Income generation activities. Programmes to enhance and commercialize agricultural products and/or bio-slurry extension programmes can help to encourage uptake of AD.

3.4 Operation of Biogas Plants

Effective management is essential for successful operation of an anaerobic digestion plant and therefore biogas production. In order to operate a biogas plant safely and highly efficient, every plant operator must have detailed knowledge about the technical equipment and the biogas process. This helps to avoid feeding errors and to correctly react in case of problems. Problems can be manifold and include interruption of gas production, insufficient gas quality, formation of scum, feeding problems or problems at the outlet, blocking of piping, and breakdown of equipment. Digestion of energy-rich and easily hydrolysable substrates in general requires more attention, while AD processes run on manure and slurry are more robust and less susceptible to failure.

Operating a biogas plant requires regular and multiple activities in different intervals (ranging from daily routine inspection, to cleaning of reactor e.g. once per year), a continuous process needs more regular attention than a batch process. The operator must be aware of the fact that in case of illness another well-informed person must be able to ensure at least the basic regular biogas plant operation.

In many cases plant operators lack knowledge and put unsuitable substrates in their digesters. Limited and expensive access to training also discourages villagers from utilising biogas, and courses sometimes are perceived as waste of money due to poor quality [22]. In particular materials containing pesticides, disinfectants, antibiotics need to be excluded from the biological process. When digesting waste materials such as food waste knowledge about necessary pre-treatment including hygienisation is imperative.

Repairs will occasionally be necessary and AD plant operators should consider repair costs when calculating economic viability of their facility. Easily available and reliable repair support can be a decisive element for trust in the technology and dissemination in a region, but this is not always the case (especially in rural areas). Often no insurance policy is available, which means that operators need to use their income to pay for repairs [22].

A Chinese adage says "A good message hardly goes beyond the gate, while bad news spreads far and wide" [22]. Failed biogas projects and bad digesters have far more impact on general perception of reliability and economic viability of the technology than good ones. This stresses the importance of a good management of a biogas plant not only for the individual operator but on more general scale. It is not sufficient to build as many AD plants as possible. It is knowledge about adequate operation, individual skills, and availability of reliable support which are crucial.

3.5 Safety Issues

Biogas from any type of digester and from any size of digestion facility is flammable, explosive (in certain mixtures with air/oxygen) and corrosive. Components exposed to AD liquids or to biogas condensation water will be subject to rapid corrosion if unsuitable materials were chosen (this includes the digester itself with various components such as stirring equipment, but also wiring systems and piping in and around digesters).

Carbon dioxide and hydrogen sulfide are both heavier than air, while methane and ammonia are lighter. Heavier components can settle and accumulate at the bottom of tanks and pits. Lighter components can accumulate especially under roofs and ceilings.

By displacing ambient air, biogas can create an environment which is oxygen deficient. Deficiency in oxygen will particularly target brain cells, resulting in judgement and coordination being hindered. At lower oxygen levels ($<10\%$) loss of consciousness will occur and deadly incidents are possible.

- Accumulating methane can create a fire or explosion.
- Ammonia typically causes irritation of eyes and the respiratory tract.
- Hydrogen sulfide is a deadly poisonous gas. In lower concentrations hydrogen sulphide can be perceived by its typical smell (rotten egg smell), which however is not the case anymore at higher concentrations (>150 ppm).

3.6 Ecological and Social Aspects of Decentralized Biogas Digesters

Biogas technology is conducive for maintaining stable power supply with renewable energy, saving fuel and protecting the environment. Through the application of biogas, most of, or even all of the traditional fuel such as fossil fuel, firewood, and straw, will be saved. If the digester produces 400 m^3 of biogas a year, 0.7 tonnes of coal or 1.2 tonnes of firewood can be saved, which is equivalent 2,100 m^2 forest area.

The CO_2 and methane air pollution due to the burning of biomass or coal could be eliminated. In the year 1991 in China more than 2 million tons of greenhouse gas has been reduced as a result of the utilization of biogas. Due to the higher number of digesters in 2010, reduction of greenhouse gas increased to 30 million tons per year [39].

Utilization of biogas offers economic, environmental as well as social benefits. Additionally working cost is substantially cut down in activities such as collecting firewood or raising crops. The time saved in the process of lighting a fire or adding firewood is also significant. It takes about 700 working hours per year in the rural areas to sustain the traditional way of life, while in the case of biogas application only about 200 working hours are needed in activities such as collecting animal manure and applying the sludge-turned fertilizer to agricultural farmland. The comparison shows that 500 working hours are saved in the case of biogas user as against the non-biogas user [7].

Application of biogas provides a solution to problems such as human and animal feces and existing parasite on the site, which together contributes to the improvement of rural sanitation. According to results of relevant tests, most biogas slurries meet the sanitary standard of China for non-hazardous treatment of night soil (96.08% of 426 biogas slurry samples) [34]. Testing further revealed that the retention rate of the total amount of nitrogen in biogas sludge has increased by 46%, while that of amino acid nitrogen has increased by more than 20%. There is almost no loss of the contents of phosphorus, potassium or other nutrients in the sludge. As a result of applying biogas slurry as the fertilizer to the soil, it has been observed that soil porosity, organic substances and pH value have increased respectively. In opposite to large scale fermentation plants, no bad odours can be recognized in the neighborhood of small scale biogas digesters.

Furthermore the fermentation process in the production of biogas is proved to be effective in eradicating parasites' eggs, which contributes to the prevention of infectious diseases and safeguard of the peasants' health. It can also effectively reduce eye syndrome, asthma or other diseases, which are caused by the smoke of traditional cooking style.

4 Utilisation of Products

Anaerobic digestion (AD) contributes to the establishment of environmentally sound waste management. It is currently the most promising way to tackle gaseous emissions from agricultural activity (climate gas emissions: CH_4, N_2O; odour nuisances). In addition, it generates digestate with improved fertiliser value, which results in better nutrient uptake by plants and fewer leaching losses. At the same time the generated energy has the potential to displace other energy sources such as fossil energy. This reduces greenhouse gas emissions and contributes towards a more sustainable energy concept.

4.1 Biogas Utilisation

Biogas has a wide variety of possible applications, the most common ones are:

- Direct use for cooking and lighting (small-scale AD plants at household level usually provide fuel to cover the demand of the household and the agricultural site; biogas burns very cleanly and causes less air pollution than other biomass fuels)
- Utilisation for heat generation
- Generation of electricity (several engine types can be fuelled with biogas; electricity generation is often accompanied by heat generation in combined heat and power plants/CHP)
- Fuel for cars/vehicles
- Feeding into the natural gas grid (after upgrading to natural gas quality; now one standard in industrialized countries when produced at large scale; different upgrading technologies exist)

Biogas can also be used to provide cooling (agricultural storage facilities, pig stables, nearby hospitals or other buildings).

When used to heat and light greenhouses, biogas will significantly increase the level of carbon dioxide, which can result in better plant growth (simple form of CO2 farming).

It is not necessary to make use of biogas directly at the production site. Local biogas grids can be an intelligent solution to provide biogas to where it can be used at highest efficiency [29].

Compared to other renewable energies, it is one advantage of the energy carrier biogas that it can be stored to be used according to fluctuating demands or to availability of alternative energies. Biogas can be a particularly advantageous choice e.g. in hybrid power systems for electricity supply in remote areas or islands [8].

4.2 Digestate

In the organic form, nitrogen must be first mineralised. Ammonia can be converted to nitrate for plant uptake, while some plants may use ammonia directly. The extent of nutrient uptake by plants depends on the time of application and there is always the possibility that nutrients will be leached from the soil when plants are unable to take them up. AD converts much of the organic N into ammonia, yielding a digestate with 60–80% of the total nitrogen content in the form of ammonia [3]. This makes it highly predictable, minimises leaching losses and is in line with the development of good agricultural practices.

The improved fertilizer value of AD digestate is to be considered as economic advantage of the AD unit. Other fertilizers are displaced and higher biomass yields are possible, as reported for napa cabbage, cauliflower [22].

Compared to granular fertiliser, digestate has the drawback that it is a pre-determined blend and its constituents cannot be altered. Once the soil requirement of the first nutrient has been met then no more should be applied to the soil. In a closely monitored agricultural system following rules of good practice other fertiliser is therefore likely to be required to top up the nutrient needs of the soil.

Since most of the nitrogen in digestate is available in the form of ammonia, digestate spreading technologies should be given special attention and priority should be given to techniques minimising ammonia losses.

Biogas can mitigate fecal-borne and parasitic diseases. Agricultural use of untreated human and animal waste is among the main pathways for transmission of serious diseases. In addition, odour emissions are significantly reduced through the anaerobic process and the system attracts less flies.

Digestate which is not fit for landspreading (e.g. due to contamination with heavy metals) must be disposed of.

5 Decentralized Biogas Technology

5.1 General

Climatic condition in most areas of Latin America and Africa is optimal for biogas plants, especially in the Caribbean and the tropical countries. The amount of available substrates for biogas digesters is very high. Nevertheless, in all of Latin

America and Africa there are only a few thousands of such plants. Biogas plants which were built in the 1980s or 1990s are almost all out of order.

Reasons for inadequate dissemination are certainly lack of support from government and lower population density. The causes for failures of existing biogas plants resulted from quite a few aspects, such as inadequate training of users, material errors, technical defects, reduced animal holdings or water problems etc.

At the present some organizations have established and started series of national and international biogas programs so as to promote development of biogas technologies as a solution to the environmental problems and covering the energy gap in Latin America and in Africa.

5.2 Biogas Technology in China

5.2.1 Overview

China is the biggest biogas producer and consumer in the world. The popularization and scientific research of biogas in China began around 1970. After more than 3 decades of development, by the end of 2007, around 26 million household biogas digesters had been counted all over the country. In China, the most basic applied type of household digester is the hydro-pressure digester with 6–10 m^3, also known as fixed-dome type digester. A large number of biogas digesters are set up in various geographic landforms and climatic areas, either in the areas of plateau in western China, or in plain areas of central China, and both in the cold climatic conditions of northern China and in the subtropical and tropical regions of southern China. Large and medium-sized farm biogas projects have increased in number from ca. 800 in 2000 to ca. 8,500 by the end of 2007, and annual biogas production has reached 10.4 billion m^3.

The Chinese Government has always attached great importance and provided strong support for rural biogas development. The biogas development enjoys a very good policy and legal environment, a relatively complete and efficient work net, which covers marketing of biogas application, technical support and maintenance outlets all together. According to the biogas development plan in 2020 it is planned that 10,000 large-scale biogas projects on livestock farms and 6,000 biogas projects utilizing industrial organic effluent will be built. About 80 million rural households (300 million people) will use biogas as their main fuel.

A relatively complete and efficient network has been developed, which covers the adoption of marketing biogas application, technical support and maintenance all together. More than 8,000 rural energy offices have been established in more than 1,900 counties and towns with 40,000 full-time staff members, who are responsible for the administration of biogas in rural areas.

The Ministry of Agriculture has also focused on education, advocacy and training: they have not only published brochures of biogas training materials,

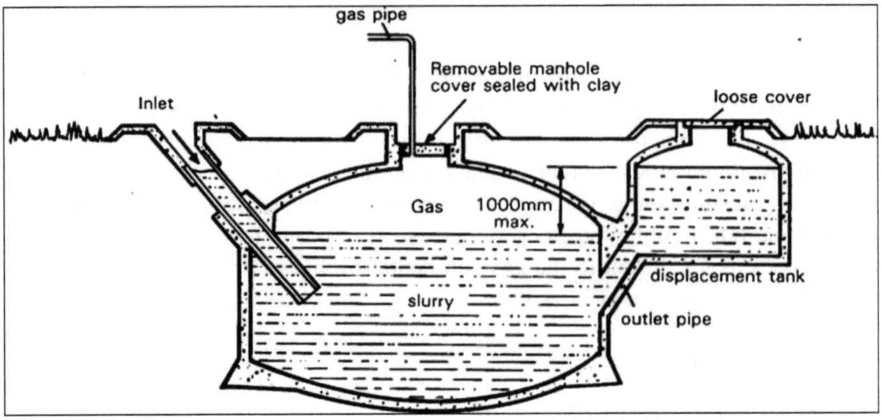

Fig. 5 Construction of fixed-dome biogas digester [17]

television and radio programs, but also organized many training courses, so that the technicians and farmers can access the knowledge of biogas technology.

5.2.2 Typical Household Biogas Plants and its Maintenance

In China, the most basic applied type of digesters is the hydro-pressure digester, which is usually an underground, circular and shallow facility with a domed top, so it is also called as fixed-dome type digester (Fig. 5). The volume of these household digesters is ca. 6–10 m^3, and the life expectance of the digester is 15 years [38].

The dynamic system is adopted for the facility, i.e., when the gas is produced and stored in the upper part of the digester, the gas pressure will increase to push manure and slurry at the bottom of the pit to flow up into the compensating tank. When the gas is used up, the gas pressure will reduce and the slurry will flow back into the digester chamber to push the gas up for usage. As a result of the constant circle of gas consumption and gas generation, the pressure is always maintained in a balanced state between the compensating tank and the fermentation room.

Construction of the reactors is carried out by trained technicians and members of the household. The construction is carried out in the following stages:

- Site selection: at least 10 ~ 15 m away from the shallow well, max. 25 m between the digester site and the biogas appliances.
- Preparation of the material (for example: brick, cement, sand etc…). The basic construction materials are concrete and bricks, which are easily available and commonly used in rural China.
- Positioning and excavation according to the layer construction
- Construction of the digester body, including bottom, walls, roof and sealed layers, as well as brushing starch to prevent leakage

- Construction of the gas transfer pipes for the biogas usage
- Quality control to ensure the construction quality and a normal gas production

These biogas plants are particularly simple to operate and maintain. Waste from the animals and toilet flows directly into the reactor. Sometimes manual fed works are also required in order to maintain the frequent input and output within the biogas digester, otherwise the gas production would drop if fresh material could not be supplied in time. About 20 kg fresh materials have to be fed into the digester and 20 kg slurry to be taken out of the plants on a daily basis. Some harmful substances are not allowed to put into the digester, which include all kinds of toxic pesticide or the crop stems with that pesticide, heavy metal compound and industrial waste water containing toxic substances etc. A regular check over pipes, joints, switches, cover of entrance, exit pipes and water column pressure is necessary. When the pressure in the digester is too high, the gas must be released immediately to protect the gas box or prevent the digester cover from bursting open, which could result in an accident.

The winter temperature in northern of China is relatively low, which is unfavorable for biogas production. A review of the development of household biogas in China shows that one of the key factors which impeded the utilization of biogas in northern China was: how a biogas digester lives through the winter. Therefore measures must be taken for rural biogas digesters to overwinter before the arrival of winter, so as to ensure a normal gas production, and insulation should be adopted such as covering plastic layer or material stack, annular ditch, pouring of hot water or integration of solar cell.

Some interrelated experiment shows that, use of solar radiation, combining with heat-proof shed, building annular ditch, covering with corn and straw could give a significant temperature raise for the biogas plant, and a long time heat preservation guarantee a normal gas production for the biogas plant even in winter.

5.2.3 Economical Assessment of Household Biogas Plants in China

The cost for a small household biogas digester is around 220 € and for a larger model (with pigpen, greenhouse, toilet, and the biogas digester) around 550 €.

After the completion of biogas plants, yearly, a total 900 € in cost-cutting and income increasing can be realized, which includes,

- 45 € for the replacement of coal and saving the cost of fuel and electricity though the application of biogas lamps
- 260 € for the saving of fertilizer application, feeds for livestock
- 610 € for the increased yield of crops and livestock

Therefore, in the first year after starting up a digester, the investment cost can be recovered.

5.3 Biogas Technology in Latin America

Climatic condition in most areas of Latin America and Africa is optimal for biogas plants, especially in the Caribbean and the tropical countries.

5.3.1 Colombia

Colombia has ideal conditions for implementation and continuation of biogas technology. It is located in northwestern South America. Colombia has 42 million inhabitants, 74% of which live in urban areas and cities. Depending on the altitude, Columbia can be divided into several climatic zones. In the lowlands the average annual temperatures is over 24°C. The amount of available substrates (the live-stock and the agricultural) is enormous, which are suitable for biogas plants. There are 2.5 million pigs and more than 25 million cattle in Columbia.

Main incentive to introduce biogas technology is the severe pollution of surface waters and thus the water resources through the deposits from animal husbandry. The first project for the construction of biogas plants started in the middle of the 80s, with the assistance of a German consulting company. Between 1985 and 1992 altogether 25 biogas plants had been built. The types of biogas plants and their sizes were very varied: a floating-drum plant, a tunnel plant, an upflow anaerobic sludge blanket and some fixed-dome plants with the size from 14 to 115 m^3. Some of the plants are equipped with plastic balloons to include biogas equipment. These plants were designed for medium and large pig and cattle breeders who had between 20 and 2,000 animals [19].

5.3.2 Jamaica

Jamaica is the third largest island in the Greater Antilles. The main island has an area of 10,991 km^2. Climate in Jamaica is tropical and temperature differences are very small throughout the year. In Kingston, the mean monthly temperature in January is 25 and 27°C in July, in the central plateau it is about three degrees lower. In Jamaica, approximately half of the 3 million inhabitants live in rural areas and 30% of the working population is employed in the agricultural sector. The history of agricultural biogas plants dates back to end of 1970s.

Back in 1978 The Scientific Research Council (SRC) started research and development activities in the area of biogas. It has always been regarded as the first and most important advocator of biogas technology in Jamaica. In the phase from 1988 to 1992, 89 biogas plants were built, mostly by Chinese fixed—dome digester type. At the end of 1992 there were only 15 of the 89 plants in operation. A new phase of the biogas plants began in 1993. In the new plants, all the manure and waste water flow by gravity into the plant. Domestic waste water, toilet effluents in particular garden and kitchen waste can be treated by the new plants. 120 biogas plants have been built in this phase.

5.3.3 Bolivia

Bolivia lies in west-central South America and has a total area of 1,098,580 km^2, of which 14,190 km^2 is water. The average annual temperatures are depending on the location and height in a range from 6°C to about 27°C. Bolivia has a population of about 7 million inhabitants, but the most of the population live on an area with only 30% of the total area of the country. Bolivia has extensive oil and gas reserves. One important motivation for biogas plants is to use the fermenters for the production of fertilizers from organic materials. Therefore the government decided to develop biogas technology as part of agricultural production, to prevent ecosystems from being destroyed by agricultural activities.

In 1986, a biogas cooperation project was started by GTZ and the Universidad Mayor de San Simon (UMSS) in Cochabamba. A total of 27 plants were constructed but in 1988 there was only one still functioning. From 1989 to 1992, around 35 plants were constructed [33].

Because of the low temperatures, operation of biogas plants at high altitude is the theoretical limit of biogas technology. In 2002 a draft of biogas digesters for the regions above 2,000 m in Bolivia was developed. In 2003 a tubular biogas digester was installed in the Altiplano, 4,100 m over sea level, with ambient temperatures under 0°C (Fig. 6). From 2002 to 2006 about 250 tubular biogas digesters were installed by departments Cochabamba and La Paz. Until now more than 1,000 biogas digesters are working in Bolivia. Most of them are the type of tubular plastic biogas digester [21].

5.3.4 Cuba

The republic of Cuba is an island country in Caribbean Sea. About 11 million people live in Cuba, of which about two million live in the capital, Havana. Cuba has rich materials, which is fit for the biogas plants:: about 75 million cubic meters of organic waste per year is available, excrement of cattle and pig and from sugar cane, remaining sugar scraps after filtering the pressed sugar juices. In Cuba, the NGO Cubasolar and church-based organization KATE are the powerhouse for the development of biogas technology, with strong support from the government. The history of biogas in Cuba dates back to 1940s, with the introduction and promotion of floating drum digesters and fixed dome digesters, the 1980s witnessed a boom in the development of biogas technologies in Cuba. In the 1980s, about 400 biogas plants were built as small plants in agriculture in Cuba. In 2000 only 50–60 plants were in operation.

At present there are 700 biogas plants in Cuba; the largest one, located in Havana, was jointly developed with United Nations Industrial Development

Fig. 6 Biogas digester in the Altiplano, 4,100 m above mean sea level; gas tank (**1**), digester (**2**)

Organization (UNIDO). Many of these plants now need an overhaul. Thus, in 2007, the repair and modernization of these systems began. From 2008, about 450 new biogas plants will be built gradually.

5.4 Biogas Technology in Africa

5.4.1 Ethiopia

Covering an area of 1,097 million square kilometers, Ethiopia is a country in northeastern Africa. The differences in climate in Ethiopia are primarily due to the altitude. Ethiopia has a population of 79.1 million, but only 16% of them are living in urban area. Ethiopia has the largest livestock population in Africa (approximately 61.5 million in 2001). Most of the households use crop residue, dung, kerosene or fire wood as fuel, but only a few people use electricity. Consequently, there is a large gap in the energy supply system and the market of biogas is enormous.

The first introduction of the biogas technology dates back to as early as 1979, with the first batch type digester being constructed at Ambo Agricultural College. Since then, around 1,000 biogas plants have been built up all over the country, sizes of which ranging from 2.5 to 200 m^3 in households, community, and governmental institutions. At present, approximately 40% of the biogas digesters are out of function. Therefore, an increasing number of people are skeptical about the development potential of biogas in Ethiopia, and the benefits for the farmers. World Vision Ethiopia has recently introduced the application of biogas under its Appropriate Agricultural Technology Promotion Initiative (AATPI) and some 150 plants have been built. As such, the total number of completed biogas plants in Ethiopia would reach 600–700 [14]. In 2007, with the help of Netherlands Development organization (SNV) a national biogas program was started in Ethiopia to build on and to further develop existing institutions and organizations for biogas technology.

5.4.2 Lesotho

The Kingdom of Lesotho has a total area of 30,000 km^2 and a population of almost 1.8 Million people. The average annual temperature in the capital Maseru is 15°C. About 84% of Lesotho's total population lives in rural areas. In Lesotho, sheep, goats, cows, oxen, and other animals are common in even the most rural areas. In rural areas, Paraffin fuel is mainly used for cooking and heating, and also biomass and dried animal dung.

Biogas technology has been used in Lesotho since the 1980s. About 80 household-size biogas digesters constructed in Lesotho were built between 1980 and 1990. But in 2002 none of them were operational. In 2004, "Technologies for Economic Development" (TED) was established with the main focus on renewable energies, sustainable sanitation and climate protection. So far, more than 300 household biogas digesters have been constructed by TED. At the beginning, the digester was used as storage tanks for the untreated wastewater to solve the sanitation problem and from this resulting high costs of operation. But now some of these digesters are also fed with organic solid waste or animal dung.

5.4.3 Tanzania

Tanzania is a country in East Africa with 945,000 km^2. There are around 41 million inhabitants. The average daily temperatures amount to 26.5–30°C. Livestock production by small farmers in the region is used to meet their own demand, as well as for commercial purposes. According to a date research, there are totally 886,500 cows, 699,300 goats and 47,500 pigs in the region of Kagera. They produce 4.3 million tons fresh cow dung and 21,400 tons pig dung per year. Considering water supply, 34% of households have an adequate water supply in the vicinity to ensure the operation of a biogas plant. Therefore it results in a plant's requirement of about 132,000 small biogas digesters in the region of Kagera.

In the early 90s several small biogas plants were constructed by the Tanzanian church in collaboration with its Danish partner organization in this region. During that period, most of the biogas plants were equipped with either domed digesters or tubular plastic digesters. Up to 2007, there were altogether 2,821 built biogas plants in Tanzania, among which 2,444 were fixed dome digester and 429 were tubular plastic digesters. However, in 2007 there were only 1900 biogas plants still in operation due to their limited service life [20]. Since 2008 Berlin regional group of Engineers Without Borders Germany and Mavuno from Tanzania work together on the project "Biogas support for Tanzania (BiogaST)" for the development and construction of decentralized small biogas plants to use of biogas as an energy source for cooking and other energy applications.

5.5 Assessment of Dissemination of Biogas Technology in Latin America and Africa

One reason for the low dissemination of biogas technologies in Latin American and Africa countries is certainly the lack of support from the government. The lower population density compared to China and India makes the environmental problems seem not quite so urgent. The other constraints for the widespread biogas technology include:

- High investment costs
- Lack of legal framework

The causes for the failure or failures of existing biogas plants are listed as follows:

- The operators/users were not sufficiently trained (information about handling and maintenance was inadequate)
- Lack of motivation of the operator/user
- Material errors and technical defects
- Reduced animal holdings
- Evacuation of ownership and water problems etc.

6 Outlook

Anaerobic digestion has potential for more widespread dissemination especially in developing countries and countries in transition. Though implementation at household level remains important, scale of operation at large-scale and municipal level will gain importance, which is closely linked to improved waste management schemes, industrialization of agriculture, and increasing energy demands.

Biogas technology qualifies as a CDM project (avoiding uncontrolled methane emissions from dumped waste or landfills). According to the Kyoto Protocol CDM allows industrialised countries with a greenhouse gas reduction commitment to invest in projects that reduce emissions in developing countries as an alternative to more expensive emission reduction in their home countries. This may further boost implementation of AD in developing countries [37].

Among the most pressing issues related to advancing sustainable dissemination of AD technology are solutions to overcome the investment barriers, educational issues, availability of help and advice (calculations on economic viability, repair services, advice on plant operation/biological process) and implementation of schemes allowing for reliable income generation via biogas (e.g. remuneration for biogas energy, digestate management schemes and programmes, favourable legal frameworks).

At the present some organizations have established and started a series of the national and international biogas programs to facilitate the biogas development in Latin-America and in Ethiopia. The "Biogas for better life" project has the goal: until 2020 there will be 2 million biogas plants installed in Africa. "IGNIS"—project committed for the waste management in Addis Ababa, Ethiopia, in which the biogas development is also an important task. GTZ, UTEC, SNV, Swiss contact, SEQUA and RELUX have initiated some cooperation with local partners in Latin-American and Africa to use and dissemination of biogas technology. They use their experience in biogas programs in a number of countries to support the setting up of biogas plants, to train more technical personnel and promote the general publicity of biogas technology. In this way, more and more people will understand the benefits of biogas and can thus be motivated to develop and disseminate the biogas technology, mobilize the enthusiasms to dissemination and development of biogas technology, so as to realize a feasible model in the rural areas of Africa and Latin America featuring stable gas supply, minimized environmental pollution and sustainable supply of resources and energy.

References

1. An BX (2005) Biogas technology development in the developing countries. J Agric Sci Technol 4(2005):75–82 (Nong Lam University)
2. Anonymus (2010) Leitfaden Biogas. http://www.gerbio.eu/neu/uploads/media/biogas-handbuch 01 pdf Accessed Dec 2010
3. Banks CJ, Salter AM, Chesshire M (2007) Potential of anaerobic digestion for mitigation of greenhouse gas emissions and production of renewable energy from agriculture: barriers and incentives to widespread adoption in Europe. Water Sci Technol 55(10):165–173
4. Batstone DJ, Keller J et al (2002) Anaerobic digestion model no. 1. IWA scientific and technical report no. 13, IWA Publishing, London
5. Bickel H et al (1995) Natura. Themenband Stoffwechsel, Klett, Stuttgart
6. Bischofsberger W et al (2005) Anaerobtechnik, 2nd edn. Springer, Berlin
7. Bo Yu (2007) The analysis of rural household energy choices and the policy of new energy extension—empirical study around Nanjing city, Jiangsu province, Nanjing Agricultural University, June 2007
8. Borges Neto MR, Carvalho PCM, Carioca JOB, Canafistula FJF (2010) Biogas/photovoltaic hybrid systems for decentralized energy supply of rural areas. Eng Policy 38:4497–4506
9. Braun R (1982) Biogas Methangärung organischer Abfallstoffe: Grundlage und Anwendungs-beispiele. Springer, Vienna
10. Buysman E (2009) Biogas for developing countries with cold climates. WECF Women in Europe for a Common Future, geres Cambodia

11. Cimatoribus C (2009) Simulation and nonlinear control of anaerobic digestion. Dissertation University of Stuttgart, Stuttgarter Berichte zur Abfallwirtschaft, vol 96. Oldenbourg, Munich
12. Cimatoribus C (2010) Vergärung. In: Kranert M, Cord-Landwehr K (eds) Einführung in die Abfallwirtschaft, 4th edn. Vieweg + Teubner, Wiesbaden
13. Deublein D, Steinhauser A (2011) Biogas from waste and renewable resources, 2nd edn. Wiley-VCH, Weinheim
14. Eshete G et al (2006) Report on the feasibility study of a national program for domestic biogas in Ethiopia, p 25
15. FNR (2006) Fachtagung für Nachwachsende Rohstoffe. Biogasgewinnung und -nutzung, 3rd edn. Gülzow
16. Fraenkel PL (1986) Water lifting devices, Rome, food and agriculture organization of the United Nations, 1986, ISBN 92-5-102515-0
17. Gehring M, Raninger B, Rundong L (2008) Derzeitiger Stand und neueste Entwicklungen der Bioabfallvergärung in China. In: Bilitewski B et al (eds) 6. Fachtagung Anaerobe biologische Abfallbehandlung, vol 57. Forum für Abfallwirtschaft und Altlasten, Dresden, pp 119–131
18. GTZ Isat (2010? report not dated, retrieved in 2010) Biogas digest, vol 1, biogas basics. Report produced for the ISAT website on the order of the GTZ project information and advisory service on appropriate technology (ISAT)
19. GTZ-GATE (1999) Biogas digest, vol 4, biogas country report, information and advisory service on appropriate technology. Eschborn, Germany
20. GTZ (2007) Feasibility study of a national domestic biogas program in Tanzania—biogas in Tanzania, p 23–24
21. Herrero JM (2008) Biodigestores familiares: guía de diseño y manual de instalación. GTZ-Energía
22. Jian L (2009) Socio-economic barriers to biogas development in rural southwest China: an ethnographic case study. Human Organiz 68(4):415–430
23. Kapp H (1984) Schlammfaulung mit hohem Feststoffgehalt. Dissertation, Stuttgarter Berichte zur Siedlungswasserwirtschaft, vol 86. Oldenbourg, Munich
24. Lehninger A (1983) Bioenergetik, 3rd edn. Thieme, Stuttgart
25. Liebetrau J (2008) Regelungsverfahren für die anaerobe Behandlung von organischen Abfällen. Dissertation, Manuskripte zur Abfallwirtschaft, vol 9. Rhombos, Berlin
26. Löffler D, Kranert M (2010) Simulation-based evaluation of control strategies for anaerobic digestion. ORBIT 2010, organic resources in the carbon economy. In: 7th international conference, 29.06.2010–03.07.2010, Heraklion Crete, Greece, proceedings p 71 and CD-ROM
27. Loll U (ed) (2002) Mechanische und biologische Verfahren der Abfallbehandlung. Ernst und Sohn, Berlin
28. Madigan MT et al (2000) Brock biology of microorganisms, 9th edn. Prentice Hall, Englewood Cliffs
29. Panic O, Hafner G, Kranert M, Kusch S (2011) Mikrogasnetze-eine innovative Lösung zur Steigerung der Energieeffizienz von Vergärungsanlagen. Energie Wasser-Praxis 2(2011): 18–23
30. Pèrez Porras J, Gebresenbet G (2003) Review of biogas development in developing countries with special emphasis in India. SLU, Department of Agricultural Engineering, rapport 252, Uppsala
31. Ratkowsky DA et al (1983) Model for bacterial culture growth rate throughout the entire biokinetic range. J Bacteriol 154(3):1222–1226
32. Sasse L, Kellner C, Kimaro A (1991) Improved biogas unit for developing countries. Deutsche Gesellschaft für Technische Zusammenarbeit (GTZ) GmbH, Eschborn
33. Schwarz W (2006) Landwirtschaftliche Nutzung der Biogas-Technologie in Lateinamerika, master thesis, University of Stuttgart. Institute of Sanitary Engineering, Water Quality and Solid Waste Management

34. Sun J et al. (2006) Functions of biogas construction on public health in rural areas. Chinese J Health Edu 22(11)
35. Suryawanshi PC, Chaudhari AB, Kothari RM (2010) Mesophilic anaerobic digestion: first option for waste treatment in tropical regions. Crit Rev Biotechnol 30(4):259–282
36. Van Lier JB et al (1997) High rate anaerobic waste water treatment under psychophilic and thermophilic conditions. Water Sci Technol 35(10):199–206
37. Vögeli Y, Zurbrügg C (2008) Biogas in cities—a new trend? Sandec News 9(2008):8–9
38. Wang H (2005) Biogas plant in China—status and development. Master thesis University of Stuttgart, Institute of Sanitary Engineering, Water Quality and Solid Waste Management
39. Zhang P, Wang G (2005) Contribution to reduction of CO_2 and SO_2 emission by household biogas construction in rural China: analysis and prediction, Transactions of the CSAE 2(12)

Use of Cement Kilns for Managing Hazardous Waste in Developing Countries

Yukari Ishikawa and Sunil Herat

Abstract Most developing countries seldom have possessed hazardous waste incinerators or decomposition technologies for treatment of hazardous waste such as PCB stockpiles, obsolete pesticides and other hazardous waste. For solving this issue, using cement kilns as a hazardous waste management option is gaining increasing favour around the world. The cement industry is an intensive energy consuming industry. Reusing the hazardous waste as the substitution of raw materials, fuel or gypsum during the cement manufacturing process provides an energy and material recovery which could be described as a great example of win–win situation. This chapter discusses the key aspects in hazardous waste management by cement kilns.

1 Production Process of Cement

Cement is the second most consumed material in the world after water [2]. It is a finely ground, inorganic and non-metallic powder, and the most important ingredient of concrete. It is the necessaries of construction industry and an essential for the development in any country [29]. Cement manufacture is a resource and energy intensive industry. Producing a tonne of cement needs 1.5–1.7 t of raw materials, 60–130 kg of fuel oil as well as around 105 kW h of electricity [2, 35]. About 5% of the global anthropogenic CO_2 emissions are originated from the cement industry [21].

Y. Ishikawa · S. Herat (✉)
Griffith School of Engineering, Griffith University, Queensland, QLD 4111, Australia
e-mail: s.herat@griffith.edu.au

Y. Ishikawa
e-mail: y.ishikawa@griffith.edu.au

A. Karagiannidis (ed.), *Waste to Energy*, Green Energy and Technology,
DOI: 10.1007/978-1-4471-2306-4_6, © Springer-Verlag London Limited 2012

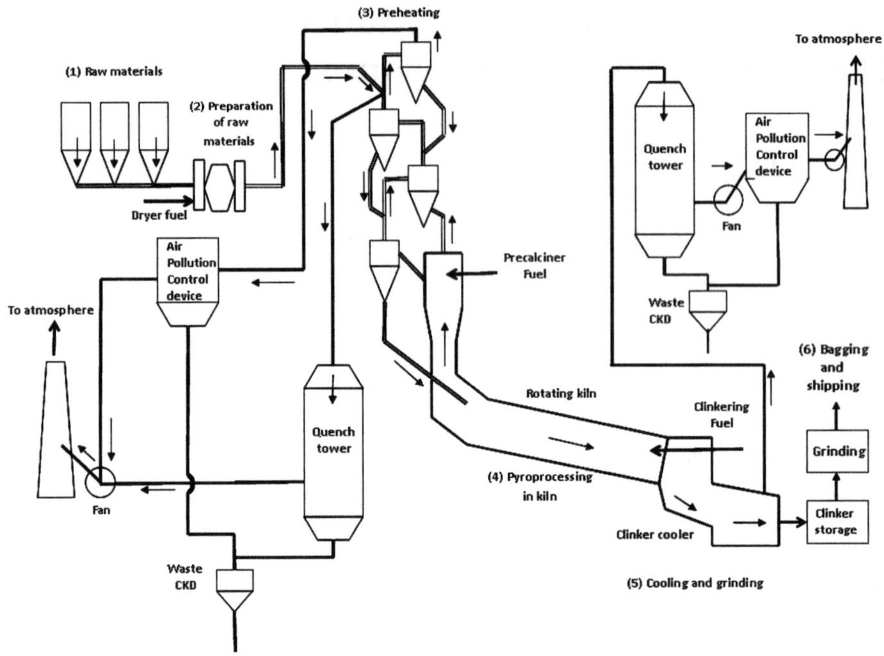

Fig. 1 Cement process

1.1 Production Steps

The production of cement usually consists of following major steps:

1. Acquisition and transportation of raw materials
2. Preparation of raw materials
3. Preheating
4. Pyroprocessing (high temperature reacting) of raw materials to form cement clinker in kiln
5. Cooling and grinding clinker into cement
6. Bagging and shipping

The principal chemical components required for the production of cement are calcium (Ca), silicon (Si), aluminium (Al) and iron (Fe). Calcium is provided by limestone, which is usually quarried near the cement plant site. Silicon and aluminium are provided by a mixture of clay, shale or sand. Iron is provided by iron ore or steel mill scale. These raw materials contain approximately 75% calcium carbonate ($CaCO_3$), 15% silicon dioxide (SiO_2), 3% aluminium oxide (AlO), 2% ferric oxide (Fe_2O_3) and 5% other minerals. The raw materials are selected, crushed, ground and proportioned so that the resulting mixture has the desired fineness and chemical composition for the pyroprocessing system (Fig. 1).

Modern cement plants have preheating system that is positioned before the kiln and preheats the mixed raw materials for saving energy. Hot exit gas from the kiln heats the mixed raw materials as they swirl through the cyclones [39]. The pyroprocessing takes place in a cylindrical rotary kiln which is an elongated steel cylinder with a length to diameter ratio between 10 and 40, ranging in size up to 250 m in length and 8 m in diameter, lined with refractory brick and inclined at angle 3–6° [14, 35]. The kiln rotates at about 70 revolutions per hour in older generation plants and about 180 revolutions per hour in newer plants. The raw materials are fed at the elevated end of kiln and are moved slowly with the rotation of the kiln down towards the firing end where heat is applied with coal, gas or oil flame, and then they become clinker. The hot clinker is tumbled to cool clinker by a great cooling system. To save energy, heat recovered from this cooling process is recirculated back to the kiln or preheater tower. The cooled clinker is kept in clinker storage until it is ready to be ground into the grey powder and packed as Portland cement [39].

1.2 Thermal Zones

There are three distinct thermal zones within an operating cement plant, as described below.

1. Drying and preheating zone (20–800°C)
2. Calcining zone (600–1,000°C)
3. Burning zone (clinkering zone) (1,250–1,500°C)

The material in the drying and preheating zone reaches a temperature of about 800°C. In this zone, all water in the material is evaporated. The calcining zone is set after preheating zone. Carbon dioxide (CO_2) is driven off from the limestone material and calcium oxide (CaO) is, thus, formed. The material temperature in this zone reaches 1,000°C. After calcination, the burning zone (also called as clinkering zone) is placed. The temperature of burning zone reaches 1,500°C that leads chemical reactions to form a clinker. The clinker is in a semi–liquid state at this stage, but cooled down by the grate cooler. The cooled clinker is referred to as cement clinker. The clinker is then finely ground with an addition of about 5% gypsum (calcium silicate: Ca_2SiO_4) to give the final cement product.

During the cement manufacturing process, it is essential that the temperature in the feed materials reaches 1,500°C in order to form the clinker. To achieve this clinkering temperature, combustion gas temperature in the burning zone of the kiln must generally exceed 1,700°C. In addition, residence time of combustion gas in the burning zone of the kiln ranges from 2 to 5 s that depends on the size of the kiln. The overall gas residence times during the process can reach 10 s.

Cement kilns operate in a counter current manner. The combustion gas and waste dust flow counter to the material, exiting at the elevated end of the kiln. Exhaust gas consists primarily of nitrogen (N_2), carbon dioxide (CO_2), water (H_2O), sulphur and nitrogen oxides (SO_x, NO_x) together with the fine particles of the feed and clinker materials. These particles are entrained in the gas flowing through the kiln, removed by electrostatic precipitators or baghouse filters fitted at the end of kiln and returned back to the kiln as the feed materials or sold as cement by–products for other industrial uses. The alkaline environment caused by the feed material in the kiln acts as a trap for sulphur dioxide (SO_2). Thus, after passing through pollution control devices, the combustible gas primarily contains carbon dioxide and water vapour and it is discharged to the atmosphere through a stack.

2 Characteristics of Cement Kiln as a Hazardous Waste Management Option

The characteristics of the cement kiln as an effective device for managing solid and hazardous waste can be summarised as follows:

1. High temperature and long residence time
2. High thermal capacity
3. Alkaline environment
4. Minimum amount of waste generated

2.1 High Temperature and Long Residence Time

The most important considering matter during the usage of cement kilns for managing hazardous waste is the by-production of persistent organic pollutants (POPs) such as polychlorinated dibenzo-*p*-dioxins (PCDD), polychlorinated dibenzofurans (PCDF), polychlorinated biphenyls (PCB) and hexachloro benzene (HCB) which are often detected from commercial waste incinerators. For preventing the by-production of POPs, the complete combustion of input materials is required. It is generally recognised that all organic compounds are adequately destroyed if they are exposed to a temperature of 1,200°C for a residence time of 2 s under oxidising conditions. For incinerating PCB, the US TSCA PCB incineration criteria states that a temperature of 1,200°C and retention time of 2 s at 3% oxygen are required [21]. Also, Council Directive [8] of the European Parliament and of the Council requires a temperature of 850°C for at least 2 s for the incineration of non-chlorinated hazardous waste and 1,100°C for 2 s retention time for organic substances containing more than 1% halogen at 2% oxygen. The conditions in the burning zone of the cement kiln exceed these requirements by a wide margin. To produce the cement clinker, the material inside the kiln must reach a temperature of 1,500°C while the combustion gas temperatures reach 1,700°C. These sustaining high combustion gas temperatures and long gas residence time (6–10 s) combined with intense turbulence inside the kiln ensure efficient destruction of even the most stable organic compounds.

2.2 High Thermal Capacity

The large amount of heated materials in the cement kiln ensures that the temperature in the kiln is stable without significant oscillations. Thus, in the case of an emergency shutdown due to operational problems, the flow of any organic waste can be halted before the temperature falls below the critical values.

2.3 Alkaline Environment

The contents inside the kiln are alkaline. Therefore, virtually all of the chlorines entering the kiln and hydrogen chlorides (HCl) formed during the combustion of chlorinated waste are neutralised by forming into calcium chloride ($CaCl_2$), sodium chloride (NaCl) and potassium chloride (KCl), relatively non–toxic compounds. Thus, emissions of hydrogen chloride from kilns are significantly lower than commercial incinerators. Most of the sulphur oxides (SO_x) are similarly trapped as calcium sulphate ($CaSO_4$).

2.4 Minimum Amount of Waste Gerenation

The combustion of waste in commercial incinerators generates ash which needs to be disposed. In contrast, there is no ash equivalent in the cement production process. The only by-product of its process is cement kiln dust (CKD), which could be recycled back to the kiln or recycled in other industrial processes. Any incombustible materials such as metal in the waste become incorporation of cement clinker that eliminates disposal problems.

3 Benefits of Managing Hazardous Waste in Cement Kilns

The benefits associated with the management of solid and hazardous waste in cement kilns can be summarised as follows:

1. Energy recovery
2. Conservation of non renewable resources
3. Reduction in cement production costs
4. Facilities already exist
5. Reduction in waste transportation costs and risks

3.1 Energy Recovery

The cement industry is an energy intensive industry as very high temperatures are required in order for the proper chemical reactions to occur in the kiln. The energy sources used by the cement industry includes coal, petroleum coke, fuel oil and natural gas and also a wide range of materials having a calorific value such as used oil, waste tyres and organic solvents. The use of secondary fuels reduces the industry's consumption of primary fossil fuel and, therefore, contributes to the principles of energy saving and sustainable development.

Burning solid and hazardous waste as supplementary fuel allows for the recovery of significant amounts of energy from these waste materials. This source of energy is one of the primary reasons for the cement industry's interest in burning solid and hazardous waste as fuels.

3.2 Conservation of Non Renewable Resources

A significant advantage of using solid and hazardous waste as a supplemental fuel in the cement kiln is the conservation of non renewable fossil fuels such as coal, gas and oil. It is possible to replace between 25 and 50% of the energy supplied by coal by using hazardous waste as fuel. For example, if only 25% of the energy used in the production of cement in the United States were replaced by hazardous waste, then 3.8 million tonnes of domestic coal or 14.4 million barrels of domestic crude oil could be saved each year [24].

3.3 Reduction in Cement Production Costs

The production of cement is an energy intensive process. The energy costs generally account for about 40% of the cost of clinker production. Thus, by using the cheaper hazardous waste as fuel, the industry can significantly reduce their manufacturing costs. The waste fuel typically has a heat value of 24 (GJ/t) which is somewhat less than coal but it is available at a fraction of the cost [5].

3.4 Facilities Already Exist

One of the key advantages of using cement kilns is that the technology and facilities are already in place. Thus, there will be significant reductions of capital expenditure instead of siting a new waste disposal plant. A small capital investment of up to $5–10 million is required to adapt a cement kiln to manage solid and hazardous waste whereas building a new incinerator facility could cost around $50 million [3]. In addition, use of a cement kiln, as opposed to the construction of a new incinerator facility, does not result in the creation of a new source of emissions. Therefore, cement kilns provide an attractive option for the burning of large volumes of waste.

3.5 Reduction in Waste Transportation Costs and Risks

Most developing countries do not have dedicated hazardous waste incinerators or any decompositional technologies for hazardous waste. Therefore, these waste are sometime abandoned and cause undesirable contamination to around environment. Otherwise, they have to be shipped to some advanced countries for adequate treatment, mainly high-temperature combustion in dedicated incinerators. However, these are usually high costs and give a pressure to governments of developing countries. Additionally, it contributes to increase the environmental risks accompanying the accidents on the way of transportation of hazardous waste. The use of cement kilns to manage solid and hazardous waste reduces the amount of waste transportation and transportation costs by having many cement plants sites close to waste generators rather than one central facility.

4 Types of Most Suitable/Not Recommended Hazardous Waste for Cement Kilns and their Fate

In the cement manufacturing process, hazardous waste could be injected to mainly three stages as described below:

1. Raw materials stage
2. Kiln burning stage
3. Clinker grinding stage

In the raw material stage, a variety of waste materials can be combined as the substitution of raw material it is important to produce the correct chemical composition. In the kiln burning stage, most of the energy produced by solid and hazardous waste can be used as supplementary fuels. At present a variety of waste fuels, both hazardous and non-hazardous are being utilised in cement kilns. Finally in the cement grinding stage, gypsum can be replaced with some suitable waste materials.

4.1 Hazardous Waste in Developing Countries

Safe management of hazardous waste is becoming a major problem in many countries around the world. While there is no doubt that the life expectancy and quality of life have increased during the past decades, it has also resulted in the production of variety of waste, which are hazardous to mankind and environment.

The UNEP specifically mentions the following categories of hazardous waste substances as cause for major concern:

- Persistent, bioaccumulative and toxic substances (PBTs)
- Chemicals that are carcinogens or mutagens or that adversely affect the reproductive, endocrine, immune, or nervous systems
- Chemicals that have immediate hazards (acutely toxic, explosives, corrosives)
- Chemicals of global concern such as POPs, greenhouse and ozone-depleting substances (ODS)
- Healthcare waste
- Electric and electronic waste (E-waste)

Past practices of dealing with these waste have resulted in tens of thousands of contaminated sites. The global cost of restoring these sites could involve vast amounts of money. In addition, we are also facing the cost of dealing with current and future production of hazardous waste in such a way that they do not impair our environment or human health now or in future.

Developing countries face number of issues with the generation, transboundary movement and management of hazardous waste. Most of these countries do not have the expertise to manage the hazardous waste in an environmentally sound manner and proper infrastructure to protect human health and environment against the adverse effects which may result from such waste. One of the main problems in developing countries is the transboundary movement of hazardous waste. Although much progress has been made on controlling transboundary movements of hazardous waste, mainly through the implementation of Basel Convention, developing countries still face major challenges in reducing the quantity of hazardous waste generated, increasing the resource recovery from such waste and managing the residual waste in an environmentally sound manner. It is common practice in developing countries to mix hazardous waste with normal municipal solid waste and dispose of it in uncontrolled landfill sites.

There is very limited data available on the generation of hazardous waste in developing countries. Most of the hazardous waste arises from the industrial, agricultural and manufacturing processes that include industries such as chemical, petroleum, metal, textiles, pulp and paper, power plants and healthcare facilities. Especially, PCB stockpiles and obsolete pesticides are emergency issues in developing countries, because of inadequate management of these hazardous waste that constitutes a threat to human health and environment [9].

A conservative estimate of hazardous waste generation in selected countries is given in *State of the Environment of Asia and Pacific* 2000 [41]. According to the state, the hazardous waste generation in countries such as China, India, Indonesia, Philippines and Thailand is expected to increase significantly.

4.2 Waste Suitable for Cement Kilns

Cement kilns accept a wide variety of hazardous waste. However, the decision on what type of hazardous waste is used has to be site-specifically considered with certain assessments. Wrong usage of hazardous waste in cement kilns could cause serious negative impact to environment. Especially, by-production of harmful compounds such as dioxins necessarily has to be prevented. The selection of waste is influenced by many factors other than the nature of the waste itself. The Draft Technical Guidelines of Basel Convention [2] shows the important matter that needs consideration as follows;

1. Kiln operation
2. Raw material and fuel compositions
3. Waste feed points
4. Gas cleaning process
5. Resulting clinker quality
6. General environmental impacts
7. Probability of formation and release of persistent organic pollutants (POPs)
8. Particular waste management problems
9. Regulatory compliance
10. Public and government acceptance.

The Draft Technical Guidelines provides a useful figure simply explaining the general decision-making process for choosing hazardous waste which is suitable for each cement kiln. We strongly recommend reviewing the figure.

4.2.1 Liquid Waste Fuel

Paint thinners, degreasing solvent, solvent wash from ink and printing industries, chemical by-products from pharmaceutical and chemical manufacturing, waste oil and other flammable, readily pumpable waste materials can be used to prepare high quality liquid waste fuels for cement kilns [6]. Furthermore, metal cleaning fluids such as metal working and machining lubricants, coolants and cutting liquids and cleaning solvents from automotive aftermarket operations such as automobile shops, paint shops and service stations and liquid petroleum and petrochemical waste can also be used as liquid waste fuels in the cement industry [11, 12]. Acid tar produced by several industries which use concentrated sulphuric acid for the treatment of hydrocarbon materials is also considered to be a useful waste fuel for the cement industry [26]. The other types of liquid waste fuels used by the cement industry include pesticides, insecticides [25, 15], PCB stockpiles and PCB contaminated solvents and coal tars [27].

4.2.2 Solid Waste Fuel

Municipal solid waste or its separated energetic fraction, commonly known as refuse derived fuel (RDF), has been used in cement industry since early 1970s [31]. Depending on the chlorine content of the waste, up to 30% of the total fuel consumption may be covered by RDF [30].

The use of waste tyres as supplementary fuel in cement manufacture is one of the most common fuel substitution used in the industry. Carbon and oxygen amounts to 88% of a tyre, which accounts for its rapid combustion and relatively high heating value [4]. Tyres contain approximately 34,900 kJ/kg, which is compared favourably with coal which, generally contains 27,900 kJ/kg. Subsequently, as substituting waste tyres for coal, it is possible to reduce coal by 1.25 kg for every kg of tyres [34].

Cement kilns also offer following advantages for the disposal of sewage sludge [32]:

- Elimination of heavy metal in sewage sludge.
- Making use of organic matter in sewage sludge as fuel.
- Making use of inorganic matter in sewage sludge as raw materials.
- Complete destruction of organic pollutants in sewage sludge.

The only reservation in using cement kilns for burning sewage sludge is mercury contamination in the sludge, which in the long term may be all emitted to atmosphere. However, the situation is same even if the sludge is utilised in agriculture.

Another type of solid waste receiving attention by the cement industry is spent pot linings from aluminium industry. Spent pot liner is a layer of carbon situated between the molten metal and refractory material inside the steel shell of an aluminium reduction cell. Generally the useful life of potliner is 3–7 years and the removal of spent potliner (SPL) is a lasting process. One of the advantages of using SPL in cement kilns is that the fluoride content of the SPL may speed up the clinkering reaction, resulting in lower operating temperatures. However, cement produced by some facilities using SPL has exhibited a relatively high alkali content due to sodium in SPL [36]. The process of incinerating SPL by fluidised bed combustion and offering the ash to cement industry is reported to be very attractive [38].

A number of cement plants is testing and/or constructing pyrolysis systems to allow organically contaminated soils to be recycled for both fuel value as well as the silicon and aluminium which present in the most soil [11]. There is an increasing volume of soil being generated from underground storage tanks removal and industrial sites that are contaminated by petroleum products. From the accepted methods of treating the contaminated soil, thermal desorption or incineration appears to be the most suitable options, although these options are costly. The cement kiln offers an economic alternative to thermal desorption in incineration plants [10]. Furthermore, cement kilns with its large capacity are capable of handling large tonnages of soil contaminated either by organic or inorganic contaminants. However, this option has only been used to a limited extent due to public perception and industry concerns [33].

In addition to the solid waste fuels mentioned above, there are number of other types of waste fuels which are presently used in the industry. These include coal waste, low grade lignite, charcoal fines, powdered graphite dust from electrode production, petroleum coke and battery casings [17]. Furthermore, wood chips and fabric dust, rice hulls and coconut shells are also used to some extent [16].

4.2.3 Gas Waste Fuel

The cement kilns are also capable of using waste gas to supplement the fuel requirements. The primary waste gas considered for fuel substitution is landfill gas. In the landfill, the organic components of the municipal solid waste decomposes producing landfill gas which contains mainly methane and carbon dioxide. The methane contents of this gas can be between 40 and 70% with the possibility of increasing up to 90% by gas separation systems. If the landfill is adjacent to the cement kiln, this gas can be utilised easily as a fuel [30].

4.2.4 Solid Waste as Raw Material Substitutes

One of the main growing trends to use cement kilns to properly and safely manage waste, which is not related to fuel substitution, is the raw material substitution. By substituting original raw materials with waste materials, it is possible to conserve the raw materials as well as reduce the expensive mining operation costs. Some sources of raw material substitution that have been demonstrated to be effective in cement making include fly ash from utilities/power plants, petroleum contaminated soils, sludge waste from paper mills, mill scale from steel production and foundry sand [7]. The other types of waste materials that can be used for raw material substitution include aluminium processing residues, contaminated soils, glass and ceramic residues, lime sludges, slags [28, 44]. Roasted pyrites, residues in the production of sulphuric acid, are also used in the raw material to supplement iron content [1].

In addition to the solid waste mentioned above for raw material substitution, large quantities of aqueous hazardous waste are currently being managed through deep well injection. Many of these wastes can be effectively managed in wet process cement kilns by substituting a percentage of water used for slurry making [6].

4.3 Waste Not Recommended for Cement Kilns

Basel Convention [2] shows the following waste as normally not recommended for cement kilns;

- Radioactive waste
- E-waste
- Whole batteries
- Corrosive waste

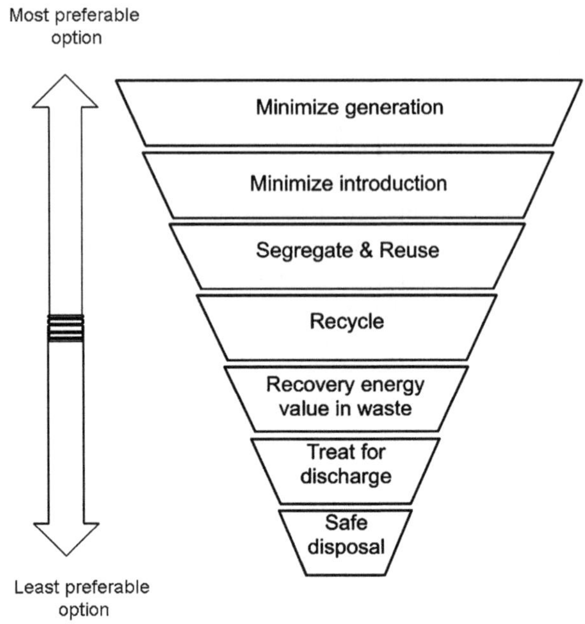

Fig. 2 Waste management hierarchy

- Reactive waste, including explosive waste, cyanide bearing waste and water-reactive waste
- Mercury waste
- Waste of unknown or unpredictable composition, including unsorted municipal waste

Generally, only waste of known composition and known energy and/or mineral value is suitable for co-processing in cement kilns. Moreover, plant-specific health and safety concerns need to be addressed as well as due consideration, which is given to the waste management hierarchy (as a general principle) presented in Fig. 2.

4.4 Emission and Fate

4.4.1 Emission of Metal

Several studies in cement kilns have attempted to study the fate and distribution of metals found in the waste fuels. Although the state of metal compound is changed during the combustion process, metal is not destroyed. Therefore, any metal, which are present in the waste fuel, fossil fuel or raw feed will be in the kiln stack emissions, the CKD or the clinker. Various test results have suggested that the cement kilns have a large potential for retaining metal elements entering with waste fuel. The tests have reported that over 99% of the metals in the waste fuel are retained in the process solids without any adverse effects to cement quality. When chlorinated waste is burnt, studies have found that metals such as lead (Pb) and cadmium (Cd) tend to appear more in the waste dust. It is also important to note that lead emissions during all the tests were lower than 10 mg/s which is approximately comparable to the emission of lead from automobiles burning leaded fuel.

4.4.2 Fate of Organic Constituents in Waste

The combustion of organic compounds produces the various end products depending on the chemical composition of original compounds. If the organic compound is composed of only carbon and hydrogen, the combustion produces simply only carbon dioxide (CO_2) and water (H_2O) as the end products. When the compound includes chlorine, the combustion could produce hydrogen chloride (HCl) or chlorine (Cl_2) as well. Further if the compound is composed of nitrogen or sulphur, the combustion could produce nitrogen oxides (NO_2) and sulphur dioxide (SO_2). A perfect decomposition of waste including organic compounds could be considered when only these end products are observed in the formed gas during the combustion process. Basel Convention defines *the values of destruction and removal efficiency (DRE)* which reveal the degree of decomposition of organic compounds. The mathematical equation for calculating DRE is as follows:

$$\mathrm{DRE} = [(\mathrm{W_{in}} - \mathrm{W_{outstack}})/\mathrm{W_{in}}] \times 100 \qquad (1)$$

By according Basel Convention, W_{in} is; the mass feed rate of one principal organic hazardous constituent (POHC) in the waste stream fed to the kiln, and W_{out} stack is; the mass emission rate of the same POHC in the exhaust emission prior to release to the atmosphere [2].

DRE indicates that whether cement kiln can achieve appropriate operation. When the cement kiln is well designed and well operated, DRE could be similar to, sometimes even better than, the DRE from commercial waste incinerators burning hazardous waste. Four nines (99.99%) removal could be reached in the most of the cases even for the most difficult to incinerate organic substances, and some test burns even could achieve DREs approaching six nines (99.9999%).

4.4.3 Emissions and Effects of Chlorine

The effect of waste combustion on particulate matter emissions has been one of interest because the earlier cement kiln tests indicated that burning chlorinated waste increased particulate emissions. The kilns equipped with electrostatic precipitators (ESPs) encountered increased particulate emissions when high amounts of chlorine were fed, especially as ESP hopper dust was recirculated to the kiln. Recirculation allows the chloride content in the dust to build up to much higher levels and eventually the resistivity of dust changes such that the ESP will cease to function properly. However, the particulate emissions from chlorinated waste combustion can be controlled by the properly operated control device and following the limitation of the chloride loading. Adjustments may be required to optimise the ESP performance to compensate for the change in dust's resistivity [13]. These adjustments have resulted in reduced particulate emissions in the recent test burns. However, the kilns equipped with a baghouse type dust control are not affected by the chlorine contents in the waste. The collection efficiencies were found to be stable for a wide range of chlorine feed rates.

When the chlorinated waste is incinerated in a cement kiln, the temperatures in the kiln favour the complete conversion of chloride in the waste to hydrogen chloride (HCl), which is absorbed by the calcium, potassium and sodium oxides in the kiln's solids and converted to their respective chlorides. These chlorides are vaporised at the hot end of the kiln and begin to condense on cooler surfaces or in the gas when they travel down the length of the kiln. The condensed chloride particles, which are entrained in the gas, are finally removed by the dust collection system. It is a routine practice to recycle a major portion of this collected dust back to the kiln effecting the dust particles to pass thorough the same cycle again. This is referred to as the 'chloride cycle'. The portion of dust which is not recycled is discarded. Thus by increasing the amount of waste dust, the chloride loading on the gas cleaning devices can be effectively decreased. Therefore, disposal of a portion of the waste dust is a common practice adopted by kiln operators and several kilns burning chlorinated waste increase their dust disposal rates to control the chloride cycle.

In most of the test burns, burning of chlorinated waste generally increased HCl emission rates. However, up to 95–99% of chlorine entering the kiln was retained by the process solids (clinker or dust), thus reducing any risk of HCl emissions to the atmosphere. The test results also indicated that high HCl removal efficiencies can be obtained by increasing the waste discard rates and optimising the chloride cycle. It was also revealed that the operation of the kiln showed no effects up to a chlorine input of 0.7% relative to clinker.

4.4.4 Considering By-Production of Dioxins in Cement Kilns

During the cement manufacturing process with using hazardous waste, by-production of POPs, especially dioxins i.e. PCDD/PCDF and coplanar PCB, needs to be seriously cautioned. It needs to mention strongly that once taking inappropriate operation, serious accident could be caused. The Stockholm Convention mentions about the potential of cement kiln industry as one of the main source of POPs emissions [37]. On the other hand, with conducting by appropriate operations, the cement kilns contribute total PCDD/PCDF emissions to air is less than 1% [21, 35]. SINTEF [35] evaluated enormous emission data of PCDD/PCDF, PCB and HCB from cement plants. It said that the emissions from modern dry preheater/precalciner kilns seem generally to be slightly lower than the emissions from wet kilns.

The mechanisms of dioxins formations have been investigated during the last two decades [18, 19, 42]. Three pathways of dioxin formation could be considerable in cement manufacturing process;

1. Dioxins formation in the high temperature zone
2. Dioxins formation from precursor compounds in the cooling zone
3. Dioxins formation via de novo synthesis in the cooling zone

Almost chlorinated aromatic compounds can be the precursors of dioxins. Especially, POPs waste has high potential to be the precursors. The dioxins formation from precursor compounds can occur under basic conditions at low temperatures around 150°C or in the presence of elemental chlorine even below 100°C [42]. De novo synthesis starts at temperatures around 200°C with a maximum formation rate between 300 and 500°C. Therefore, quick cooling of the kiln exhaust gas down to lower then 200°C at the inlet of the air pollution devices such as electrostatic precipitators (EP) and placements of bag filters (BF) are strongly required for preventing dioxins formation.

The studies indicate that the appropriateness of operation leads the cement industry into a very different position whether "POPs emitter" or "cleaner production". Therefore, establishing a procedure of environmental assessment, which allows us to assess the appropriateness of waste, potential impacts on environment, qualities of operation and product, is a matter of consequence.

4.5 Test Burns Conducted in Cement Kilns

The waste materials used by the cement industry contain organic materials as well as various amounts of metal components. In order to determine whether the cement kilns can manage these waste effectively without any adverse effects, especially to the environment, the fate of both organic and metal constituents must be determined. This task is achieved by conducting a study, commonly known as a test burn, in a cement kiln with various operating conditions and different types of waste. A summary of test burns conducted in cement kilns is given by [20].

4.5.1 Case Study

China

In 2008, the world's cement production was estimated to be 2.9 billion tonnes [2]. Chinese cement industry produces about half of them [2, 43]. The usage of cement kilns for managing hazardous waste has just started in China. Successes of popularising the appropriate management of hazardous waste in cement kilns in China would make enormous amounts of saving raw materials and fuel, and also hold down the emissions of global warming gas. As next step, establishing appropriate regulatory framework, legal requirement and standards in China is required. Yan et al. [43] referred that draft technical guidelines and background documents have recently been developed in China, however, modification of the guidelines and documents for Chinese conditions is still necessary. In this paper, a representative overview of Chinese cement industry was reported. It was based on the collection data from 12 different cement companies with 18 production lines in China. As this report, there were 6,846 cement enterprises in China in 2005. The most predominant size of them was small, 31.07% proportion in total national output, each of which annually produces cement less than 20,000 t. The vertical-shaft kilns and the precalciner rotary kilns were mainly used in 2006. They produced 50 and 44% of total national output in 2006, respectively. Only a few plants will have the capacity and knowledge of managing hazardous waste. The existence of general lack of co-processing knowledge, capabilities, and infrastructure was recognized. To establish guidelines and standards for the management is strongly required.

Sri Lanka

The first test burn of PCB waste oil by a cement kiln in a developing country was carried out in Sri Lanka by Karstensen et al. [23]. The assessment results of the feasibility and destruction performance showed that the Sri Lanka cement kiln had the ability to decompose PCB-oil safely without causing any new formation of PCDD/PCDF or HCB. The DRE was better than 99.9999% at the highest PCB feeding rate.

Thailand

The emissions of PCDD/PCDF from Thai cement plant during burning liquid hazardous waste and/or tires were investigated. The Pollution Control Department of the Ministry of Science, Technology, and Environment of the Government of Thailand with the financial support of GTZ, UNEP Chemicals, and Euro Chlor as a part of a dioxin program for making an emission inventory for PCDD/PCDF from Thai industries in 2001. All test results of PCDD/PCDF concentrations in stark gas were far below the orientation value of 0.1 ng I-TEQ/m^3. The results reveal that the addition of the hazardous waste had no effect on the emission results. Annotation: The cement plant is furnished the best state-of-the-art technology [40].

Vietnam

A test burn with obsolete insecticides, Fenobucarb and Fipronil, were conducted at a cement plant in Vietnam [22]. The obsolete insecticides, which were mixed as a solvent-based insecticide mixture, were fed to the main burner flame of the cement kiln via stainless steel pipes through a calibrated flow meter. Then they were burned with coal. The both DREs of Fenobucarb and Fipronil were better than >99.9999%. Any by-productions of PCDD/PCDF, HCB or PCBs were not recognized.

Others

Karstensen [21] referred to a test burn with used industrial solvents which was carried out in Egypt. The baseline tests were carried out before and after the test burn. All three results showed a PCDD/PCDF concentration less than 0.001 ng TEQ/m^3.

A test burn carried out in Colombia. 900 tones of pesticide contaminated soil fed to a kiln inlet of a 58 m long five-stage preheater kiln. The result showed a DRE of 99.9999% for all the introduced pesticides [21, 35].

SINTEF reported several case studies of POPs emission from cement kilns in developing countries such as Venezuela, South Africa, Chile, Philippines, Vietnam, Hong Kong, Taiwan etc. [35].

5 Conclusions

In this chapter, the hazardous waste management in cement kilns in developing countries was discussed. Burning of hazardous waste in cement kilns would be to the benefit of both the producer of hazardous waste and the cement industry.

Several case studies, which were carried by appropriate operations and assessments, demonstrated that the hazardous waste had been destroyed in an irreversible and environmental sound manner without any hazardous by-products such as POPs.

On the other hand, the existence of general lack of co-processing knowledge, capabilities, and infrastructure in the developing countries' cement industry was reported. It shows possibilities that serious environmental contaminations could be occurred any time. For the further progress of implementation and popularisation of the use of cement kilns for managing hazardous waste in developing countries, establishing guidelines and standards could be mentioned as the first priority.

References

1. Ahling B (1987) Alternative raw materials for the production of cement–are waste materials a resource. Paper prepared for CEMENTA AB, Sweden
2. Basel Convention (2010) Draft technical guidelines on co-processing of hazardous waste in cement kilns. http://www.basel.int/techmatters/index.html
3. Baxendale T (1990) Co-processing waste derived fuels help solve compliance problems and produce useable materials. Hazmat World 3(1):57–58

4. Blumenthal M (1992) The rationale for using whole tyres. Rock Prod 95(7):48–50
5. Canadian Portland Cement Association (1989) Resource recovery: the cement kiln solution. Report published by the canadian portland cement association, Canada
6. Chadbourne JF, Freeman HM (1988) Hazardous waste as fuel in cement kilns. In: Abbou R (ed) Hazardous waste: detection, control, treatment, Part B. Elsevier, New York, pp 1315–1324
7. Dawson B (1992) Emerging technologies for utilising waste in cement production. In: Proceedings of international conference on the role of cement kilns in waste management—KILNBURN'92, Brisbane, Australia, pp 109–115
8. Directive 2000/76/EC of the European Parliament and of the Council (2000) Council directive 2000/76/EC of the European parliament and of the council of 4 Dec 2000 on the incineration of waste. J Eur Communities, L 332/91. http://www.central2013.eu/fileadmin/user_upload/Downloads/Document_Centre/OP_Resources/Incineration_Directive_2000_76.pdf
9. FAO (1997) FAO pesticide disposal series 5—prevention and disposal of obsolete and unwanted pesticide stocks in Africa and the near east, Second consultation meeting. http://www.fao.org/ag/AGP/AGPP/Pesticid/Disposal/pdf/w3338e.pdf
10. Ferdo AM, Hawks RL (1990) Processing of multimedia waste in cement kilns. In: Proceedings of air and waste management association international specialty conference on waste combustion in boilers and industrial furnaces, Missouri, USA
11. Gabbard WD, Gossman D (1990) Hazardous waste fuels and the cement kiln. ASTM Stand News 18(9):38–41
12. Gossman D (1992) Reuse of petroleum and petrochemical waste in cement kilns. Environ Prog 11(1):1–6
13. Hazelwood D, Smith F, Gartner E (1982) Assessment of waste fuel use in cement kilns. United States Environmental Protection Agency, USA (Report No. EPA–600/2–82–103)
14. Herat S (1997) Protecting the environment from waste disposal: the cement kiln option. Environ Prot Eng 23:25–34
15. Huden GH (1989) Opportunities and constraints for hazardous waste disposal in developing countries. Paper presented at 2nd international symposium of Pacific basin consortium for hazardous waste, Singapore
16. Huhta RS (1985) Waste fuel survey report–II. Rock Prod 88(5):46–49
17. Huhta RS (1986) Another look at waste fuels. Rock Prod 89:55–59
18. Ishikawa Y, Noma Y, Mori Y, Sakai S (2007) Congener profiles of PCB and a proposed new set of indicator congeners. Chemosphere 67:1838–1851
19. Ishikawa Y, Noma Y, Yamamoto T, Mori Y, Sakai S (2007) PCB decomposition and formation in thermal treatment plant equipment. Chemosphere 67:1383–1393
20. Jones PH, Herat S (1992) Incineration of intractable wastes in cement kilns. In: Proceedings of 1st national hazardous and solid waste convention, Sydney, Australia, pp 181–188
21. Karstensen KH (2008) Formation, release and control of dioxins in cement kilns. Chemosphere 70:543–560
22. Karstensen KH, Kinh NK, Thang LB, Viet PH, Tuan ND, Toi DT, Hung NH, Quan TM, Hanh LD, Thang DH (2006) Environmental science and policy, vol 9, pp 577–586
23. Karstensen KH, Mubarak AM, Gunadasa HN, Wijagunasekara B, Ratnayake N, Alwis AD, Fernando J (2010) Test burn with PCB-oil in a local cement kiln in Sri Lanka. Chemosphere 78:717–723
24. Mantus EK (1992) All fired up. Report published by environmental toxicology international, Seattle, Washington
25. Marcil AG (1990) Cement kiln incineration of out–dated agricultural chemicals in Pakistan. Waste Mange Res 8(2):171–174
26. Milne DD, Clark AI, Perry R (1986) Acid tars: their production, treatment and disposal in the UK. Waste Manage Res 4(4):407–418
27. Mourighan RE (1992) An overview of the disposal of solid and hazardous wastes in cement kilns. Technical information document prepared by US environmental protection agency, Kansas, USA

28. Mullick AK, Aluwalia SC (1992) Utilisation of wastes in Indian cement industry. In: Proceedings of 1st international symposium on cement industry solutions to waste management, Alberta, Canada, pp 489–512
29. Nemerow NL, Agardy FJ, Sullivan P, Salvato JA (2009) Environmental engineering, sixth edition–environmental health and safety for municipal infrastructure, land use and planning, and industry. Wiley, Hoboken, NJ
30. Neumann E (1992) Energy alternatives—the substitution of fossil fuels in cement kilns. Int Cement Rev (May) pp 61–67
31. Neumann E, Duerr M, Kreft W (1990) The substitution of fossil fuels in cement kilns. World Cem 21(3):80–88
32. Obrist A, Lang T (1987) Incineration of sewage sludge in cement kilns. Paper presented at 1987 IEEE cement industry technical conference, California, USA, pp 369–380
33. Parker A (1987) Modern methods of treatment of contaminated land and waste disposal. Paper presented at international pollution abatement conference, Birmingham, UK
34. Serumgard JR, Bluementhal MH (1992) The use of scrap tyres in cement rotary kilns. In: Paper presented at proceedings of 1st international symposium on cement industry solutions to waste management, Alberta, Canada
35. SINTEF (2006) Formation and release of POPs in the cement industry, Second edition. In: World business council for sustainable development. Cement Sustainability Initiative. http://www.wbcsd.org/DocRoot/piF5rKj2ulwpFpYRMI8K/formation_release_pops_second_edition.pdf. Accessed 23 Jan 2006
36. Spiegel SJ, Pelis TK (1990) Regulations and practices for the disposal of spent potliner by the aluminium industry. J Metals 42:70–73
37. Stockholm Convention (2004) Article 5 and annex C of the stockholm convention on persistent organic pollutants. Available at http://www.pops.int/documents/meetings/bat_bep/2nd_session/egb2_followup/draftguide/1BArticle5andAnnexC.pdf
38. Tabery RS, Dangtran K (1990) Fluidised bed combustion of aluminium smelting waste. Environ Prog 9(1):61–67
39. The Portland Cement Association. http://www.cement.org/
40. UNEP/IOMC (2001) Thailand–PCDD/DF sampling and analysis program—report UNEP Chemicals, Geneva, Switzerland. http://www.chem.unep.ch/Pops/pdf/thdioxsamprog.pdf
41. UNESCAP (2000) State of the environment of Asia and the Pacific
42. Weber R (2007) Relevance of PCDD/PCDF formation for the evaluation of POPs destruction technologies—review on current status and assessment gaps. Chemosphere 67:S109–S117
43. Yan D, Karstensen KH, Huang Q, Wang Q, Cai M (2010) Coprocessing of industrial and hazardous wastes in cement kilns: a review of current status and future needs in China. Envi Eng Sci 27(1):37–45
44. Zimmerman L, Coles C (1992) The utilisation of processed waste by–products for cement manufacturing. In: Proceedings of 1st international symposium on cement industry solutions to waste management, Alberta, Canada, pp 533–545

Thermodynamic Approach to Design and Optimization of Biomass Gasifier Utilizing Agro-Residues

Buljit Buragohain, Pinakeswar Mahanta and Vijayanand Suryakant Moholkar

Abstract In recent years, biomass gasification has emerged as a viable option for decentralized power generation, especially in developing countries like India. In this chapter, we have given an overview of the energy scenario in India with statistics of electricity generation through various sources along with data on coal deposits and agro-residues, which are potential feedstock for electricity production through thermal route. Moreover, using the non-stoichiometric equilibrium and semi-equilibrium models, we have assessed the outcome of gasification process for different combinations of operating conditions and various biomass as feedstock, either single biomass or their mixtures. Four key parameters have been used for optimization, viz. biomass type, air or equivalence ratio, temperature of gasification, and gasification medium. The gasification medium has been taken to be either pure air or air–steam mixture. It is revealed that the biomass mixtures also form potential feedstock for gasifiers, and the performance of gasifier is very similar to that with single biomass as feedstock. On the basis of simulations results, we have also attempted to optimize the operating conditions for biomass gasifier operation. The optimum sets of operating conditions for gasifier for application of decentralized power generation are: Air Ratio = 0.3–0.4, Temperature = 700–800°C with gasification medium being air.

B. Buragohain · P. Mahanta · V. S. Moholkar (✉)
Center for Energy, Indian Institute of Technology,
Guwahati, 781 039, Assam, India
e-mail: vmoholkar@iitg.ernet.in

A. Karagiannidis (ed.), *Waste to Energy*, Green Energy and Technology,
DOI: 10.1007/978-1-4471-2306-4_7,

1 Introduction

Electricity is a vital form of energy input for overall growth of a developing country and transitional economy like India. It is one of the main infrastructural requirements for socio-economic development, and also development in various other sectors such as agriculture and industry. In the last 6 decades (after independence in 1947), India's energy use has increased 16 times with population rise from 328 million in 1947 to 1.165 billion in 2010. The installed capacity of electricity generation has increased more than 100 times from a meager 1,362 MW in 1947 to 167 GW in 2010. Although the present installed capacity of electricity generation in India is fifth largest in the world, it is far insufficient to meet the demands. Table 1 gives list of top 10 electricity producing countries in world [20].

Even with 167 GW of installed capacity, per capita electricity consumption in India is very low (less than 1/4th of the world average of 2,646 kW h). Greater details of Indian power sector are given in next section. The average peak hour shortage of electricity in India is $\sim$10% with the highest shortage of 17% in the western part of India and least (3.3%) in the southern part [8]. The current rate of growth of economy is $\sim$8–9% per annum, and with rapid urbanization and improvement in living standards of population, the electricity demand is bound to grow very rapidly. Integrated Energy Policy Report [15] has predicted 3–4-fold increase in primary energy demand and insisted on 5–6-fold increase in generation capacity (considering 2003–2004 year as basis) to sustain growth rate of 8% through 2031–2032 [15].

Table 2 gives the past, present and future scenario of electricity demand and generation in India for desired annual GDP growth rate of 8 or 9%, as foreseen by expert committee for energy policy [15]. According to this forecast, the power generation capacity in India must increase to $\sim$800 GW by 2031–2032 to meet the energy demands. Meeting this challenge to provide energy security is a daunting task before India to achieve economic growth targets and raising living standards of entire (urban as well as rural) population. Before proceeding to main text of this chapter, we would like to ponder over definition of "Energy Security" in the context of a developing country like India [15].

2 Energy Security in the Indian Context

World Energy Assessment Report has defined energy security as: The continuous availability of energy in various forms in sufficient quantities at reasonable prices. However, the Integrated Energy Policy of India has suggested modification of this definition as follows: "Supply of life line energy to all citizens irrespective of their ability to pay for it as well as meet their effective demand for safe and convenient energy to satisfy their various needs at competitive prices at all times, and with a prescribed confidence level considering shocks and disruptions that can be

Table 1 Global electricity production scenario [20]—data holds for year 2005

Sr. No.	Country	Annual electricity generation (MW h)	Population	Per capita electricity consumption (kW h)
1.	China	4,190,000,000	1,315,844,000	3,184
2.	USA	3,741,485,000	298,213,000	12,546
3.	Japan	963,852,000	128,085,000	7,525
4.	Russia	857,617,000	141,927,297	6,043
5.	India	600,649,000	1,103,371,000	544
6.	Brazil	600,029,000	186,405,000	3,219
7.	Canada	549,476,000	32,268,000	17,029
8.	Germany	544,467,000	82,329,758	6,613
9.	France	460,944,000	60,496,000	7,619
10.	South Korea	386,169,000	47,817,000	8,076
11.	World	17,109,665,000	6,464,750,000	2,647

reasonably expected". This modification of definition in inspired by several facts and circumstances that need to be considered in the context of a transitional and diverse economy like India.

The motivations for the modification of the definition are outlined below briefly: (1) Energy security makes sense only when energy needs of vast population in diverse regional, social and economic sectors are met. (2) Life line energy or basic minimal energy needs of all citizens must be fulfilled. If any section of society is not financially capable of meeting energy needs at market price, proper subsidies need to be given. (3) Meeting needs of one form of energy through other form (e.g. use of electricity for cooking than LPG or kerosene) results in loss of energy. (4) Interruption of energy supply can cause significant damage or loss of economy. (5) Probable disturbance and disruption in energy generation/supply should be well anticipated and accounted for. In case of 100% capacity utilization for generation and effective transmission and distribution is not possible; certain confidence levels should be prescribed.

Roadmap to Energy Security

Principal measures for the energy security can be defined in two categories depending on level of uncertainties or risks involved. First category is reduction of risks by various means such as:

1. reduction in energy requirement by increasing efficiency of energy production and use;
2. greater use of local fuels, cutting down on imports;
3. increasing fuel flexibility and supply (or in other words diversification of fuel choices);
4. exploration of new and local energy sources.

The second category involves dealing with risks. Measures in this category include:

Table 2 Past and future of electricity requirement in India [15]

Year	Population (billion)	Total energy requirement (billion kW h)		Energy required at bus bar (billion kW h)		Peak energy demand (GW)		Required installed capacity (GW)	
		GDP growth rate 8%	GDP growth rate 9%	GDP growth rate 8%	GDP growth rate 9%	GDP growth rate 8%	GDP growth rate 9%	GDP growth rate 8%	GDP growth rate 9%
2003–2004	1.065	633	633	592	592	89	89	131	131
2006–2007	1.114	761	774	712	724	107	109	153	155
2011–2012	1.197	1,097	1,167	1,026	1,091	158	168	220	233
2016–2017	1.275	1,524	1,687	1,425	1,577	226	250	306	337
2021–2022	1.347	2,118	2,438	1,980	2,280	323	372	425	488
2026–2027	1.411	2,866	3,423	2,680	3,201	437	522	575	685
2031–2032	1.468	3,880	4,806	3,628	4,493	592	733	778	960

1. increase endurance for unforeseen interruption in fuel supply;
2. increase ability to cope up with market fluctuations and risks;
3. increase technical standards and support to face technical risks.

In addition to these, maintenance of strategic fuel reserves and obtaining equity oil or gas abroad could help tackle supply and market risks. In the context of subject matter of this chapter, we shall elaborate two important measures of energy security in the first category, viz. (1) local or domestic alternatives for imported fuels, and (2) development of alternate fuel sources.

2.1 Local or Domestic Alternatives

Substituting conventional (or imported) energy by locally available substitutes would augment self-reliance of country in energy. It would also bring down risk factors and associated economic penalties on supply of imported energy. Nonetheless, this would increase the load on local energy supplies. If the substitute energy source is of renewable kind, this risk factor is lessened. Some examples in this category are electrification of railways or wood plantation on wasteland. Electrification of railways would, of course, necessitate additional electricity generation through coal-thermal route, but at the same time would bring down burden on diesel. Wasteland plantation could yield up to 20 tons of (dry) wood, which could be used for electricity generation through combustion or gasification route. Some other examples of alternate energy sources are liquid fuels such as biodiesel or ethanol, or synthetic fuels from coal liquefaction or Fischer–Tropsch synthesis coupled to biomass gasification.

2.2 Development of Alternate Energy Sources

This route will mainly involve research-based development for new energy sources, in addition to enhancing recoveries from existing sources. Better mine design and use of advanced technologies could enhance oil, gas as well as coal recovery. For coal deposits available at great depths (from which mining is not feasible), in-situ gasification of coal is a viable solution giving higher recoveries at economic prices. Methane adsorbed on coal seams could also be recovered. This technique is already in practice in USA and Australia. The estimated potential of CBM (Coal Bed Methane) is in the range of 1,400–2,600 billion m^3 equivalent to 1,260–2,340 million tons of oil equivalent. Identifying potential of this energy resource, Government of India has formulated a policy for CBM in 1997 and development of CBM has been a joint venture of Ministry of Petroleum and Natural Gas and Ministry of Coal. In addition to these, effective utilization of renewable energy sources is also vital to development of alternate energy sources. We shall deal with this topic in greater details in Sect. 4.

3 Indian Power Sector: Facts and Figures

As noted earlier, the total installed generation capacity of India by December 2010 stands at 167 GW [8]. However, the resource as well as regional distribution of this capacity is uneven. Table 3(A&B) gives the installed capacity in various states and union territories of India through various routes (thermal, hydro, nuclear and renewables) [8, 25]. Highest capacity is installed in the western region of countries with industrialized states. On the other hand, northeastern region comprising of seven states with hilly and mountainous terrains and very low population has least installed capacity for generation. Secondly, the generation is dominated by coal-thermal route ($\sim$54% of total capacity). The total contribution of thermal route (coal + gas + diesel) stands at $\sim$65%. This is followed by hydro ($\sim$22%), renewables ($\sim$10%) and nuclear ($\sim$2.7%). Major cause leading to dominance of coal based generation is that India has over 200 billion tons of coal deposits. However, these coal deposits are not uniformly spread over the country, but mainly concentrated in northern and northeastern states.

Table 3(C) gives an account of some important statistical figures of Indian power sector [8]. Although the plant load factor has remained more-or-less uniform over past 1 decade, the consumption of coal has increased over 50%. The transmission and distribution losses are enormous—almost 1/3rd of power generated at source.

Per capita electricity consumption has shown a steady growth. An important aspect of the economy of electricity sector is represented by the unit prices of electricity supply and purchase. Due to large subsidy given Government of India to rural and agricultural sector, the cost of supply of electricity is not completely recovered by the electricity boards. The electricity boards of the states are thus running in huge losses.

Rural Electrification: *Rajiv Gandhi Grameen Vidyutikaran Yojana*—Electrification of rural areas, especially located in the remote and hilly regions where extension of grid is not feasible, remains a major hurdle in India's infrastructural development. Providing access to electricity to all rural households has been a major component of the common minimum program of Government of India. To meet this objective, a scheme named "Rajiv Gandhi Grameen Vidyutikaran Yojana (RGGVY)" (Rajiv Gandhi Rural Electrification Program) was launched in April 2005 [25]. Rural Electricity Corporation (REC) is the nodal agency of this scheme, and the total budget allocated for this scheme in the 11th plan is Rs. 280 billion. The principal aim of this scheme is to electrify the 125,000 unelectrified villages and to extend electrical connection to an estimated 23.4 million households recognized as "below poverty line (BPL)". In addition, strengthening and augmentation of electricity supply infrastructure in currently electrified 462,000 villages by 2010 is also an important component of this program. At this juncture we would like to specifically mention the definition of an "electrified village". In October 1997, Ministry of Power, Government of India defined electrified village as: "a village will be deemed to be electrified if electricity is used in the

Table 3 (**A**) Power scenario in various states of India (installed capacity in MW of power utilities including allocated shares in joint & central sector utilities). (**B**) Summary of the generation installed capacity in various parts of India [8, 25]. (**C**) Some representative statistics of Indian electricity sector

State	Mode wise breakup				Nuclear	Hydro (renewable)	Renewable energy sources	Grand total
	Thermal			Total thermal				
	Coal	Gas	Diesel					
(A)								
States in northern region								
Delhi	3,293.96	1,008.01	0.00	4,301.97	122.08	597.62	1.05	5,022.72
Haryana	3,812.99	560.29	3.92	4,377.20	109.16	1,334.68	105.90	5,926.94
Himachal	118.30	61.88	0.13	180.31	34.08	1,731.94	337.82	2,284.15
Jammu & Kashmir	263.70	304.14	8.94	576.78	77.00	1,503.53	129.33	2,286.64
Punjab	3,208.19	288.92	0.00	3,497.11	208.04	2,972.89	295.63	6,973.67
Rajasthan	4,659.48	665.03	0.00	5,324.51	573.00	1,467.80	1,162.45	8,527.76
Uttar Pradesh	7,261.84	549.97	0.00	7,811.81	335.72	1,624.42	612.18	10,384.13
Uttarakhand	261.26	69.35	0.00	330.61	22.28	1,924.18	132.97	2,410.04
Chandigarh	27.09	15.32	0.00	42.41	8.84	47.74	0.00	98.99
Central (unallocated)	713.19	290.35	0.00	1,003.54	129.80	417.95	0.00	1,551.29
Grand total for northern region	23,620.00	3,813.26	12.99	27,446.25	1,620.00	13,622.75	2,777.32	45,466.32
States in western region								
Goa	277.03	48.00	0.00	325.03	25.80	0.00	30.05	380.88
Daman & Diu	19.04	4.20	0.00	23.24	7.38	0.00	0.00	30.62
Gujarat	7,588.89	3,894.49	17.48	11,500.86	559.32	772.00	1,851.04	14,683.22
Madhya Pradesh	4,282.10	257.18	0.00	4,539.28	273.24	3,223.66	244.36	8,280.54
Chhattisgarh	4,383.00	0.00	0.00	4,383.00	47.52	120.00	218.95	4,769.47
Maharashtra	11,768.05	3,475.93	0.00	15,243.98	690.14	3,331.84	2,573.88	21,839.84
Dadra & Nagar Haveli	22.04	27.10	0.00	49.14	8.46	0.00	0.00	57.60
Central (unallocated)	950.35	196.91	0.00	1,147.26	228.14	0.00	0.00	1,375.40
Grand total for western region	29,290.50	7,903.81	17.48	37,211.79	1,840.00	7,447.50	4,918.28	51,417.57
States in southern region								
Andhra Pradesh	6,759.88	2,878.40	36.80	9,675.08	214.28	3,695.53	721.59	14,306.48

(continued)

Table 3 (continued)

State	Mode wise breakup				Nuclear	Hydro (renewable)	Renewable energy sources	Grand total
	Thermal			Total thermal				
	Coal	Gas	Diesel					
Karnataka	4,752.67	220.00	234.42	5,207.09	195.36	3,599.80	2,437.49	11,439.74
Kerala	765.38	533.58	256.44	1,555.40	78.10	1,881.50	142.86	3,657.86
Tamil Nadu	5,519.81	1,026.30	411.66	6,957.77	478.50	2,122.20	5,215.59	14,774.06
NLC	100.17	0.00					0.00	100.17
Pondicherry	207.01	32.50	0.00	239.51	16.28	0.00	0.03	255.82
Central (unallocated)	1,067.58	0.00	0.00	1,067.58	117.48	0.00	0.00	1,185.06
Grand total for southern region	19,172.50	4,690.78	939.32	24,802.60	1,100.00	11,299.03	8,517.55	45,719.18
States in eastern region								
Bihar	1,661.70	0.00	0.00	1,661.70	0.00	129.43	64.10	1,855.23
Jharkhand	1,737.88	0.00	0.00	1,737.88	0.00	200.93	4.05	1,942.86
West Bengal	6,756.34	100.00	12.20	6,868.54	0.00	1,116.30	160.55	8,145.39
DVC	3,703.10	90.00	0.00	3,793.10	0.00	193.26	0.00	3,986.36
Orissa	2,428.10	0.00	0.00	2,428.10	0.00	2,166.93	68.60	4,663.63
Sikkim	68.10	0.00	5.00	73.10	0.00	75.27	47.11	195.48
Central (unallocated)	1,280.16	0.00	0.00	1,280.16	0.00	0.00	0.00	1,280.16
Grand total for eastern region	17,635.38	190.00	17.20	17,842.58	0.00	3,882.12	344.41	22,069.11
States in northeastern region								
Assam	60.00	441.32	20.69	522.01	0.00	429.72	27.11	978.84
Arunachal Pradesh	0.00	21.05	15.88	36.93				
Meghalaya	0.00	25.96	2.05	28.01	0.00	230.58	31.03	289.62
Tripura	0.00	181.84	4.85	186.69	0.00	62.37	16.01	265.07
Manipur	0.00	25.96	45.41	71.37	0.00	80.98	5.45	157.80
Nagaland	0.00	19.19	2.00	21.19	0.00	53.32	28.67	103.18
Mizoram	0.00	16.28	51.86	68.14	0.00	34.31	36.47	138.92
Central (unallocated)	0.00	55.40	0.00	55.40	0.00	127.15	0.00	182.55
Grand total for northeastern region	60.00	787.00	142.74	989.74	0.00	1,116.00	223.32	2,329.06

(continued)

Table 3 (continued)

State	Mode wise breakup				Nuclear	Hydro (renewable)	Renewable energy sources	Grand total
	Thermal			Total thermal				
	Coal	Gas	Diesel					
Islands								
Andaman & Nicobar	0.00	0.00	60.05	60.05	0.00	0.00	5.35	65.40
Lakshadweep	0.00	0.00	9.97	9.97	0.00	0.00	0.75	10.72
Grand total for Islands	0.00	0.00	70.02	70.02	0.00	0.00	6.10	76.12
(B)								
Northern region	23,620.00	3,813.26	12.99	27,446.25	1,620.00	13,622.75	2,777.32	45,466.32
Western region	29,290.50	7,903.81	17.48	37,211.79	1,840.00	7,447.50	4,918.28	51,417.57
Southern region	19,172.50	4,690.78	939.32	24,802.60	1,100.00	11,299.03	8,517.55	45,719.18
Eastern region	17,635.38	190.00	17.20	17,842.58	0.00	3,882.12	344.41	22,069.11
Northeastern region	60.00	787.00	142.74	989.74	0.00	1,116.00	223.32	2,329.06
Islands	0.00	0.00	70.02	70.02	0.00	0.00	6.10	76.12
Grand total for all India	89,778.38	17,384.85	1,199.75	108,362.98	4,560.00	37,367.40	16,786.98	167,077.36

Year	T&D losses (%)	Plant load factor (%)	Coal consumption (MMTPA)	Per capita electricity consumption (kWÑ h)	Cost of electricity supply (Rs.)	Cost of electricity charged (Rs.)*
(C)						
2001–2002	–	69.9	240	–	2.46	1.81
2002–2003	32.54	72.2	253	566.7	2.38	1.95
2003–2004	32.53	72.7	263	592.0	2.39	2.03
2004–2005	31.25	74.8	278	612.5	2.54	2.09
2005–2006	30.42	73.6	280	631.5	2.58	2.21
2006–2007	28.65	76.8	302	671.9	2.76	2.27
2007–2008	27.2	78.6	330	717.1	–	–
2008–2009	25.47	77.2	355	733.5	–	–

* Average cost paid by Indian households (including the subsidy for the agricultural use of electricity)

inhabited locality within the revenue boundary of the village for any purpose whatsoever" [25]. In 2004–2005, this definition was modified with following conditions imposed [25]:

1. Basic infrastructure such as distribution transformer and distribution lines are provided in the inhabited locality as well as the Dalit Basti/hamlets where it exists (no necessity of transformer for electrification through non-conventional energy sources).
2. Provision of electricity to public places like schools, Gram Panchayat (Village Council) offices, health centers etc.
3. Number of households electrified should be at least 10% of the total number of households in the village.
4. Compulsory certification from the Gram Panchayat regarding completion of the village electrification.

Under the RGGVY scheme, BPL households will be connected to electricity free of cost, while rest of the program receives 90% subsidy from Government of India. However, an estimated 54.6 unelectrified households above poverty line are expected to get electricity connection on their own, without any subsidy. This has not been realized, unfortunately. Despite an ambitious program with huge investments, an estimated 40% households still remain without electricity in the villages which have been certified as "electrified" as per the new definition stated earlier. The latest statistics of the RGGVY program is given in Table 4 [25].

Successful implementation of RGGVY scheme requires expansion in connectivity with an equivalent expansion in generation and supply of electricity. Given the current shortage in supply and excessive loads on state electricity boards, it is practically impossible to meet these demands through grid-supplied electricity. Thus, successful implementation of RGGVY program necessitates special efforts in installation of decentralized generation systems, mainly through use of renewable energy sources. Perhaps, this remains the primary driver for the renewable energy efforts in India.

4 Renewable Energy Endeavors of India

India is blessed with all forms of renewable energy sources, viz. biomass, solar, small hydro and wind. Effective utilization of these renewable energy sources has always been an important component of India's energy policy and planning. For this, Government of India established Department of Non-conventional Energy Sources in 1982, which was transformed in full-fledge ministry after a decade as Ministry of Non-Conventional Energy Sources (MNES) which was later renamed as Ministry of New and Renewable Energy (MNRE), [24]. This has been a nodal agency of Government of India for implementation of broad spectrum of program related to new and renewable energy such a installation of plants and infrastructure for effective utilization of renewable power, promotion of use of renewable energy

Table 4 Statistics of the Rajiv Gandhi Grameen Vidyutikaran Yojana as on January 31, 2011 [25]

Sr. no.	State/union territory name (total no. of districts)	Electrification of un-/de-electrified villages		No. of connections to rural households including BPL		No. of connections to BPL households	
		Coverage in no.	Achievement in no. (%)	Coverage in no.	Achievement in no. (%)	Coverage in no.	Achievement in no. (%)
1	Andhra Pradesh (23)	0	0 (0.0)	3,954,128	3,092,410 (78.2)	2,592,140	2,576,129 (99.4)
2	Arunachal Pradesh (16)	2,129	500 (23.5)	76,407	10,648 (13.9)	40,810	8,170 (20.0)
3	Assam (23)	8,525	4,969 (58.3)	1,414,828	520,626 (36.8)	991,656	520,626 (52.5)
4	Bihar (38)	23,211	21,596 (93.0)	6,022,036	1,589,893 (26.4)	2,762,455	1,589,893 (57.6)
5	Chhattisgarh (16)	1,132	136 (12.0)	1,285,545	379,269 (29.5)	777,165	379,269 (48.8)
6	Gujarat (25)	0	0 (0.0)	1,595,853	645,521 (40.4)	955,150	645,521 (67.6)
7	Haryana (20)	0	0 (0.0)	569,686	177,192 (31.1)	224,073	177,192 (79.1)
8	Himachal Pradesh (12)	93	13 (14.0)	36,479	3,278 (9.0)	12,448	1,587 (12.7)
9	Jammu & Kashmir (14)	283	107 (37.8)	295,221	27,063 (9.2)	136,730	27,063 (19.8)
10	Jharkhand (22)	19,737	16,035 (81.2)	2,926,260	1,013,555 (34.6)	1,691,797	1,013,555 (59.9)
11	Karnataka (27)	132	59 (44.7)	1,932,797	894,021 (46.3)	891,939	771,261 (86.5)
12	Kerala (14)	0	0 (0.0)	92,736	17,238 (18.6)	56,351	17,238 (30.6)
13	Madhya Pradesh (48)	806	243 (30.1)	2,653,536	415,136 (15.6)	1,376,242	299,059 (21.7)
14	Maharashtra (36)	6	0 (0.0)	2,633,742	966,995 (36.7)	1,876,391	954,012 (50.8)
15	Manipur (9)	882	230 (26.1)	192,148	8,844 (4.6)	107,369	8,203 (7.6)
16	Meghalaya (7)	1,943	150 (7.7)	188,648	28,295 (15.0)	116,447	28,295 (24.3)
17	Mizoram (8)	137	31 (22.6)	44,334	7,185 (16.2)	27,417	7,185 (26.2)
18	Nagaland (11)	105	52 (49.5)	142,992	22,568 (15.8)	69,900	16,061 (23.0)
19	Orissa (30)	17,895	11,569 (64.6)	4,858,292	1,842,404 (37.9)	3,185,863	1,842,404 (57.8)
20	Punjab (17)	0	0 (0.0)	405,023	48,144 (11.9)	148,860	48,144 (32.3)
21	Rajasthan (32)	4,454	3,652 (82.0)	2,229,442	1,427,151 (64.0)	1,750,118	880,042 (50.3)
22	Sikkim (4)	25	14 (56.0)	28,166	5,450 (19.3)	11,458	5,221 (45.6)
23	Tamil Nadu (30)	0	0 (0.0)	1,692,235	498,873 (29.5)	545,511	498,873 (91.5)

(continued)

Table 4 (continued)

Sr. no.	State/union territory name (total no. of districts)	Electrification of un-/de-electrified villages		No. of connections to rural households including BPL		No. of connections to BPL households	
		Coverage in no.	Achievement in no. (%)	Coverage in no.	Achievement in no. (%)	Coverage in no.	Achievement in no. (%)
24	Tripura (4)	160	64 (40.0)	228,759	54,695 (23.9)	194,730	54,695 (28.1)
25	Uttar Pradesh (70)	30,802	27,759 (90.1)	1,694,075	872,993 (51.5)	1,120,648	872,993 (77.9)
26	Uttarakhand (13)	1,469	1,505 (102.5)	357,309	223,470 (62.5)	281,615	223,470 (79.4)
27	West Bengal (18)	4,573	4,169 (91.2)	3,974,005	1,188,216 (29.9)	2,699,734	1,170,717 (43.4)
	Total of all states (587)	118,499	92,853 (78.4)	41,524,682	15,981,133 (38.5)	24,645,017	14,636,878 (59.4)

in rural areas such as domestic/street lighting, cooking and agriculture, use of renewable energy in urban, commercial and industrial sectors, and research, design and development of new and renewable energy technologies, products and services. Nodal agencies of the MNRE have also been established in various states to mobilize state level government and non-government organization (NGOs), infrastructure and machinery for implementation of various programs related to renewable energy. The financial wing of MNRE known as Indian Renewable Energy Development Agency (IREDA) is also established with aim of market development and financing renewable energy projects. Till March 31, 2010, IREDA has financed 1,921 projects with loan commitment of Rs. 121.8 billion [28]. An installed power generation capacity of 4.38 GW has resulted out of this investment. A profile of renewable energy development in India over 10, 11, 12 and 13th plan is given in Table 5 [28]. A total capacity of 72.4 GW is envisaged by end of 2022 with major contributions from wind (38.5 GW) and solar (20 GW) power. Table 6 gives the latest scenario of various renewable energy sources in India [1]. As of October 2010, the total installed capacity of renewable power stands at impressive 18,780 MW, which is approximately 10% of the total installed capacity [1]. By end of 11th plan in 2012, the total targeted installed capacity of renewable is 22,700 MW with net generation of 60 billion kW h, which is $\sim$4.4% of total electricity mix [28].

An unfortunate yet true fact is that even after successful achievement of the projected expansion of grid interactive renewable power to capacity of 72.4 GW with net generation of 173 billion kW h, the actual contribution of renewable sources to the total generation (projected at 2,516 billion kW h) will stand at meager 5–6%. This projection, which is rather in consistence with projections for other countries in the world, shows that India would still be mainly dependent on fossil fuels for another decade or so [15].

Electricity Act 2003 of Government of India has special provisions and resolutions for development of grid interactive renewable power projects [25]. This act mandates State Electricity Regulatory Commission (SERCs) to promote co-generation and generation of electricity from renewable sources of energy by providing suitable measures for connectivity with grid and sale of electricity to any person, and fix certain minimum percentages for purchase of renewable power in the area of each distributor licensee. The terms and conditions for determination of tariff shall be guided by (1) principles and methodologies specified by central commission, (2) generation and transmission, distribution and supply of electricity on commercial principles, (3) encouragement of competition, efficiency, economic use of resources, good performance and optimum investments, (4) safeguarding of consumers' interest and recovery of cost of electricity in reasonable manner. Tariff Policy 2006 has authorized SERCs to purchase electricity from renewable sources taking into account availability of such sources in the region, and its impact on retail tariff, and procurement by distribution companies at preferential tariffs determined by SERCs. In addition to these, several financial and fiscal incentives have been offered by MNRE to promote private investment in setting up of projects for power generation from renewable sources such as relaxation and/or

Table 5 Profile of renewable energy efforts in India [28]

Resource	Capacity at the end of 10th plan (MW)	11th Plan Target (MW)	Total targeted capacity (MW) by end of 11th plan (2012)	12th plan projection (MW)	Total targeted capacity (MW) by end of 12th plan (2017)	13th plan Projection (MW)	Total targeted capacity (MW) by end of 13th plan (2022)
1. Wind power	7,094	9,000	~16,100	11,200	27,300	11,200	38,500
2. Small hydro power	1,976	1,400	~3,400	1,600	5,000	1,600	6,600
3. Biomass power							
A. Biomass gasification	525	500	1,025	500	1,525	1,000	2,525
B. Bagasse co-generation	616	1,200	1,816	1,400	3,216	700	3,916
C. Waste to energy	44	80	124	200	324	500	824
Total	1,185	1,780	~3,000	2,100	~5,100	2,200	~7,300
4. Solar power	3	200	~200	3,800	4,000	16,000	20,000
Total capacity of renewable power	10,258	12,500	22,700	18,700	41,400	31,000	72,400
Net generation through renewable sources (billion kW h)	27	–	60	–	106	–	173
Capacity addition through conventional sources	122,071	78,700	~200,000	83,000	283,000	100,000	383,000
Net generation through conventional sources (billion kW h)	662	–	1,314	–	1,859	–	2,516
Grand total (conventional + renewable)	132,329	91,200	222,700	101,700	32,430	130,800	455,100
Share of renewable power (installed capacity)	7.7%	–	10.2%	–	12.8%	–	15.9
Share of renewable power (net generation)	3.5%	–	4.4%	–	5.4%	–	6.4

Table 6 Renewable energy in India at a glance (data as of October 30, 2010) [1]

Sl. no.	Source/system	Estimated potential (MW)	Cumulative installed capacity (MW)/number
I.	**Power From Renewables**		
A	*Grid interactive renewable power*		
1.	Bio power (agro-residues and plantations)	16,881	979.1
2.	Wind power	45,195	12,906.73
3.	Small hydropower (≤25 MW)	15,000	2,850.25
4.	Co-generation bagasse	5,000	1,494.53
5.	Waste to energy (urban)	2,700	72.46
6.	Solar power	–	17.82
	SUBTOTAL (MW)	**84,776**	**18,320.89**
B	*Captive/combined heat and power/distributed renewable power*		
7.	Biomass/co-generation (bagasse)	–	267.08
8.	Biomass gasifier	–	128.16
9.	Energy recovery from waste (sum total of urban, rural and industrial)	–	61.23
10.	Solar PV power plants and street lights (<1 kW)	–	2.39
11.	Aero generator/hybrid power systems	–	1.07
SUBTOTAL (MW)		–	**459.93**
TOTAL (A + B)			**18,780.82**
II.	**Remote Village Electrification**		
1.	Villages	–	5,329
2.	Hamlets	–	1,538
III.	**Decentralized Energy Systems**		
1.	Family type biogas plants	12 million	4.28 million
2.	Solar photovoltaic programme	20 MW/km^2	
	(i) Street lighting systems (no.)		121,634
	(ii) Home lighting systems (no.)		619,428
	(iii) Solar lanterns (no.)		813,380
	(iv) Solar pumps		7,495
3.	Wind pumps (no.)	–	1,352
4.	Solar cookers (no.)	–	0.67 million
5.	Solar water heating system (m^2 collector area)	140 million	3.53 million

cancellation of excise and custom duties, capital/interest subsidy, and accelerated depreciation policy. The range of capital subsidy is 10–90% of total project cost, and is mainly a function of region of project implementation and type of renewable source. Special subsidies are also in place for remote regions such as northeastern states or special category states. Generation based incentives for wind power and feed in tariff for solar power have also been introduced. As a result of these policies, a capacity addition of 6,761 MW has been achieved in the 10th plan, and another 12,500 MW of generation capacity is envisaged in the 11th plan.

4.1 Renewable Energy Options for India and Their Feasibility

Table 6 depicts the estimated potential and cumulative achievement/installations as of October 30, 2010 of various renewable energy sources in India [1]. Among all renewable energy options, biomass-based generation has special merits in the Indian context such as relatively low capital investments, simple and proven technology needing unskilled/semi-skilled labor, low operational costs and abundant as well as uniform availability of biomass throughout the year in the country at reasonable rates. The word biomass in the context of gasification refers to all organic matter originating from land and/or water based vegetation, which includes algae, trees and crop and forest residues. In order to quantify this resource, National Productivity Council of India undertook a comprehensive survey. The basis of this survey was to get a ratio of crop residue to crop production. On the basis of state wise production of various crops, the net generation of crop residue was estimated. This data has been updated regularly. In 2006–2007, the total annual biomass residue generation was estimated at 500 million tons [24]. However, not all of these residues are available for power generation. Domestic use as cooking fuel, and other conventional uses such as cattle feeds, fuel for paper manufacture, brick factories etc would take up approx. 350 million tons and net biomass surplus available for power production is 150 million tons, with power production potential of 18.7 GW [24].

Another notable attempt to quantify the biomass resource in India was made by Indian Institute of Science Bangalore by creating a comprehensive biomass atlas of India [4]. This atlas gives both state and district wise land for agriculture and forests, the total crop production, total biomass production and total biomass surplus. Table 7 give the detailed account of biomass production in India for the year 2008–2009 through both agro-residues and forest & wasteland [4]. For the year 2008–2009 the total surplus biomass through agro-residues is estimated at 145 millions tons with another 104 million tons biomass through forest & wastelands. Total power generation capacity of this biomass is 333 GW. In addition, the co-generation capacity of sugar mills through biogases is estimated at 6–7 GW, equivalent to approx. 45 million tons of coal equivalent [24]. Another argument that can be made in support of biomass-based generation is on the basis of comparative HHV (higher heating values) of coal and biomass. The chemical composition of coal and biomass is different. Coal contains $\sim$70% fixed carbon and only 20–30% volatiles as against 60–80% volatiles and 15–25% fixed carbon in biomass. However, most of Indian coal contains large ($\sim$30%) amount of ash, and its HHV is in the range of 20–22 MJ/kg, which is close to the HHV range of biomass (16–18 MJ/kg) [6]. Table 8 gives a listing of typical capacity ranges, capital cost per kW of installed capacity, and level zed unit cost of electricity generation from various renewable sources [27]. It could be inferred from Table 8 that biomass gasification beats competition among all renewable energy sources for small scale ($\sim$50 kW or less) decentralized power generation.

Table 7 (**A**) State-wise biomass production (in the form of both agro-residues and forest & wasteland) for the year 2008–2009 in India [27]. (**B**) Summary of biomass atlas of India (Biomass Atlas)

State	Biomass class	Area (kHa)	Crop production (kT/Yr)	Biomass generation (kT/Yr)	Biomass surplus (kT/Yr)	Power potential (MWe)
(A)						
Andhra Pradesh	Forest & wasteland	3,623.9	–	5,151.6	3,484.4	487.8
	Agro	9,983.2	21,167.1	43,893.2	6,956.4	863.3
Arunachal Pradesh	Agro	208.5	251.1	400.4	74.5	9.2
	Forest & wasteland	5,467.4	–	8,313.1	6,045.3	846.3
Assam	Forest & wasteland	2,676.8	–	3,674.0	2,424.2	339.4
	Agro	3,460.3	8,250.6	11,443.6	2,346.9	283.9
Bihar	Forest & wasteland	906.0	–	1,248.3	831.9	116.5
	Agro	7,348.7	18,817.6	25,756.9	5,147.2	641.1
Chhattisgarh	Agro	4,758.2	6,636.6	11,272.8	2,127.9	248.5
	Forest & wasteland	8,762.1	–	13,592.3	9,065.8	1,269.2
Goa	Forest & wasteland	153.4	–	180.7	119.3	16.7
	Agro	154.2	489.5	668.5	161.4	20.9
Gujarat	Agro	8,007.6	23,895.7	29,001.0	9,085.5	1,224.8
	Forest & wasteland	9,030.3	–	12,196.3	8,251.8	1,155.2
Haryana	Forest & wasteland	294.7	–	393.3	259.6	36.3
	Agro	5,707.3	15,226.2	29,034.7	11,342.9	1,456.9
Himachal Pradesh	Agro	788.3	1,504.0	2,896.9	1,034.7	132.6
	Forest & wasteland	2,259.8	–	3,054.6	2,016.0	282.2
Jammu & Kashmir	Agro	749.4	773.8	1,591.3	279.6	37.1
	Forest & wasteland	9,838.0	–	11,461.7	7,564.7	1,059.1
Jharkhand	Agro	1,850.3	2,459.5	3,644.9	890.0	106.7
	Forest & wasteland	3,506.8	–	4,876.6	3,249.8	455.0
Karnataka	Forest & wasteland	6,993.7	–	10,001.3	6,600.8	924.1
	Agro	9,683.6	43,139.6	34,167.3	9,027.2	1,195.7

(continued)

Table 7 (continued)

State	Biomass class	Area (kHa)	Crop production (kT/Yr)	Biomass generation (kT/Yr)	Biomass surplus (kT/Yr)	Power potential (MWe)
Kerala	Forest & wasteland	1,235.4	–	2,122.1	1,429.1	200.1
	Agro	2,306.8	5,561.0	11,644.3	6,352.1	864.4
Madhya Pradesh	Forest & wasteland	12,802.2	–	18,398.2	12,271.2	1,718.0
	Agro	13,167.3	17,951.7	33,344.8	10,329.3	1,373.3
Maharashtra	Forest & wasteland	13,177.4	–	18,407.1	12,440.4	1,741.7
	Agro	18,851.5	64,336.1	47,624.8	14,789.6	1,983.7
Manipur	Agro	340.8	435.1	909.4	114.4	14.3
	Forest & wasteland	1,260.9	–	1,264.0	834.2	116.8
Meghalaya	Agro	174.4	284.2	511.1	91.6	11.3
	Forest & wasteland	1,532.6	–	1,705.9	1,125.6	157.6
Mizoram	Agro	19.0	33.3	61.1	8.5	1.12
	Forest & wasteland	1,638.8	–	1,590.9	1,050.0	147.0
Nagaland	Agro	179.6	276.1	492.2	85.2	10.0
	Forest & wasteland	786.4	–	843.8	556.9	78.0
Orissa	Forest & wasteland	6,265.0	–	9,370.2	6,084.8	851.9
	Agro	6,667.6	12,262.7	20,069.5	3,676.8	429.3
Punjab	Forest & wasteland	229.1	–	398.5	263.0	36.8
	Agro	6,993.5	35,934.0	50,847.6	24,842.9	3,172.2
Rajasthan	Forest & wasteland	14,135.0	–	9,541.6	6,297.5	881.6
	Agro	14,851.4	16,135.5	29,851.3	8,645.7	1,126.7
Sikkim	Agro	58.0	69.1	149.5	17.8	2.29
	Forest & wasteland	372.8	–	531.5	350.8	49.1
Tamil Nadu	Forest & wasteland	3,187.2	–	4,652.4	3,070.6	429.9
	Agro	4,165.1	30,415.4	22,507.6	8,900.0	1,160.0

(continued)

Table 7 (continued)

State	Biomass class	Area (kHa)	Crop production (kT/Yr)	Biomass generation (kT/Yr)	Biomass surplus (kT/Yr)	Power potential (MWe)
Tripura	Agro	9.5	3.70	40.9	21.1	2.94
	Forest & wasteland	831.0	–	1,035.5	683.4	95.7
Uttar Pradesh	Forest & wasteland	3,856.5	–	5,478.4	3,672.0	514.1
	Agro	15,950.9	138,945.4	60,322.2	13,737.9	1,746.2
Uttaranchal	Agro	1,015.7	7,783.3	2,903.2	638.4	80.9
	Forest & wasteland	2,885.5	–	4,559.2	3,055.3	427.7
West Bengal	Forest & wasteland	1,113.9	–	1,430.7	949.0	132.9
	Agro	6,090.2	22,807.8	35,989.9	4,301.5	529.3
(B)						
	Agro-total	143,540.9	495,845.6	511,041.0	145,026.6	18,728.7
	Forest & wasteland total	118,822.9	0.000	155,474.0	104,047.4	14,566.6
	Total	262,363.8	495,845.6	666,515.0	249,074.0	33,295.4

Table 8 Comparative evaluation of unit capital cost and levelized unit cost of generation of different renewable energy sources in Indian context [27]

Sr. no.	Renewable energy source	Typical range of unit capacity available (kW)	Range of unit capital cost (Rs/kW)	Range of unit cost of electricity generated (Rs/kW h)
1.	Biomass gasifiers			
	(a) Dual fuel engine system	5–40	122,000–44,000	25.00–13.14[a]
	(b) 100% Producer gas engine	9–40	95,000–75,000	18.53–15.02[a]
2.	Diesel generating set	5–40	35,000–16,500	21.38–13.51[a]
3.	Small hydro power	10–100	124,000–216,000	14.56–8.31[b]
4.	Solar photovoltaic systems	2.5–25	308,000–279,000	32.32–29.26[c]
5.	Small wind generators	1–50	203,000–67,000	44.17–6.30

[a] Plant load factor assumed to be 25%
[b] Plant load factor assumed to be 40%
[c] Solar radiation 4.89 kW h/m^2

Essentially, there are two options for electricity generation from biomass on commercial level, viz. (1) biomass gasification and (2) boiler-steam turbine route. As far as second technology is concerned it is immensely popular in sugar industries. In this technology, high pressure steam generated from biogases fired boiler is used for driving turbine for electricity generator, and later meeting the energy demands of the process. The overall efficiency of the process is about 60%, with typical investment of Rs. 30–40 million/Mew. However, these units are economically feasible only for capacities higher than 5 MW or so. These are not suitable for decentralized generation in remote locations [5].

Biomass gasifier coupled to either dual fuel or 100% producer gas engine is a relatively much simpler and low-cost technology. Moreover, the biomass gasifier units are available in low to moderate capacities of 100–500 kW, which are sufficient to meet the electricity requirement of villages with population 2,000–5,000 (approx. 250–500 households). Biomass gasifiers are available in various designs such as downdraft, updraft, cross-draft, bubbling fluidized bed and circulating fluidized bed. Of these, the most popular designs are updraft and downdraft type operating at atmospheric pressure with air as gasification medium. However, fuel specificity is a major drawback of these designs. Moreover, the calorific value of gas produced from these units is moderate (3–4 MJ/N m^3) and tar/particulate content is also high. Thus, an efficient gas cleaning system must precede the engine–generator in which this gas is fired. Frequent cleaning and maintenance of the engine is also required. Typical capacity of these designs does not exceed 250 kW. Fluidized bed gasifiers overcome most of these demerits. Principal merit of these designs is fuel flexibility (in terms of both type and size), high carbon conversion, low tar content and use of in-bed catalyst for tar cracking as well as methane reforming. These designs also have potential of scale up to MW level. The calorific value of gas resulting from these units is moderate to high (4–6 MJ/N m^3) [5].

In addition to basic mechanical design, the performance of gasifier also depends on operating parameters, type of biomass and gasification medium. We have done an extensive thermodynamic study of performance of gasifiers at various operating conditions. We have considered biomass mixtures as feedstock in our analysis so as to assess the performance of gasifier with respect to fuel flexibility. In the next section we have presented a gist of this study [7].

5 Thermodynamic Assessment of Gasifier Performance with Biomass Mixtures as Feedstock

Most of these gasifiers mainly use wood chips as fuel, although several new designs have been developed that can use alternative fuels such as coconut shells and briquettes. Typical capacity of these gasifiers ranges from 5 to 250 kW [13, 23]. For capacities higher than 1 MW, fluidized bed gasifier is the most feasible design [3]. These gasifiers have the merits of fuel flexibility, uniformity of temperature over reactor volume, low tar content of producer gas and high overall carbon conversion. Typical specific consumption of biomass fuel in these gasifiers is 1–1.2 kg/kW h [26]. Thus, the annual biomass requirement of a typical 5 MW gasifier plant (with capacity utilization factor of 70%) is more than 35,000 tons. It is rather unlikely, in any region of the country, that a single biomass would be available through out the year in such large quantities for meeting the fuel demands of the plant, and thus, mixtures of different biomasses that are available in different seasons would have to be used. This necessitates a thorough study of the performance of gasifier in terms of fuel flexibility, i.e. variation in the quality and quantity of the producer gas resulting from gasification of biomass mixtures of different compositions. Such a study would provide important guidelines for design and scale-up of fluidized bed gasifiers with biomass mixtures as fuel input.

In this study, we have assessed gasification characteristics of mixtures three biomass, which are available in abundance in the northeastern states of India [4], viz. rice husk, saw dust and bamboo dust. In addition, we also evaluate the gasification process with semi-equilibrium models, which take into consideration partial conversion of carbon in biomass.

For decentralized power generation, principal performance parameters of gasifier are: Net gas yield and LHV of the producer gas resulting from biomass gasification. The ultimate analyses of the individual biomasses considered in this study are given in Table 9(A). Major operational parameters that influence performance of gasifier are: (1) temperature of gasification and (2) air or equivalence ratio, which is the ratio of actual oxygen supplied for gasification to the oxygen required for complete combustion of biomass [9, 14]. From a previous study [5, 6], we have established that the most suitable ranges of these parameters are: temperature = 700–1,000°C and air ratio = 0.2–0.4.

We, therefore, have chosen four representative temperatures, viz. 700, 800, 900 and 1,000°C and three air ratios, viz. 0.2, 0.3 and 0.4 for the simulations. Binary biomass mixtures have been considered for analysis. We combine these two individual biomasses in three proportions in weight percent as 25–75, 50–50 and 75–25%. Thus, we have nine combinations of biomass mixtures. The elemental analysis of these mixtures of biomasses along with a representative molecular formula for the biomass mixture is given in Table 9(B). The elemental compositions (or elemental vector input) for the 27 mixtures for gasification process are given in Table 10.

5.1 Incomplete Carbon Conversion

In a fluidized bed biomass gasifier, the residence time of the biomass mixture is small, as it is carried out of the riser section with the gasification air. Therefore, kinetics of various chemical reactions in the gasification process comes into the picture. The major result of short residence time of biomass is incomplete conversion (oxidation) of carbon in it. This incomplete conversion of carbon leads to reduction in the quality as well as quantity of producer gas. We have also tried to

Table 9 Elemental compositions (**A**) Individual biomasses (ultimate analysis). (**B**) Biomass mixtures (basis: 100 g of total biomass mixture) [7], reproduced with permission from International Energy and Environment Foundation, Al-Najaf, Iraq

Biomass	Composition in weight percent (dry basis)					Molecular formula
	Carbon	Hydrogen	Nitrogen	Oxygen	Ash	
(A)						
Saw dust	52.28	5.2	0.47	40.85	1.2	$CH_{1.193}N_{0.007}O_{0.585}$
Rice husk	37.03	5.25	0.09	40.94	16.69	$CH_{1.699}N_{0.003}O_{0.828}$
Bamboo dust	39.88	5.5	0.89	47.92	5.81	$CH_{1.657}N_{0.018}O_{0.904}$

Biomass components	Mixture composition	Elemental composition (g atoms)				Molecular formula*	Net energy content (kJ per 100 g)
		C	H	N	O		
(B)							
Rice husk (*RH*) Saw dust (*SD*)	RH = 25%, SD = 75%	4.039	5.213	0.027	2.555	$CH_{1.291}N_{0.007}O_{0.633}$	1,867
	RH = 50%, SD = 50%	3.721	5.225	0.02	2.556	$CH_{1.404}N_{0.005}O_{0.687}$	1,729
	RH = 75%, SD = 25%	3.404	5.238	0.013	2.557	$CH_{1.539}N_{0.004}O_{0.751}$	1,590
Bamboo bust (*BD*) Rice husk (*RH*)	BD = 25%, RH = 75%	3.145	5.313	0.021	2.668	$CH_{1.689}N_{0.007}O_{0.848}$	1,468
	BD = 50%, RH = 50%	3.205	5.375	0.035	2.777	$CH_{1.677}N_{0.011}O_{0.866}$	1,485
	BD = 75%, RH = 25%	3.264	5.438	0.049	2.886	$CH_{1.666}N_{0.015}O_{0.884}$	1,502
Bamboo dust (*BD*) Saw Dust (*SD*)	BD = 25%, SD = 75%	4.098	5.275	0.041	2.664	$CH_{1.287}N_{0.010}O_{0.650}$	1,884
	BD = 50%, SD = 50%	3.84	5.35	0.049	2.774	$CH_{1.393}N_{0.013}O_{0.722}$	1,762
	BD = 75%, SD = 25%	3.582	5.425	0.056	2.885	$CH_{1.515}N_{0.016}O_{0.805}$	1,641

* The molecular formula represents the mixture of two biomasses as a single entity. All biomasses are assumed to contain 10% w/w moisture

Table 10 Elemental vector input (in gatom) for simulations (basis: 100 g of total biomass mixture + air for gasification)

Element/elemental ratio	AR 0.2	0.3	0.4	Element/elemental ratio	AR 0.2	0.3	0.4	Element/elemental ratio	AR 0.2	0.3	0.4
Biomass: RH (25%) + SD (75%)				**Biomass: RH (50%) + SD (50%)**				**Biomass: RH (75%) + SD (25%)**			
C	4.039	4.039	4.039	C	3.721	3.721	3.721	C	3.404	3.404	3.404
H	6.324	6.324	6.324	H	6.336	6.336	6.336	H	6.349	6.349	6.349
N	6.180	9.257	12.334	N	5.696	8.534	11.373	N	5.212	7.812	10.411
O	4.736	5.549	6.362	O	4.611	5.361	6.111	O	4.487	5.173	5.860
H/C	1.566	1.566	1.566	H/C	1.703	1.703	1.703	H/C	1.865	1.865	1.865
O/C	1.173	1.374	1.575	O/C	1.239	1.441	1.642	O/C	1.318	1.520	1.722
Biomass: BD (25%) + RH (75%)				**Biomass: BD (50%) + RH (50%)**				**Biomass: BD (75%) + RH (25%)**			
C	3.145	3.145	3.145	C	3.205	3.205	3.205	C	3.264	3.264	3.264
H	6.424	6.424	6.424	H	6.486	6.486	6.486	H	6.549	6.549	6.549
N	4.773	7.150	9.526	N	4.819	7.211	9.602	N	4.864	7.271	9.679
O	4.479	5.107	5.735	O	4.596	5.228	5.860	O	4.714	5.350	5.986
H/C	2.043	2.043	2.043	H/C	2.024	2.024	2.024	H/C	2.006	2.006	2.006
O/C	1.424	1.624	1.824	O/C	1.434	1.631	1.828	O/C	1.444	1.639	1.834
Biomass: BD (25%) + SD (75%)				**Biomass: BD (50%) + SD (50%)**				**Biomass: BD (75%) + SD (25%)**			
	0.2	0.3	0.4		0.2	0.3	0.4		0.2	0.3	0.4
C	4.098	4.098	4.098	C	3.840	3.840	3.840	C	3.582	3.582	3.582
H	6.386	6.386	6.386	H	6.461	6.461	6.461	H	6.536	6.536	6.536
N	6.226	9.318	12.410	N	5.787	8.656	11.525	N	5.348	7.994	10.640
O	4.863	5.670	6.487	O	4.846	5.604	6.362	O	4.838	5.537	6.237
H/C	1.558	1.558	1.558	H/C	1.683	1.683	1.683	H/C	1.825	1.825	1.825
O/C	1.187	1.384	1.583	O/C	1.262	1.459	1.657	O/C	1.351	1.546	1.741

Note: All biomasses are assumed to contain 10% w/w moisture. The elements C, H, N and O are given in g atom while the elemental ratio is dimensionless.
Source [7] reproduced with permission from International Energy and Environment Foundation, Al-Najaf, Iraq

assess this effect with approach of semi-equilibrium model. In this approach, we reduce the moles of carbon in the elemental vector input (given in Table 10).

The number of input moles of other three elements, viz. H, N and O are kept unchanged, or in other words, conversion of these elements is assumed to be complete. This approach is known as semi (or quasi)-equilibrium model [17–19]. In this category, however, we have considered binary biomass mixtures with even composition only (i.e. 50–50% w/w fraction of two biomasses in the mixture). Moreover, we have considered only one air ratio (=0.3) and temperature of gasification (800°C or 1,073 K). We have chosen three representative values for carbon conversion in our simulations, viz. 60, 70 and 80% (or 0.6, 0.7 and 0.8). These parameters have been chosen in view of practical values of carbon conversions (CC) observed in fluidized bed gasification [21]. The g atom of carbon in the elemental vector input in the semi-equilibrium model are CC × C, where C are g atom of carbon in the biomass mixtures as given in Table 10 (i.e. elemental vector input in the equilibrium model). The balance carbon, i.e. (1− CC) × C, is assumed to be remain unconverted, and appears as elemental carbon (C) species in the products of biomass gasification.

6 Mathematical Model

For the simulations, we have used software FACTSAGE [2, 12]. This software employs the algorithm SOLGASMIX proposed by Eriksson [10] for calculation of thermodynamic equilibrium using Gibbs energy minimization of the system. We give below the main equations of this model. These equations can be solved using method of LaGrange multipliers for calculation of equilibrium composition of a chemical system, i.e. mole numbers and fractions of gas/condensed phase species at equilibrium that could result from reactant species (or set of elements with predetermined g atoms) at a specific temperature and pressure, for which the total free energy of the system is at its minimum (with constraint of mass balance equations). The input to the thermodynamic model is given in terms of elemental vector, which could be determined from the ultimate analysis of biomass, and given air or equivalence ratio.

6.1 Algorithm for Gibbs Energy Minimization

For a system comprising of mixture of I species, the total Gibbs free energy (G) is:

$$G = \sum_i x_i g_i \tag{1}$$

x_i—Mole number of a substance or species in the mixture. g_i is the chemical potential written as:

$$g_i = g_i^0 + RT \ln a_i \tag{2}$$

For gaseous species, the activities a_i are equal to the partial pressure p_i (assuming ideal behavior):

$$a_i = p_i = (x_i/X)P \tag{3}$$

X—Total number of moles in the gas phase; P—total pressure of the system. Activity of the condensed substances is taken as one under assumption of purity. A new dimensionless quantity (G/RT) is defined as:

$$G/RT = \sum_{i=1}^{m} x_i^g[(g^0/RT)_i^g + \ln P + \ln(x_i^g/X)] + \sum_{i=1}^{s} x_i^c(g^0/RT)_i^c \tag{4}$$

R—Ideal gas constant. Superscripts g and c represent gas phase and condensed phase, respectively. m and s represent the total number of substances in the gas phase and condensed phase, respectively, at equilibrium. The value of (go/RT) for a certain substance is calculated using the expression:

$$(g^o/RT) = (1/R)[G^o - H^o_{298})/T + \Delta_f H^o_{298}/RT \tag{5}$$

Superscript o refers to the thermodynamic standard state; subscript 298 refers to the reference temperature (25°C= 298.15 K); subscript f denotes the formation of a compound from the elements in their standard states. The mass balance among various species can be written as:

$$\sum_{i=1}^{m} a_{ij}^g x_i^g + \sum_{i=1}^{s} a_{ij}^c x_i^c = b_j (j = 1, 2, \ldots, l) \tag{6}$$

a_{ij}—number of atoms of the jth element in a molecule of the ith substance; b_j—total number of moles of the jth element; l—total number of elements. The method involves a search for a minimum value of the free energy G of a system (or equivalently G/RT as given in Eq. 6) subject to the mass balance relation as subsidiary conditions. For solution of this system of equations, Lagrange's methods of undetermined multipliers can be used.

7 Results of Simulations

7.1 Trends in Simulation Results (Equilibrium Model)

7.1.1 Net Gas Yield (Fig. 1)

For all biomass mixtures, the net gas yield increases with air ratio; however, the temperature rise from 700 to 1,000°C does not affect the gas yield. For biomass mixtures containing saw dust (which has higher carbon content than other two biomasses), the gas yield slightly reduces as the proportion of the saw dust in the mixture reduces. However, for mixtures of rice husk and bamboo dust, the gas yield is practically independent of the mixture composition.

7.1.2 Hydrogen Content of Producer Gas (Fig. 2)

For a given gasification temperature, the hydrogen content of the producer gas decreases with air ratio for all biomass mixtures. On the other hand, for a given air ratio, the hydrogen content does not show a common trend with gasification temperature. For AR = 0.2, the hydrogen content rises till 900°C and thereafter decreases, whereas for AR = 0.3 and 0.4, the hydrogen content reduces continuously with rising gasification temperature. No particular trend can be seen for hydrogen with constituents of the biomass mixture. For a given combination of temperature and air ratio, the hydrogen content of producer gas varies by less than ±10% with composition of the mixtures. Even among mixtures of different biomasses (RH + SD; SD + BD or RH + BD), the hydrogen content of producer gas shows insignificant variation.

7.1.3 Carbon Monoxide Content of Producer Gas (Fig. 3)

For a given gasification temperature, the CO content of the producer gas reduces with air ratio. This trend is consistent for all nine mixtures of biomasses. Similarly, for all nine mixtures, the CO content of producer gas increases with temperature at a constant air ratio. For biomass mixtures constituting saw dust, at any combination of air ratio and gasification temperature, the CO content is significantly higher (by about 25–40%) than the corresponding value for mixtures of rice husk and bamboo dust. Moreover, for biomass mixtures comprising saw dust, the CO content reduces with the proportion of saw dust in the mixture. This effect is clearly attributed to higher carbon content of saw dust than rice husk and bamboo dust.

7.1.4 LHV of Producer Gas (Fig. 4)

The major combustible components of the producer gas are CO and H_2. It is thus obvious that the trend in the LHV of the producer gas is similar to that of hydrogen and carbon monoxide. For a given gasification temperature, the LHV reduces with increasing air ratio; whereas, for a given air ratio, the LHV increases with gasification temperature. For biomass mixtures comprising saw dust, the LHV values for any combination of air ratio and gasification temperature are higher than the corresponding values for mixtures of rice husk and bamboo dust. Moreover, for mixtures of saw dust, the LHV at any air ratio and temperature reduces with

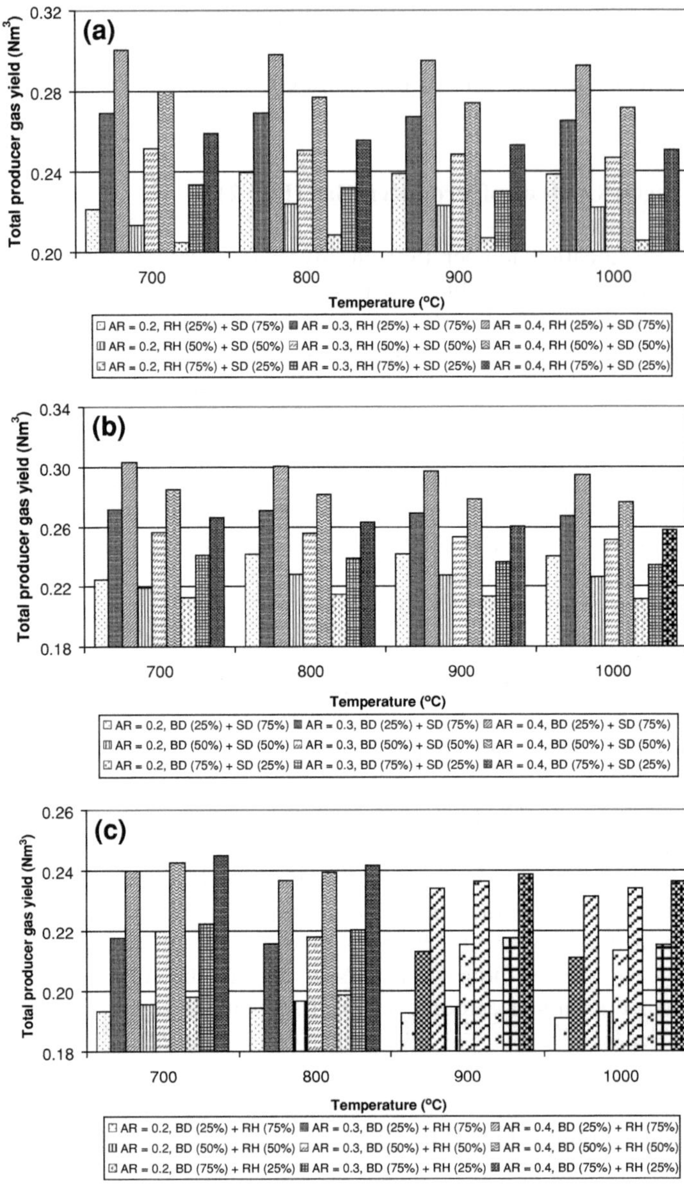

Fig. 1 Simulations results for the gasification of biomass mixtures (basis: 100 g of total biomass mixture). Variation in total producer gas yield for different biomass mixtures with air ratio and temperature. **a** Mixtures of rice husk and saw dust. **b** Mixtures of bamboo dust and saw dust. **c** Mixtures of bamboo dust and rice husk. *Source* [7]—reproduced with permission from International Energy and Environment Foundation, Al-Najaf, Iraq

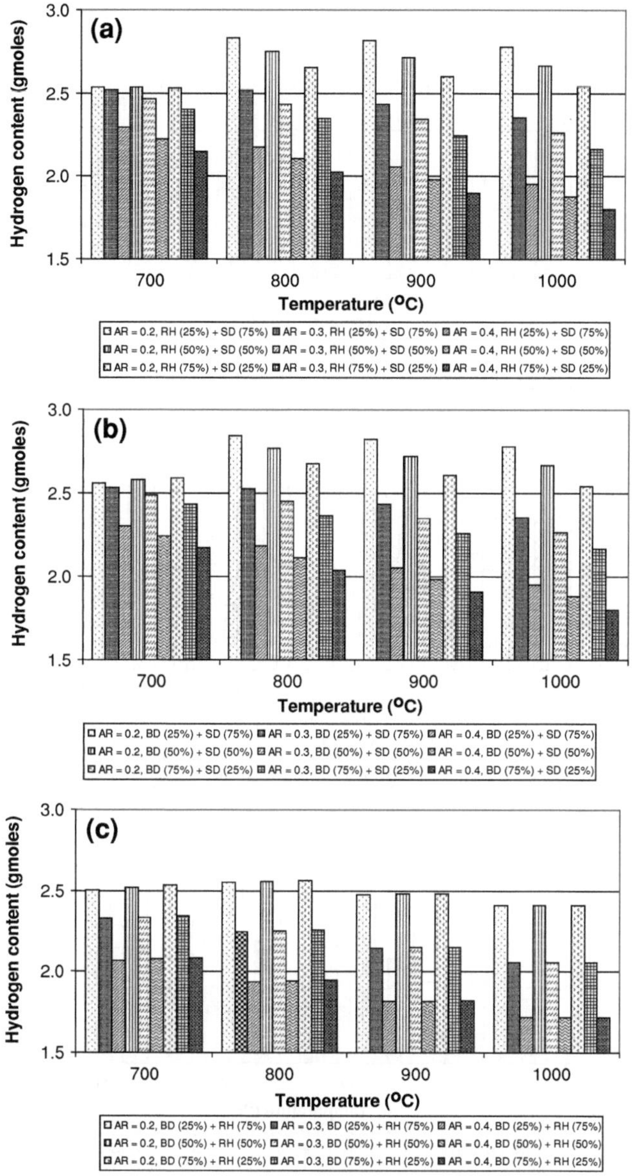

Fig. 2 Simulations results for the gasification of biomass mixtures (basis: 100 g of total biomass mixture). Variation in hydrogen content in producer gas for different biomass mixtures with air ratio and temperature. **a** Mixture of rice husk and saw dust. **b** Mixture of bamboo dust and saw dust. **c** Mixture of bamboo dust and rice husk. *Source* [7]—reproduced with permission from International Energy and Environment Foundation, Al-Najaf, Iraq

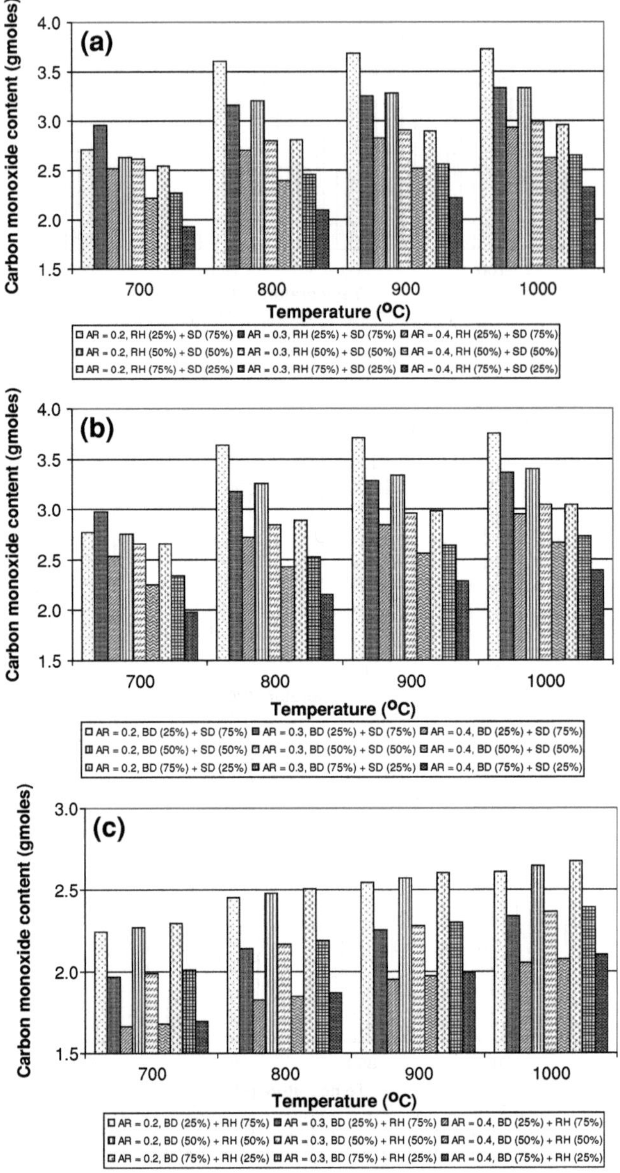

Fig. 3 Simulations results for the gasification of biomass mixtures (basis: 100 g of total biomass mixture). Variation in carbon monoxide content in producer gas for different biomass mixtures with air ratio and temperature. **a** Mixtures of rice husk and saw dust. **b** Mixtures of bamboo dust and saw dust. **c** Mixtures of bamboo dust and rice husk. *Source* [7]—Reproduced with permission from International Energy and Environment Foundation, Al-Najaf, Iraq

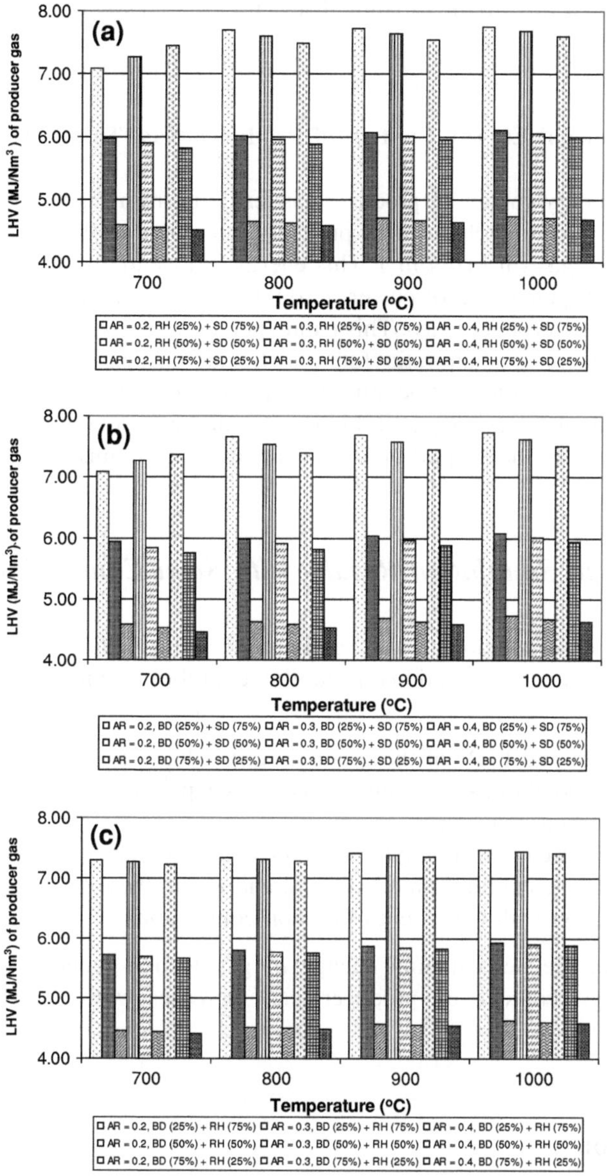

Fig. 4 Simulations results for the gasification of biomass mixtures (basis: 100 g of total biomass mixture). Variation in LHV of producer gas for different biomass mixtures with air ratio and temperature. **a** Mixtures of rice husk and saw dust. **b** Mixtures of bamboo dust and saw dust. **c** Mixtures of bamboo dust and rice husk. *Source* [7]—reproduced with permission from International Energy and Environment Foundation, Al-Najaf, Iraq

proportion of saw dust in the mixture. These trends are essentially same as that of CO content.

7.1.5 Net Thermal Energy Content of Producer Gas (Table 11)

The thermal energy content of producer gas resulting from gasification of 100 g of biomass mixture can be obtained by product of net yield of the gas (in N m^3) and the LHV of the gas (in MJ/N m^3). This energy is essentially the potential of the producer gas for generation of power (through engine–generator sets operating on dual fuel or 100% producer gas). This parameter shows same trends as the LHV. It reduces with increasing air ratio for a particular gasification temperature, and increases with gasification temperature for a particular air ratio. Moreover, thermal energy content of producer gas is higher for mixtures containing saw dust and varies directly with the proportion of the saw dust in the mixture.

7.2 Trends in Simulation Results with Semi-Equilibrium Model

The results of simulations with semi-equilibrium models are presented in Table 12. Two major deviations from the equilibrium model are evident as follows:

1. The net yield and LHV of the producer gas reduces as compared to the total equilibrium conditions. Obviously, these two parameters vary directly with extent of carbon conversion.
2. Hydrogen and carbon monoxide content of the producer gas also reduces as compared to the equilibrium conditions, and again, these two parameters show direct variation with the extent of carbon conversion.

In addition, the net enthalpy change of gasification reduces, while the net thermal energy content of producer gas increases with increasing carbon conversion.

8 Discussion

Major result of our simulations is that the quality and quantity of producer gas resulting from gasification of biomass mixtures has high potential for power generation through dual fuel or 100% producer gas engines. This potential can be quantified as follows: if a gasifier consumes 100 g of 50–50% w/w mixture of rice husk and saw dust per second (corresponding to 360 kg/h of gross consumption of mixture) for air ratio of 0.3 and gasification temperature of 800°C, the maximum thermal energy available (as seen from Table 11) in the producer gas after

Table 11 Net thermal energy (ΔHth, P, kJ) content of producer gas (basis: 100 g of total biomass mixture) source: [7]—reproduced with permission from International Energy and Environment Foundation, Al-Najaf, Iraq

Temp. (°C)	Air ratio			Temp. (°C)	Air ratio			Temp. (°C)	Air ratio		
	0.2	0.3	0.4		0.2	0.3	0.4		0.2	0.3	0.4
(A) Biomass: RH (25%) + SD (75%)				(A) Biomass: BD (25%) + SD (75%)				(A) Biomass: BD (25%) + RH (75%)			
700	1,570	1,604	1,382	700	1,594	1,612	1,389	700	1,411	1,244	1,069
800	1,844	1,616	1,386	800	1,853	1,624	1,393	800	1,427	1,249	1,071
900	1,848	1,617	1,386	900	1,857	1,625	1,393	900	1,428	1,249	1,071
1,000	1,848	1,617	1,386	1,000	1,857	1,625	1,392	1,000	1,428	1,249	1,071
(B) Biomass: RH (50%) + SD (50%)				(B) Biomass: BD (50%) + SD (50%)				(B) Biomass: BD (50%) + RH (50%)			
700	1,548	1,481	1,275	700	1,596	1,498	1,290	700	1,421	1,252	1,077
800	1,702	1,491	1,278	800	1,721	1,507	1,292	800	1,437	1,258	1,078
900	1,704	1,491	1,278	900	1,723	1,508	1,292	900	1,437	1,258	1,078
1,000	1,705	1,491	1,278	1,000	1,723	1,508	1,292	1,000	1,437	1,258	1,078
(C) Biomass: RH (75%) + SD (25%)				(C) Biomass: BD (75%) + SD (25%)				(C) Biomass: BD (75%) + RH (25%)			
700	1,527	1,359	1,169	700	1,569	1,384	1,190	700	1,431	1,260	1,083
800	1,560	1,367	1,172	800	1,588	1,391	1,192	800	1,446	1,266	1,085
900	1,562	1,367	1,172	900	1,590	1,391	1,192	900	1,446	1,266	1,085
1,000	1,562	1,367	1,171	1,000	1,590	1,391	1,192	1,000	1,446	1,265	1,085

Table 12 Simulation results for gasification of biomass mixtures with semi-equilibrium model (incomplete carbon conversion; basis: 100 g of biomass mixture) *source* [7]—reproduced with permission from International Energy and Environment Foundation, Al-Najaf, Iraq

CC (%)	H_2 moles	CO moles	ΔH (kJ)	Gas yield (N m^3)	LHV (MJ/N m^3)	ΔH_P (kJ)
(A) Biomass mixture: RH (50%) + SD (50%)						
60	1.36	0.91	−800.83	0.19	3.42	647
70	1.69	1.33	−743.78	0.21	4.17	858
80	1.97	1.79	−685.35	0.22	4.82	1,069
(B) Biomass mixture: BD (50%) + RH (50%)						
60	1.19	0.67	−814.70	0.16	3.28	530
70	1.51	0.99	−767.87	0.18	4.02	712
80	1.79	1.35	−730.35	0.19	4.67	894
(C) Biomass mixture: BD (50%) + SD (50%)						
60	1.33	0.90	−832.36	0.19	3.32	636
70	1.67	1.33	−775.67	0.21	4.09	854
80	1.97	1.80	−728.28	0.23	4.76	1,072

Abbreviations CC carbon conversion, LHV Lower heating value, ΔH net enthalpy change of the biomass gasification process, ΔHth, *P* net available thermal energy for power generation

attainment of total equilibrium in the gasification system is 1,491 kJ/s or kW t h. Typical efficiency of dual fuel engine–generator sets available in the market is ~30%. Thus, maximum electrical power generated with the producer gas is 1,491 × 0.3 = 450 kWe. However, practically the conversion of biomass in gasifier is not complete. As inferred from Table 12, the net thermal energy in the producer gas for 60, 70 and 80% conversion (at similar biomass consumption rate of 100 g/s or 360 kg/h) is 647, 858 and 1,069 kW t h. The net power generation through this producer gas is 190, 255 and 320 kWe for engine–generator efficiency of 30%. These figures also help in calculation of specific biomass consumption for electricity generation. For example, for 50–50% w/w mixture of rice husk and saw dust, the minimum possible specific fuel consumption (for total equilibrium conditions) is 360/450 = 0.8 kg/kW h. This consumption varies inversely with carbon conversion. For 60, 70 and 80% carbon conversion, the typical consumption of biomass mixture is 360/190, 360/255 and 360/320, i.e. 1.89, 1.41 and 1.13 kg/kW h, respectively. It should be noted however, that these calculations do not include recycled unconverted biomass, as in case of circulating fluidized bed gasifier. With that taken into account, requirement of fresh biomass reduces.

The values of specific fuels consumption reported in literature are higher than the calculated values in the present study by ±5–20%. For gasification of rice husk alone, Lv et al. [21] have reported carbon conversion in the range of 0.7–0.9, while Mansaray et al. [22] have reported slightly lower values of 0.6–0.8. Yin et al. [29] have reported specific fuel consumption of 1.7–1.9 kg/kW h for rice husk gasification for electricity generation capacity of 800 kW or higher. However, for low capacity (~200 kW), the specific fuel consumption is reported to be as high as 3.5 kg/kW h. Values of specific fuel consumption reported by Mansaray et al. [22] are in the same range (1.91 kg/kW h). It should be noted that in addition to overall carbon conversion in the gasifier, the specific fuel consumption also depends overall efficiencies of gasifier, duel fuel or 100% producer gas engine and the generator set. Typical values of these efficiencies are 55, 33 and 88%, respectively [16]. Kapur et al. [16] have given following formula for estimation of specific fuel consumption for electricity generation through rice husk gasification using duel fuel engine with generator set:

$$Q_h = \left[\frac{1/DF - (1 - RF)}{\eta_g \times CV_h \times \eta_d \times \eta_a}\right] \times 3.6 \tag{7}$$

where DF—derating factor; RF—diesel replacement factor; CV_h—calorific value of rice husk; η_d—efficiency of diesel engine; η_a—efficiency of the duel fuel generator; η_g—efficiency of gasification. With representative values of various factors, as given by Kapur et al. [16], DF = 0.75, RF = 0.7, CV_h = 13.4 MJ/kg, $\eta_g = 0.55$, $\eta_d = 0.33$ and $\eta_a = 0.875$, Q_h is calculated as 1.74 kg/kW h. Specific fuel consumption also depends on gasifier capacity and plant load factor. The range of specific fuel consumption, as reported by Nouni et al. [26], is in the range 1.1–1.68 kg/kW h for fixed bed downdraft gasifier of capacity 20–40 kWe employing either dual fuel or producer gas engine, and plant load factor ranging between 50 and 75%. It should be noted that "economy of scale" is another dominant aspect determining specific fuel consumption, in addition to various factors mention above. For large scale gasifiers (circulating fluidized bed type) in the range of 30 MWe or higher, the specific fuel consumption (either demolition wood or clean wood) could be as low as 0.9 kg/kW h [11].

Another important factor is the net enthalpy change of gasification process (or the energy released in gasification). Part of this energy is absorbed by the gasification system itself and part is carried out of the gasifier by the producer gas. The absorbed heat helps maintaining the temperature of the gasifier. Moreover, the heat in the producer gas can be recovered through various means such as preheating of gasification air or drying of the biomass feed.

It can be seen from Table 11 that thermal energy content of producer gas essentially stays constant after 800°C. Therefore, at first impression it appears that operation of gasifier at higher temperatures (900 or 1,000°C) may not fetch additional advantage. However, for a fluidized bed system where residence time of biomass is limited, higher gasification temperature can enhance single-pass conversion of biomass. Air ratio, on the other hand, has high influence on the process. For all biomass mixtures, rise of air ratio from 0.2 to 0.4 has been found to reduce the thermal energy content of producer gas by 30–40%. This result points out that low air ratios favor better performance of gasifier, but one must also take into consideration incomplete conversion of carbon even under total equilibrium conditions for air ratio of 0.2. This effect is more pronounced for biomass mixtures containing saw dust, which has higher carbon content than rice husk or bamboo dust.

The power generation capacity of gasifier is higher for biomass mixtures containing saw dust. Among all nine biomass mixtures, the least power generation is seen for biomass mixtures containing higher proportions of rice husk. This effect is clearly attributed to high ash content of rice husk.

9 Conclusion

In the first part of this chapter, we have given an overview and state-of-the-art of the electricity sector of India. Rapid economic growth in India, with development of industrial an agriculture sector, with fast urbanization leading to ever increasing energy demand, and increasing concerns over greenhouse gas emission leading to climate change have made quest for alternate and renewable energy sources as urgent need of the hour. Among the various renewable energy sources, biomass gasification has special importance for a developing country like India, with agriculture forming the base of the economy. Vast biomass reserves in India have a potential of 20 GW of power generation, which will play vital role in rural electrification through decentralized power generation, as well as reducing peak hour power shortages through grid interactive power. Effective utilization of this resource will pave way for a new economy that is not only inclusive and sustainable but also helps in reduction of green house gas emissions and risks of climate changes. Indian government has been promoting implementation of renewable energy systems through special policies, financial incentives and promotional programs since past 3 decades, with commendable success with more than 18 GW of total installed capacity of generation. Despite these efforts, the vast biomass resource remains highly underutilized. The principal causes leading to this effect are technology development and adaptation, innovation induction and strategies to up scaling and field implementation. The renewable energy program of India has a long way ahead. Following strategies could be recommended for enhancement of effective utilization of biomass energy [5, 6]: (1) Establishment of standard design guidelines, performance standards with testing and certification; (2) Development of an "energy technology package" by the manufacturers of gasifier that includes manufacturing, installation, operation and maintenance of gasifiers of wide range of capacities; (3) Subsidies on fixed as well as working capital required for gasifier; (4) Increasing the capacity utilization factor of gasifier along with plant load factor; (5) Setting up of local energy service companies (ESCOs) with help of NGOs and community based organizations; (6) Additional measures for cost reduction such as direct purchase of biomass from villagers and conduction of information and awareness program for promotion of gasifier based generation.

In the second part of this chapter, we have given a gist of our work on the feasibility of use of biomass mixtures of common biomasses found in northeastern states of India such as rice husk, bamboo dust and saw dust as fuel in biomass gasifiers for decentralized power generation. The results of our simulations have revealed interesting trends in performance of gasifiers with operating parameters such as air ratio, temperature of gasification and composition of the biomass mixture. For all biomass mixtures, the optimum air ratio is $\sim$0.3 with gasification temperature of 800°C. Under total equilibrium conditions, and for engine–generator efficiency of 30%, the least possible fuel consumption is found to be 0.8 kg/kW h. This parameter shows an inverse variation with the extent of carbon conversion (or oxidation) achieved in the gasifier, which in turn depends on parameters such as air ratio, temperature of gasification and residence time of biomass in the gasifier. For low carbon conversions ($\sim$60% or so), the specific fuel consumption could be as high as 1.5 kg/kW h. On a whole, this study confirms that performance of gasification process for decentralized electricity generation stays essentially same after replacement of single biomass (either saw dust or rice husk) by mixtures of these biomasses with bamboo dust in different proportions. This feature, obviously, adds to the flexibility of operations of gasifiers in different locations under different operating conditions.

References

1. Akshay U (2010) Ministry of new and renewable energy, New Delhi. Available online at: http://mnre.gov.in. Nov–Dec 2010
2. Bale CW, Chartrand P, Degterov SA, Eriksson G, Hack K, Mahfoud K, Melancon J, Pelton AD, Petersen S (2002) Factsage thermochemical software and databases. Calphad 26(2):189–228
3. Bharadwaj A (2002) Gasification and combustion technologies of agro–residues and their application to rural electric power systems in India. Ph.D. Dissertation, Carnegie Mellon University, Pittsburgh, USA
4. Biomass atlas of India, Combustion gasification and propulsion laboratory, Indian Institute of Science, Bangalore. http://cgpl.iisc.ernet.in. Accessed Jan 2011
5. Buragohain B, Mahanta P, Moholkar VS (2010) Biomass gasification for decentralized power generation: the Indian perspective. Renewable Sustainable Energy Rev 14(1):73–92
6. Buragohain B, Mahanta P, Moholkar VS (2010) Thermodynamic optimization of biomass gasification for decentralized power generation and Fischer–Tropsch synthesis. Energy 35:2557–2579
7. Buragohain B, Mahanta P, Moholkar VS (2011) Investigations in gasification of biomass mixtures using thermodynamic equilibrium and semi–equilibrium models. Int J Energy Environ 2(3):551–578
8. Central electricity authority, Ministry of power, Government of India. http://www.cea.nic.in. Accessed Jan 2011
9. Desrosiers R (1981) Thermodynamics of gas-char reactions. In: Reed TB (ed) Biomass gasification—principles and technology. Noyes Data Corporation, Park Ridge
10. Eriksson G (1975) Thermodynamic studies of high temperature equilibria–XII: SOLGAMIX, a computer program for calculation of equilibrium composition in multiphase systems. Chem Script 8:100–103
11. Faaij A, van Ree R, Waldheim L, Olsson E, Oudhuis A, van Wijk A, Daey–Ouwens C, Turkenburg W (1997) Gasification of biomass wastes and residues for electricity production. Biomass Bioenergy 12:387–407
12. FactWeb. http://www.factsage.com. Accessed Nov 2009
13. Ghosh D, Sagar A, Kishore VVN (2004) Scaling up biomass gasifier use: applications, barriers and interventions paper no. 103 (climate change series). World Bank Environment Department, Washington
14. Gumz W (1950) Gas producers and blast furnaces: theory and methods of calculations. Wiley, New York
15. Integrated energy policy: report of the expert committee. Planning commission, Government of India, New Delhi. Available online at: http://planningcommission.nic.in/reports. Aug 2006
16. Kapur T, Kandpal TC, Garg HP (1998) Electricity generation from rice husk in Indian rice mills: potential and financial viability. Biomass Bioenergy 14:573–583
17. Kersten SRA (2002) Biomass gasification in circulating fluidized beds. Ph.D. Dissertation, Twente University Press, Enschede, The Netherlands
18. Li X (2002) Biomass gasification in circulating fluidized bed. Ph.D. Dissertation, University of British Columbia, Vancouver, Canada
19. Li X, Grace JR, Watkinson AP, Lim CJ, Ergudenler A (2001) Equilibrium modeling of gasification: a free energy minimization approach and its application to a circulating fluidized bed coal gasifier. Fuel 80:195–207
20. List of countries by electricity consumption. Wikipedia, http://en.wikipedia.org/w/index.php?oldid=412694038. Accessed Jan 2011
21. Lv PM, Xiong ZH, Chang J, Wu CZ, Chen Y, Zhu JX (2004) An experimental study on biomass air–steam gasification in a fluidized bed. Bioresour Technol 95:95–101

22. Mansaray KG, Ghaly AE, Al–Taweel AM, Hamdullahpur F, Ugursal VI (1999) Air gasification of rice husk in a dual distributor type fluidized bed reactor. Biomass Bioenergy 17:315–332
23. McKendry P (2002) Energy production from biomass (part 3): gasification technologies. Bioresour Technol 83:55–63
24. Meshram JR, Mohan S (2007) Biomass power and its role in distributed power generation in India. In: 25 years of renewable energy in India, Ministry of new and renewable energy, New Delhi, pp 109–134
25. Ministry of power, Government of India, New Delhi. http://powermin.nic.in. Accessed Jan 2011
26. Nouni MR, Mullick SC, Kandpal TC (2007) Biomass gasifier projects for decentralized power generation in India: a financial evaluation. Energy Policy 35:373–1385
27. Nouni MR, Mullick SC, Kandpal TC (2008) Providing electricity access to remote areas in rural India: An approach towards identifying potential areas for decentralized power supply. Renewable Sustainable Energy Rev 12:1187–1220
28. Renewable energy in India: progress, vision and strategy. Paper presented by ministry of new and renewable energy at Delhi international renewable energy conference (DIREC), New Delhi. Available online at: http://mnre.gov.in. Oct 2010
29. Yin XL, Wu CZ, Zheng SP, Chen Y (2002) Design and operation of a CFB gasification and power generation system for rice husk. Biomass Bioenergy 23:181–187

Decisions Under Uncertainty in Municipal Solid Waste Cogeneration Investments

Athanasios Tolis, Athanasios Rentizelas, Konstantin Aravossis and Ilias Tatsiopoulos

Abstract The issue of Municipal Solid Waste (MSW) management is an ever increasing problem for all countries. Developed countries face the problem of dealing with very large amounts of MSW per capita, forcing them to develop new technologies and systems. On the other hand, countries with developing or transitional economies may generate lower amounts of MSW per capita, but the rate of increase is high and the current practices of MSW management are not as advanced as those of developed countries. Therefore, countries with developing or transitional economies may benefit from adopting MSW management technologies used by developed economies. One aspect of MSW management in developed economies is the energy recovery from MSW. The advantages of this type of technologies are mainly the significantly reduced waste volume for landfilling, the reduction of total greenhouse gas emissions, the potential for generating electricity or co-generation of electricity and heat. In this work, a comparative study of the most prominent co-generation technologies using MSW as a fuel source is presented, focusing on the evolution of their economical performance over time. An algorithm based on real-options has been applied for four technologies of MSW energy recovery: (1) incineration, (2) gasification, (3) landfill biogas exploitation using a pipeline system and (4) anaerobic digestion facilities. The financial contributors are identified and the impact of greenhouse gas trading is analyzed in terms of financial yields, considering landfilling as the baseline scenario. The greenhouse gas trading system presents an opportunity for investing in environmentally friendly technologies for MSW energy recovery, through the Clean Development Mechanism (CDM), in most developing countries. The results of this

A. Tolis (✉) · A. Rentizelas · K. Aravossis · I. Tatsiopoulos
School of Mechanical Engineering, Industrial Engineering Laboratory
National Technical University of Athens, Iroon Polytechniou 9 Street
15780 Athens, Greece
e-mail: atol@central.ntua.gr

A. Karagiannidis (ed.), *Waste to Energy*, Green Energy and Technology,
DOI: 10.1007/978-1-4471-2306-4_8,

work indicate an advantage of combined heat and power over solely electricity generation. The most attractive technology among the ones examined proves to be incineration, mainly due to its higher power production efficiency, lower investment costs and lower emission rates. Despite the fact that these characteristics may not drastically change over time, either immediate or irreversible investment decisions might be reconsidered under the current selling prices of heat, power and CO_2 allowances.

1 Introduction

The management of Municipal Solid Waste is a major issue worldwide, though its characteristics vary among developed countries and developing or transitional ones. Developed countries face mainly the problem of dealing with very large amounts of MSW per capita, forcing them to develop new technologies and systems. On the other hand, countries with developing or transitional economies may currently generate lower amounts of MSW per capita, but the rate of increase is high and the current practices of MSW management are not as advanced as those used in developed countries. Therefore, countries with developing or transitional economies may benefit from adopting MSW management technologies used by developed economies. The application of appropriate MSW management techniques constitutes an important component of sustainability and environmental protection for every country. The most important issues confronted in planned or operational waste management projects span among social acceptance, economic efficiency, organizational matters and water, soil and air pollution.

Various policies for Municipal Solid Waste management are implemented world-wide like recycling, composting and low enthalpy treatments, which are characterized by eco-friendly properties. Despite their proven environmental benefits though, not much evidence has been available regarding their efficiency and social adoption in big cities with high population density and rate of increase. On the other side, environmental experts agree that the goals set for the waste utilization rate would never be achieved without energy recovery [22]. The advantages of energy recovery from waste are mainly the significantly reduced waste volume for landfilling, the reduction of total greenhouse gas emissions, the potential for generating electricity or co-generation of electricity and heat. Furthermore, waste exists in all countries, societies and communities. This fact implies that if waste could be used to generate electricity—and potential heat—it is mostly the communities in developing countries that do not have currently access to electricity grid that would benefit from this application and would ameliorate their living conditions.

Innovative Waste-to-Energy (WtE) technologies have recently emerged, showing interesting characteristics compared to older but proven ones. However, the risk of investing in such innovative technologies might lead to the postponement of similar projects funded by private funds unless safer fiscal conditions are ensured. Moreover, interventions on environmental policies may change the relevant legal status thus further increasing uncertainty and complicating future strategies and decision making.

The management of waste is responsible for a significant amount of carbon emissions worldwide, due mainly to logistics and waste fermentation. The Kyoto protocol and the associated directives of European Union have recently led to various tools for the reduction of carbon emissions. The emissions trading market is one of these tools through which carbon intensive industries should pay a penalty for their production activities unless they take some measures for the mitigation of CO_2 emissions, having the option to act in another country to reduce CO_2 emissions. The Clean Development Mechanism has been established, allowing some flexibility for Annex I parties (developed economies) to reduce their carbon emissions, by performing environmentally friendly investments in developing countries. Within this framework, investments that allow environmentally friendlier waste management in developing countries may be eligible for CDM funding, thus ensuring en extra revenue stream for these projects.

Markets have been established for trading the CO_2 allowances and consequently, their corresponding prices acquire a non-stationary, volatile path over time. The prices of electricity sold to the grid as well as the electricity demand may also present a volatile behavior. Moreover, the prices of fuels may induce additional uncertainties in energy markets: On the one hand they constitute volatile cost factors, but on the other hand they may induce volatility on the revenues of co-generation projects, as long as the revenues from heat production depend on the volatile prices of fossil fuels. Co-generation plants may have additional revenues from trading the CO_2 allowances generated by the displacement of conventional, domestic boilers (fired by oil or natural gas), thus introducing more uncertainties to their economy related with the volatile CO_2 allowance prices. From the above described rationale, it may be seen that the context of the classical investment analysis investigating immediate and irreversible decisions becomes no longer optimal in energy markets. Optimal investment entry times should rather be inquired for investments under multiple uncertainties. Project planning should thus focus not only on logistical or production-related considerations but also on strategic decisions like the selection of the most profitable energy conversion method over time, the measures for the mitigation of CO_2 emissions and the optimal investment decision timing.

Within the frame of the traditional Discounted Cash Flow (DCF) methodology, many parameters such as the energy product prices, the fuel prices and the discounting factor (i.e. the interest rates) were usually assumed to be constant throughout the projects' duration. With the introduction of the real-options concept during the last two decades, the decision-making process has been drastically affected. Modern business plans have acquired time–dependent characteristics, which may allow optimization processes in respect of time. Optimal decisions in WtE market may not be limited to the selection of an appropriate technology but they may be extended to the optimization of investing time according to the varying fiscal conditions and the volatile prices of fuels, electricity and CO_2 allowances.

The starting point of the present study is a large city with high population density and increasing rate of MSW disposal. The inputs of the case study presented come from the city of Athens, Greece. Despite the fact that Greece belongs to the developed countries group, its waste management system is mainly based on landfilling with low rates of recycling and no energy recovery from waste, thus having different structure from most West-European countries. Therefore, the results obtained from this study may be similar to applications in developing or emerging economies, where waste management is mainly or entirely based on landfilling.

The scope of the study is to compare from an economic point of view four competing methods of combined heat and power (CHP) production based on MSW: (1) incineration, (2) gasification, (3) landfill biogas combustion considering gas supply through a pipeline system and (4) anaerobic digestion. The major milestones of the study are to analyze the cost structure and identify the impact of greenhouse gas trading on MSW-CHP projects. The baseline scenario used for comparing the investigated WtE options is assumed to be the landfilling of the entire MSW quantity. The objective of the study is the determination of the optimal investment entry times for each one of the competing technologies, and the identification of the most promising technology among the ones examined.

The rest of the work is organized as follows: In Sect. 2, a literature survey is given. In Sect. 3, a description of the case study is provided, followed by the mathematical formulation of time dependent CHP investments. The description of the model inputs and parameters is provided in paragraph 4. Section 5 includes the results of the model as well as analytical, explanatory comments. In Sect. 6 the sensitivity of the model proposed is investigated in respect of the assumed MSW price profile over time. Finally, in Sect. 7 the conclusions of the study are summarized.

2 Relevant Studies in Recent Literature

2.1 Waste Management in Developing Countries and Transitional Economies

Many researchers report on the status of waste management in developing countries. Despite the fact that every country has its own particular conditions, it is a common finding that in the majority of developing countries, waste management systems suffer from lack of appropriate infrastructure, which results in low rate of MSW collection and environmental hazards. According to Onu [28], solid waste management in developing countries is characterized by highly inefficient waste collection practices, variable and inadequate levels of service due to limited resources, lack of environmental control systems, indiscriminate dumping, littering and scavenging and poor environmental and waste awareness of the general public. In the work of Parrot et al. [30], the MSW management in Yaoundé, Cameroon is analyzed. The main characteristics of the system are the lack of even basic infrastructure (bins), the high population growth—and consequently growth of the MSW quantities–, the very low recycling rate, reported as about 5%, and the use of dump sites as a disposal facility, without any type of treatment for the MSW. It is also interesting that the authors acknowledge a low rate of about 40% of MSW collection, which is even lower in a large number of neighboring countries. In a similar vein, Troschinetz and Mihelcic [39] report a recovery rate from 5 to 40% for a study in 23 developing countries. The relationship between MSW generation and income varies with respect to the developmental stage of a nation. As a country develops, its waste generation rate increases. In contrast, a weak correlation exists between income and waste generation for middle- and upper-income countries, and waste generation actually decreases in the wealthiest countries [24].

The Clean Development Mechanism has already been used for funding projects for improving Municipal Solid Waste management in developing countries. According to the work of [41], it is interesting to note there were already 119 energy recovery from MSW projects examined in the frames of the CDM mechanism, out of which 88 projects involved generation of electricity that is supplied to the grid, which is also the case examined in this work. Furthermore, the authors acknowledge the very low standard of landfills in India and the need to improve it. Similarly, Barton et al. [3] examine the options for funding MSW management projects in developing countries, through the CDM mechanism. In their work, they evaluate the greenhouse gas emissions reduction achieved by applying several MSW management methods, such as landfilling (passive venting, gas capture with flaring) and composting of the digestate, together with two waste-to-energy options: landfill gas capture with electricity generation and composting and anaerobic digestion with electricity production. The authors conclude that there is a significant opportunity for developing CDM projects to attract investments in developing countries for improving waste management infrastructure. Energy exploitation of waste has been also examined in the past, i.e. in [7, 9, 23], but mainly in areas with lack of space for landfills, such as in the work of Kathirvale et al. [19] for Malaysia.

2.2 The Competing Waste-to-Energy Technologies

Higher efficiencies and lower emission levels are the main targets of the technological innovations in power generation. These benefits characterize emerging technologies, which compete with older but proven ones. In the present study four different technologies will be investigated: (1) MSW Incineration, (2) MSW gasification, (3) landfill biogas (LFG) exploitation through pipelines and (4) anaerobic digestion. Moreover, two energy product scenarios will be compared:

(a) Only electricity is produced. (b) Combined Head and Power production. It is emphasized that a district heating infrastructure is not available in the case study city (Athens), but CHP will be investigated in order to reveal its potential benefits over electricity production. For this reason it is assumed that a suitable district heating (or district cooling) infrastructure has been already installed. It is also assumed that a pre-sorting facility has been installed in order to separate the recyclable from the non-recyclable MSW.

Incineration is perhaps the oldest method for recovering the energy stored in MSW. The newly built projects for electricity production seem to be more efficient, compared to older installations: WtE plant MKW Bremen with efficiency of 30.5%, EVI Laar 30.5%, AEC Amsterdam 34.5%, AZM Moerdijk 32.5%. In the case of CHP production the net electrical efficiency is close to 23% whilst its thermal efficiency is approximately 45%, which is technically possible by using the back pressure turbine technology. The prevailing technology of MSW incineration is the moving grate, which is designed to handle large volumes of MSW with no pre-treatment. This type engages large-scale combustion in a single-stage chamber unit where complete combustion or oxidation occurs [42]. In the so-called Mass Burn Incinerators (MBI), the thermal energy generates electricity through steam turbines. When Combined Heat and Power is the case, the residual heat is recovered for district heating, hot water supply etc. [29].

Gasification may theoretically produce electricity at an efficiency of about 27% and heat at about 24% [26]. This would suggest that gasification of MSW is competing with incineration. However, in practice, gasification has not been proven and only recently has been realized in some WtE applications. In large-scale systems, combined cycle gas turbines may increase electrical efficiency but they may also reduce the temperature of the residual heat in the steam. Therefore, thermal energy production is significantly lower than that produced by incineration. Moreover, some installations in Europe have faced technical problems, whilst the average electrical efficiency noticed in Japanese installations is not more than 10% [12]. In the report of the Thermoselect project in Karlsruhe [15] it is stated that no more than 0.56 MWel/tMSW may be achieved even in optimized future realizations (by assuming highly efficient gas engines). This performance indicates an electrical efficiency of about 20%, which has to be proved in practice.

Biogas may be generated by digesting the organic fraction of MSW. The produced biogas may be utilized for either electricity or CHP production. Biogas exploitation requires significantly less investment costs than the thermal conversion technologies (incineration and gasification). Anaerobic digestion with biogas recovery is one treatment option for urban organic waste. Several systems for source separation, collection and pre-treatment of the municipal organic waste prior to treatment in biogas plants are available [14]. In the present case-study, the methane-enriched stream is utilized for either electricity conversion or CHP production by natural gas engines. The case of anaerobic digestion is also investigated assuming multiple decentralized installations being able to handle the entire annual MSW quantities.

2.3 Time-Optimal Energy Investments

Real options theory aims to replace traditional models of irreversible investments, since it may handle the uncertain, volatile pattern of multiple stochastic variables. Thus the potential investor may be able to select the most interesting investment using advanced time-dependent criteria and moreover to optimize the investment entry time based on the forecasts of stochastic variables like demand and prices. Among the various contributions on real-options theory, one may distinguish the studies of Brennan and Schwartz [4], Dixit and Pindyck [6] and Trigeorgis [37].

The effects of combined uncertainties in climate policy interventions have been investigated in Fuss et al. [8] and Laurika and Koljonen [21] and optimal investment timing decisions were sought. In the above mentioned works, the variables under uncertainty were: fuel and electricity prices, CO_2 allowance prices as well as demand of electricity. The time evolution of these variables was represented by Geometric Brownian Motion (GBM) models. In the present work the heating-energy market is also considered as stochastically evolving. This means that apart from the above mentioned variables, the savings due to the potential displacement of conventional boilers are represented by GBM models too, as long as they rely on the stochastic projection of oil prices. Additionally, interest and inflation rates are assumed as stochastically evolving according to mean-reverting processes. The stochastic differential equations (SDE) of these models resemble to the GBM models as they are characterized by normally distributed samples of Brownian differentials [27, 35]. However, their behavior is mean-reverting according to the Ingersoll-Ross models through which positive projections are ensured [17]. The solution of the above mentioned stochastic evolution models is based on Euler simulators [20] but subsequently a Monte-Carlo algorithm [11] is used to produce multiple solution sets and average them to a final projection output.

3 Methodological Approach

3.1 The Case Study

The present study investigates the economy of WtE alternatives as a function of time. A long-term estimation of MSW adequacy should therefore be conducted prior to any other techno-economical consideration in order to ensure MSW availability for the entire operational life-time of a potential WtE project. The basic MSW quantitative data for the case study region are comprised of the MSW disposal rate in Athens, which is currently estimated to be close to 6,500 t/day, with a current -annually increasing- rate of approximately 3% as recorded by ACMAR [1] and a relatively low percentage (13%) of MSW, which is recycled on source. The recycled percentage of disposed MSW is currently increasing by 1.5% each year [10].

In order to ensure long-term availability of the fuel source, a small portion of the totally available quantity of MSW will be used for energy exploitation, to account for potentially successful recycling campaigns in the future. It is therefore assumed that for the entire examined time horizon (50 years) an amount of 1.300.000 t/a will be available for WtE projects, as if the current increasing rate of the recycled MSW portion would hold for 50 years. Therefore, in the present case-study, this MSW supply rate determines the annual energy production of the hypothetical WtE plant. As stated before, four different WtE technologies will be investigated: incineration, gasification, biogas exploitation from landfills using pipelines and anaerobic digestion units. Two scenarios of energy production will be examined, i.e. electricity production and alternatively CHP production. A pre-sorting facility is assumed to separate recyclable materials from the non-recyclable portion of MSW, which is utilized for energy conversion. The baseline scenario considers landfilling of the entire MSW quantity. In that case, significant CH_4 quantities would be released in the atmosphere, which correspond to significant CO_2-equivalent emissions.

Uncertainty has been introduced for the following stochastic variables: Electricity prices, oil prices, CO_2 allowance prices, interest rates and inflation rates. MSW price and running costs were considered to follow the evolution of inflation rate, since only current estimations were available instead of historical time-series. The determination of the statistical parameters (drift, volatility and correlation) needed for the GBM representation of the stochastic variables' evolution [5] was based on recent historical data. An Euler solver and a Monte-Carlo simulation sub-routine were used to produce multiple SDE solutions and average them, thus providing the requested time paths. The investment costs were calculated as a function of time too, through appropriate learning curves, thus considering the experience acquired by previous installations of the same technologies [18, 34]. The above forecasts were introduced as inputs to a real-options algorithm which in turn determined the Net Present Values (NPV) of the project. This process was performed using an iterative procedure. The NPV numerical calculation was repeatedly shifted by one-year steps, meaning that the decision for investment may be postponed for as many years as needed for the investment to be more profitable. Arrays of project NPVs are therefore created as a function of time. The optimality was determined numerically by selecting the maximum NPV from the oncoming decision period.

3.2 Problem Formulation

The experience accumulated during the last decades on the construction of power production units is reflected in the investment costs, which may be mathematically formulated through global learning curves according to Eq. 1:

$$I_{i,t} = I_{i,0}\left(\frac{GQ_{i,t}}{GQ_{i,0}}\right)^{\log_2(LR_i)} \quad \forall i \tag{1}$$

where $LR_i = 1 - b_i \; \forall i$

b_i is an appropriate learning rate used for each technology i, $I_{i,t}$ is the capital cost needed for realizing an investment (i) at time-point (t). $GQ_{i,t}$ denotes the globally installed capacity of technology (i) at the time point (t).

The problem includes the following two agents:

- a developing or transitional electricity market in which fuel, CO_2 allowance and electricity prices follow a GBM generated path. The SDE that describes this process is represented by Eq. 2:

$$d\mathbf{Y}_t = \mu(t) \cdot \mathbf{Y}_t dt + D(t, \mathbf{Y}_t) \cdot \mathbf{V}(t) d\mathbf{W}_t \tag{2}$$

In the above equations, $\mathbf{Y}_t$ denotes the vector of the stochastic processes (variables), $\mu(t)$ denotes the drift vector as a function of time (t), $\mathbf{V}(t)$ denotes the volatility vector function of time (t), $D(t, \mathbf{Y}_t)$ denote the diffusion vector function of time (t) and $d\mathbf{W}_t$ denotes the Brownian Motion vector differential. The variables are given in vector form thus corresponding to any stochastic variable they may represent.

- potential WtE investors, planning to engage in WtE projects.

The financial balance of the plant is calculated on a day-by-day basis. By integrating for each year (z) of the operational life-time, the annual financial balances are obtained. The time differential (dt) is assumed to be equal to one-day interval. The carbon allowances, generated by replacing conventional energy sources with MSW, contribute to the annual revenues. The above mentioned economic terms are described using the following Eq. 3, which represents the annual financial balance $E(z)$:

$$E(z) = P_{el} \cdot C \int_0^{365} F(t)dt + P_{th} \cdot H \cdot \int_0^{365} F_{th}(t)dt + \int_0^{365} F_{CO_2}(t)dt \qquad \forall z \in \left[v + C_{t,i}, v + C_{t,i} + O_{t,i}\right] \tag{3}$$

where (P_{el}) and (P_{th}) denote the electricity and heat capacity of the planned energy conversion system, while (C) and (H) denote the power and thermal capacity coefficients, which are the percentage of operational time within a year respectively. The cost-terms inside the two first integrals of Eq. 3 are expressed in Euros per energy unit thus justifying the external multiplication with the plant capacity (either power or thermal). The operational life and the construction lead time for each technology (i) are denoted by ($O_{t,i}$) and ($C_{t,i}$) respectively while (v) denotes the investment decision time. $F(t)$ denotes the unitary algebraic balance of the daily cash-flows due to electricity production. In the case of CHP production, it is assumed that the conventional domestic burners may be displaced while the produced heat may be distributed using a pre-installed district heating network, thus allowing significant fossil fuel savings. Therefore, a second integral is included in Eq. 3 corresponding to the revenues from the heat sales ($F_{th}(t)$).

Obviously the second integral is accounted only in the CHP case whilst it is omitted when solely electricity production is considered. The unitary algebraic balance of the daily cash-flows is calculated by subtracting the unitary operational expenses of the power plant (MSW costs f_{MSW} and other running costs f_r) from the electricity selling incomes (f_{el}):

$$F(t) = (f_{el} - f_{MSW} - f_r)(t) \quad \forall t \in [0, 365], \forall z \tag{4}$$

The F_{co_2} term in Eq. 3 represents the daily revenues from the greenhouse gas emission trading:

$$F_{CO_2}(t) = f_{CO_2}(t) \cdot Ef \cdot \dot{Q}_{MSW}(t) \qquad \forall t \in [0, 365], \forall z \tag{5}$$

Where $F_{CO_2}(t)$ denotes the daily CO_2 allowance prices, simulated by Eq. 2 (shown in Fig. 2). $\dot{Q}_{MSW}(t)$ denotes the daily MSW supply rate, which in the present case-study correspond to 1.300.000 t/a or equivalently 3,560 t/day. The differential time (dt) equals to one-day interval. The utilized emissions factor, denoted by Ef is explained in detail in paragraph 4.1 (Eq. 8). By accounting Eqs. 3, 4 and 5 becomes:

$$E(z) = P_{el} \cdot C \cdot \int_{0_t}^{365} (f_{el} - f_{MSW} - f_r)(t)dt + P_t h \cdot H \cdot \int_{0}^{365} F_{th}(t)dt + \int_{0}^{365} f_{CO_2}(t) \cdot |Ef| \cdot \dot{Q}_{MSW}(t)dt \, \forall z \in [v + C_{t,i}, \, v + C_{t,i} + O_{t,i}] \tag{6}$$

The cost terms inside the integrals represent the evolution of stochastic variables (prices of MSW and electricity as well as the heat production revenues) which are endogenously modeled by the stochastic differential Eq. 2. Especially for the heat production revenues $F_{th}(t)$, it was assumed that an attractive pricing strategy has been adopted (equal to 75% of the simulated heating oil prices per energy unit). The urgency for smooth penetration of MSW-based district heating in the domestic heating energy market and the need for the social acceptance of this method might justify the above mentioned pricing policy assumption.

The annual integrals of Eq. 6 are given in nominal prices, but they are further converted to present values (PV), using the stochastically evolving interest rates modeled by a mean reverting derivative of the SDE described in Eq. 2. The cash-flow PVs are summed up, thus resulting to an aggregate NPV, which accounts for the entire operational life-time of each technology (plant). The above procedure is described in the following Eq. 7:

$$NPV_{i,v} = \sum_{z=v+C_t}^{v+C_t+O_t} \left[\frac{E(z)}{(1 + r_z)^z} \right] - I_{i,v} \tag{7}$$

where ($I_{i,v}$) denotes the capital cost needed for realizing an investment (i) at time-point (v), calculated using Eq. 1, whilst (r_z) denotes the stochastic interest rates. It is noted that the stochastic rates are averaged on a yearly basis in order to produce annual NPV results. The entire process is iterated for every year (v) of a 15-year period within which an optimal investment entry time-point should be decided. Optimality is achieved for the year (v) and technology (i) with the maximum value of the project's $NPV_{i,v}$ [max($NPV_{i,v}$)].

4 Setup of the Numerical Algorithm

4.1 Input Data of the Model

The historical data of actual loads and electricity system marginal prices (SMP) were acquired by the Hellenic Transmission System Operator [16]. The historical data were available on an hourly basis for the time-period 2001–2009, but a mean daily average was finally used. The historical data of inflation and central bank interest rates were acquired by the Hellenic Statistical Service [10]. CO_2 allowance prices were retrieved by Point Carbon [31] whilst heating oil prices were acquired by the Greek Ministry of Development [13].

The net calorific value of the non-recyclable portion of the MSW used for energy conversion is assumed to be 10 GJ/tMSW or 2.8 MW_{th}/tMSW [32], which is assumed to remain constant over time. The complete set of techno-economical inputs is presented in the following Table 1. The data correspond to 1.300.000 t MSW on a yearly basis. This quantity determines the specification of power production for each technology, based on recorded electrical and thermal efficiencies per MSW unit, which have been retrieved by Gohlke [12], Hesseling [15], Murphy and McKeogh [26]. The investment and operational costs (either running or fixed costs) were retrieved by the study of Tsilemou and Panagiotakopoulos [38].

The emission factors correspond to the CO_2 emission savings obtained by exploiting the entire MSW quantity for a WtE project instead of landfilling them (baseline scenario). The CO_2 savings were calculated by considering the replacement of the current conventional mix of electricity generation plants and a corresponding emission savings factor. Additional CO_2 emission savings are considered through the displacement of conventional (fossil-fuelled) heat generation plants. The endogenous CO_2 emissions from the energy conversion process are the only positive pollutant contributors. This rationale is analytically formulated in Eq. 8, which provides the emission saving factors of Table 1:

$$Ef = -e \cdot Ef_e - h \cdot Ef_h + Ef_p - Ef_{lf} \tag{8}$$

where, e and h denote the electricity and thermal production per fuel unit respectively, whilst Ef_e and Ef_h denote the emissions savings due to fossil power and thermal plants' displacement respectively. The CO_2 emissions of the process and the landfill emissions are denoted by Ef_p and Ef_{lf} respectively.

Table 1 Model inputs for electricity and CHP production

WtE process of MSW	Power generation capacity (MW_{el})	Investment costs (for 2009) (€/tMSW/a)	Efficiency El. only or El.//Th. (%)	CO_2 emissions ($tnCO_2$/tMSW)	Fixed costs (€/kW/a)	Running costs (€/tMSW)	Learning rate
Incineration (electricity only)	135	500	30	−2.02	4.5	42	0.01
Incineration (CHP)	102		23//45	−2.17	5		
Gasification (electricity only)	90	730	20	−1.78	3.2	60	0.02
Gasification (CHP)	56		12//26	−1.79	4		
Landfill biogas (electricity only)	26	180	6	−1.39	1.4	15	0.05
Landfill biogas (CHP)	25		5//9	−1.46	2		
Anaerobic digestion (electricity only)	95	125	22	−1.69	1.2	40	0.05
Anaerobic digestion (CHP)	60		13//19	−1.74	2.1		

The above computed emissions factor (*Ef*) is utilized in the calculation of the annual integrals of the greenhouse gas trading revenues (Eqs. 5 and 6). The notation, the units and the numerical values of each variable shown in Eq. 8 are presented in the following Table 2.

The source for the Ef_e, Ef_h values was the study of Rentizelas et al. [33]. *e*, *h* and Ef_p values were acquired by processing the numerical data reported in Gohlke [12], Hesseling [15], Moller et al. [25], Murphy and McKeogh [26] and Papageorgiou et al. [29]. Finally, the Ef_{lf} data have been retrieved by Tuhkanen et al. [40].

4.2 Stochastic Analysis

The simulation of the stochastic variables resulted to the MSW and oil prices evolution shown in Fig. 1 as well as to the CO_2 allowance and electricity price forecasts shown in Fig. 2.

The stochastic differential equations representing the evolution of the relevant stochastic variables are solved with an Euler solver. A Monte-Carlo algorithm is used in order to produce multiple results based on past data and normally distributed samples of Brownian differentials (noise). These are further averaged thus contributing to the reduction of noisy variations. From the SDE solution it is shown that increasing gate fees may be anticipated whilst on the other hand the evolutions of oil and electricity prices are mean-reverting, despite their GBM modeling. This behavior is in line with past relevant studies [2]. The results of the CO_2 allowance price representation are based on recent data and therefore, not enough experience has been gathered concerning its behavior within this newly born market. Also, it has to be noted that the future projections shown in Figs. 1 and 2 may not be considered as safe forecasts. They are rather based on historical data and represented through GBM stochastic processes, thus constituting modeled evolution paths.

Concerning the evolution path of MSW gate-fees, a starting point is required. This is based on its current (2009) value, derived from a holistic reverse-logistics algorithm [36] which in turn utilizes activity based costing methods. The entire supply chain is taken into account, including collection, transportation, warehousing, handling and treatment activities. The resulting range was approximated between 21 and 24 Euros/tMSW, which is close to gate-fee calculations retrieved from the literature [26, 29]. The evolution of MSW price (gate fee) over time has been assumed to follow the inflation rate, which in turn has been represented by an appropriate mean-reverting derivative of the stochastic differential Eq. 2. The same assumption holds for the running costs of each technology for which, only current values were available and retrieved by the studies of Murphy and Mc Keogh [26] and Tsilemou and Panagiotakopoulos [38]. Due to the uncertainty of the logistical—activity based—calculation of gate-fee, a sensitivity analysis is conducted in order to investigate the sensitivity of the model in this crucial parameter. The lower and upper price bounds, shown in the MSW graph (Fig. 1-up) indicate the limits of the above mentioned analysis.

Table 2 Emission factors

WtE process of MSW	Electricity production per fuel unit	Thermal production per fuel unit	Emission savings due to fossil power-plant displacement	Emission savings due to fossil thermal plant displacement	CO_2 emissions of the process	Landfill emissions
(Symbol) Unit	(e) $MWh_{el}/tMSW$	(h) $MWh_{th}/tMSW$	(Ef_e) tCO_2/MWh_{el}	(Ef_h) tCO_2/MWh_{th}	(Ef_p) $tCO_2/tMSW$	(Ef_{lf}) $tCO_2/tMSW$
Incineration (Electricity only)	0.8	0	0.876	0	0.28	1.6
Incineration (CHP)	0.6	1.2		0.27	0.28	
Gasification (Electricity only)	0.53	0		0	0.28	
Gasification (CHP)	0.33	0.7		0.27	0.28	
Landfill biogas (Electricity only)	0.2	0		0	0.35	
Landfill biogas (CHP)	0.2	0.3		0.27	0.35	
Anaerobic digestion (Electricity only)	0.5	0		0	0.31	
Anaerobic digestion (CHP)	0.35	0.5		0.27	0.31	

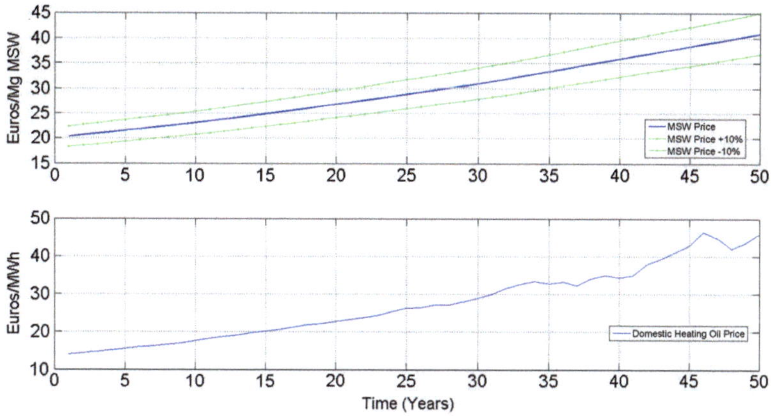

Fig. 1 Forecasting of MSW (*up*), and heating oil (*down*) prices

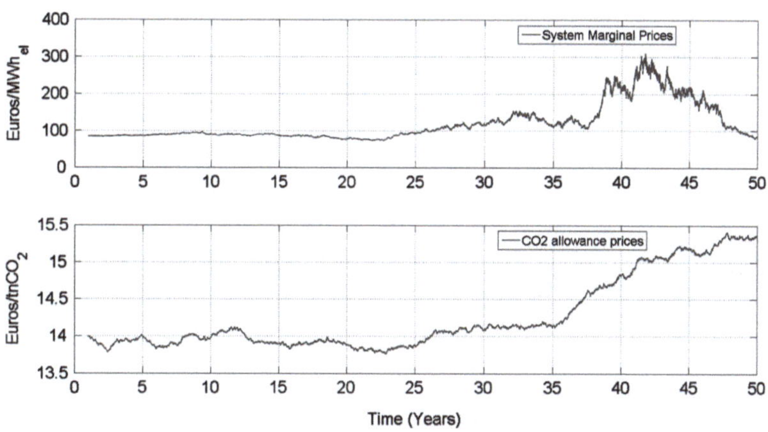

Fig. 2 Forecasting of electricity (*up*) and CO_2 allowance (*down*) prices

5 Model Results

The NPV comparison of the investigated technologies for the scenario of electricity production is presented in Fig. 3 while the corresponding NPV comparison for the CHP scenario is shown in Fig. 4.

Investing on proven CHP-incineration constitutes the optimal strategy in terms of economic efficiency. Higher power generation efficiency and lower emission rates render it the most promising method. Gasification, on the other hand, is not yet a mature technology despite the long lasting research, and does not seem to be able to compete with the other options. In the case of electricity production, the incineration technology also proves to be the most interesting WtE option due to

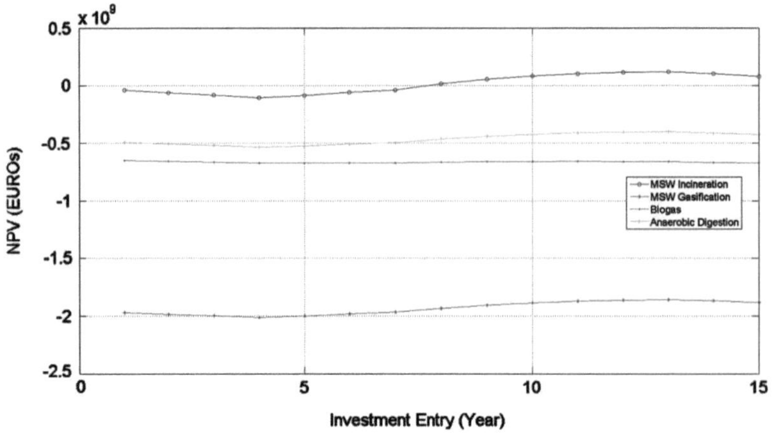

Fig. 3 NPVs for electricity production

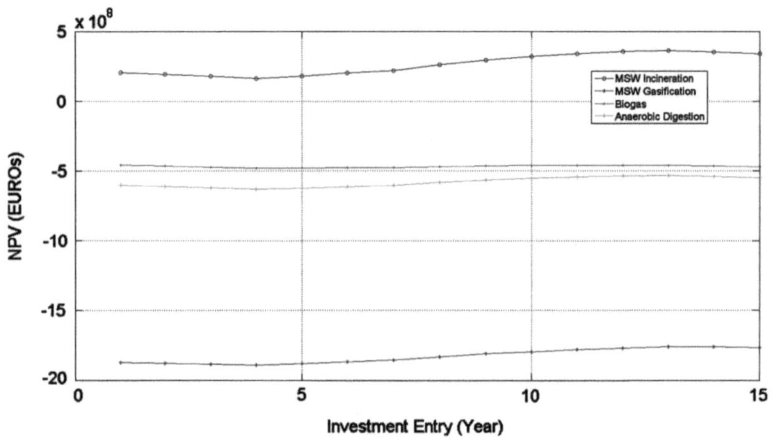

Fig. 4 NPV for CHP production

its higher electrical efficiency and lower investment costs and emission rates. On the other hand the gasification technology comprises of negative NPVs over time and therefore will probably fail to gain a considerable market share in the next years. Landfill biogas exploitation (pipeline gas supply) does not seem to be able to follow the energy market trends, despite its low running and capital costs.

The resulting NPVs are also negative in both scenarios of energy production—either power or CHP production independently of the investing entry time. Low efficiencies of power and heat production per input unit of MSW fuel are responsible for this poor performance. It is emphasized that biogas exploitation is an environmentally friendly activity that ensures efficient controlling of methane gas generated by landfill reactions. It is believed that oncoming improvements in power (and/or heat) production per input MSW-unit, will lead to much more efficient biogas projects in the near future. The anticipated yields of anaerobic digestion and landfill biogas exploitation through pipelines, are quite similar. The low efficiencies of power and heat production render anaerobic digestion a non profitable WtE option. Negative NPVs (close to the NPVs of landfill gas exploitation) characterize this technology too. Moreover, the heat production efficiency is very low thus resulting to slightly lower NPVs compared to those of solely power production. This result characterizes anaerobic digestion whilst the economic performance of the remaining technologies may be significantly improved by considering CHP production. District heating networks based on MSW fuel, may contribute to additional revenues for WtE-CHP plants. As stated before, an attractive pricing is a pre-requisite for the acceptance of MSW-fired district heating and for the subsequent displacement of the conventional domestic burners.

From the above results an optimal time of investment entry may be identified, based on the stochastic evolution of incomes and expenses. One would expect that the anticipated increasing of electricity prices in the distant future (Fig. 2) might necessitate the postponement of the investment decision for more than a decade. From the results obtained, it is concluded that this may be the case for all the examined WtE technologies except from the exploitation of landfill biogas. Its lower power-generation efficiency leads to lower sensitivity in electricity price variations thus resulting to almost constant NPV time-paths. Of particular interest for a potential investor may be the option of immediate investments, which should not be easily rejected. Although the optimal NPVs correspond to investments that may be decided in almost 13 years from today—as indicated by the model and the analysis of the results—the NPVs of immediate investment entries are expected to be slightly lower than the optimal ones. The business strategy of potential investors, the environmental policies, as well as State/EU interventions are among the factors that may necessitate the realization of WtE project plans and may finally determine the time-point of investment decision. It is emphasized that the time-dependent NPVs shown in the Figs. 3 and 4 are solely based on the stochastic representation of variables under uncertainty (fuel, CO_2 and electricity prices, interest rates and inflation rates) which in turn depend on their historical data and on their respective statistical parameters.

In the chart of Fig. 5 the financial break-down of a WtE project is presented. These results have been obtained for: (1) the optimal investment entry time, (2) the optimal technology selection (MSW incineration and CHP production) and (3) by assuming a 33-year period of operational life-time. The most important income and expense contributors may be identified. It is noted that the revenues from electricity selling to the grid exceed the respective heat-selling revenues (fossil fuel savings) despite that the electrical efficiency has been assumed to be lower than that of heat production. This may be attributed to the higher electricity (MWh_{el}) prices, compared to the anticipated unitary oil prices (€/MWh_{th}) shown in Figs. 1 and 2.

It is reminded that stochastic modeling might not be considered to represent the real future evolution of the corresponding stochastic variables. It rather reflects their

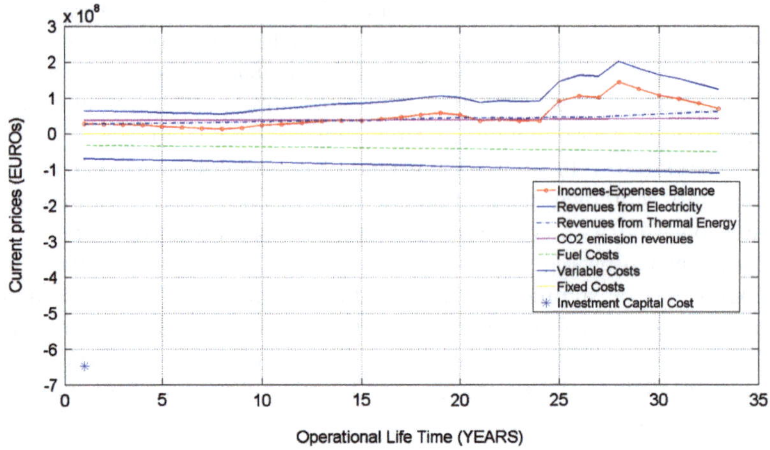

Fig. 5 Financial break-down for the project's operational life time for optimal investment entry

past behavior by sampling the induced uncertainties through appropriate probability distributions, determined by recent history. This inherent limitation of stochastic modeling should definitely be accounted during any decision making process.

6 Sensitivity Analysis

The model has been analyzed in respect of its sensitivity in the MSW price variations. More specifically, the optimal NPV, the optimal investment entry time as well as the payback period are calculated for various MSW projection paths in the range (−10, +10%) compared to the original MSW price path. The results are shown in the following Fig. 6.

Concerning the anticipated yields (NPV) the model is sensitive enough but—on the contrary—the optimal investment entry and the payback periods are slightly influenced by MSW price variations. The investment entry is clearly insensitive as the imposed quantitative variations of MSW price modify homogeneously its projection profile over time. The payback periods are slightly reduced as the MSW price reduces, thus indicating a weak sensitivity which in turn may be explained by the small proportion of MSW logistical costs as compared to the other expense streams.

7 Conclusions

An investigation of four different WtE options has been conducted in respect of their long-term economical efficiency. MSW incineration, gasification, landfill biogas exploitation and anaerobic digestion have been compared, either by solely

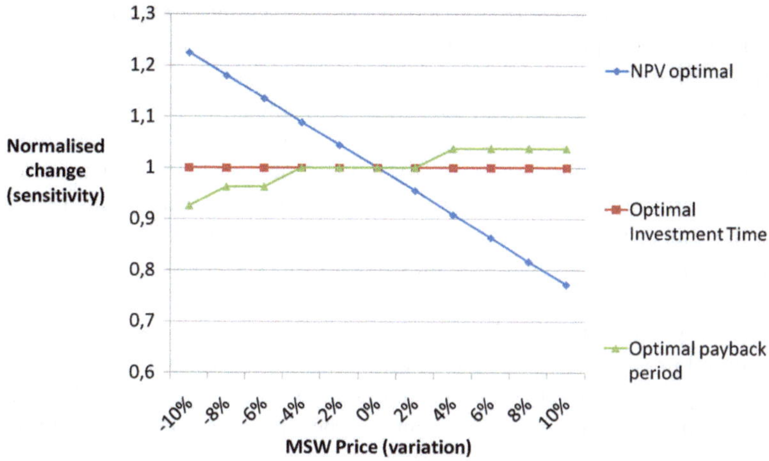

Fig. 6 Sensitivity of the model in the MSW price

considering electricity production or by assuming combined heat and power production. The comparison was based on a modern investment analysis tool, namely the real options theory, thus forcing the determination of optimal investment strategies over time. Prior to the investment analysis, the stochastic modeling of the introduced uncertainties allowed the simulation of the participating volatile variables: heat production revenues, electricity and CO_2 allowance prices as well as interest and inflation rates, which were used for representing the evolution of running costs and gate fees. The current gate fee has been externally derived in the range of 21–24 Euros/tMSW. The conclusions from the analysis may be summarized as follows:

The traditional but proven MSW incineration remains the most interesting method of energy recovery from waste in terms of financial yield, for either electricity or CHP production. The results of the analysis indicated that gasification may not constitute a profitable WtE choice under the assumptions made. Moreover, it is not yet a reliable method of MSW energy recovery; several gasification plant failures have been recently experienced in Europe, despite the intensive research focusing on that technology during the last decades. The energetic exploitation of landfill biogas-either with pipeline systems or by using anaerobic digesters- fails to prove its efficiency as long as they present negative financial yields over time. Nonetheless, the environmental benefits of biogas exploitation render it a crucial requirement for any existing landfill. It should be reminded though, that according to the European environmental policy, landfilling is not considered a sustainable waste treatment option. Therefore, in the proposed model, the landfilling option has been assumed to be the baseline scenario, thus taking into account its significant environmental issues (methane emissions, CO_2 equivalent emissions, leachates etc.).

CHP is economically a superior option but an existing infrastructure of district heating network is a prerequisite. The higher surplus of anticipated yields might probably be invested for promoting such infrastructure.

Under the current conditions and prices, immediate investments might be reconsidered in favor of future—potentially more profitable—opportunities. If immediate investments are required, the above mentioned classification of WtE technologies still holds; actually the ranking of the WtE technologies remains the same in the short and medium terms. The incineration technology may be the most attractive technology, but is rather sensitive in the variations of fiscal conditions over time. The gasification is significantly less competitive than incineration but simultaneously it is equivalently sensitive over time.

The model presented is sensitive in the variations of MSW price, provided that the sensitivity criterion relies on the anticipated NPVs over time. The payback period and the optimal investment entry times present either a slight or zero sensitivity respectively, thus indicating a weak dependence on MSW price variations.

The gas trading revenues constitute an important profit factor. The CO_2 allowances generated by assuming landfilling as the baseline scenario, contribute significantly to the financial yields of WtE-CHP projects. The analysis of the incomes through the entire operational life of such projects renders electricity selling revenues as the most important income source followed by CO_2 trading revenues, and district heating incomes respectively.

Further research is required for investigating additional emerging technologies possibly interesting for WtE projects, like: thermal depolymerisation, plasma arc gasification etc. The real options algorithm described in the present work may contribute to the investment analysis of such planned projects over time, thus leading to interesting policies and strategic WtE interventions.

References

1. ACMAR (2009) Association of Communities and Municipalities of the Attica Region, 11.09.09 www.esdkna.gr
2. Barlow MT (2002) A diffusion model for electricity prices. Math Finan 12(4):287–298
3. Barton JR, Issaias I, Stentiford EI (2008) Carbon—making the right choice for waste management in developing countries. Waste Manag 28:690–698
4. Brennan MJ, Schwartz E (1985) Evaluating natural resource investments. J Bus 58:135–157
5. Clewlow L, Strickland C (2000) Energy derivatives: pricing and risk management. Lacima Publications, London
6. Dixit A, Pindyck R (1994) Investment under uncertainty. Princeton University Press, Princeton
7. Eleftheriou P (2007) Energy from waste: a possible alternative energy source for large size municipalities. Waste Manag Res 25:483–486
8. Fuss S, Szolgayova J, Obersteiner M, Gusti M (2008) Investment under market and climate policy uncertainty. Appl Energy 85:708–721

9. Garg A, Smith R, Hill D, Longhurst PJ, Pollard SJT, Simms NJ (2009) An integrated appraisal of energy recovery options in the United Kingdom using solid recovered fuel derived from municipal solid waste. Waste Manag 29:2289–2297
10. General Secretariat of National Statistical Service of Greece 10.09.2009, www.statistics.gr
11. Glasserman G (2004) Monte-Carlo methods in financial engineering. Springer-Verlag, NY
12. Gohlke O (2009) Efficiency of energy recovery from municipal solid waste and the resultant effect on the greenhouse gas balance. Waste Manag Res 27:894–906
13. Greek Ministry of Development (2009) 13.09.09 www.ypan.gr
14. Hansen TL, Jansen J, Davidson A, Christensen TH (2007) Effects of pre-treatment technologies on quantity and quality of source-sorted municipal organic waste for biogas recovery. Waste Manag 27:398–405
15. Hesseling WFM (2002) Thermoselect Facility report, TNO R2002/126
16. HTSO SA (2009) Hellenic transmission system operator. 08.08.09, www.desmie.gr
17. Ingersoll JE, Ross SA (1992) Waiting to invest: investment and uncertainty. J Bus 65(1):1–29
18. Junginger M, Faaij A, Turkenburg WC (2005) Global experience curves for wind farms. Energy Policy 33(2):133–150
19. Kathirvale S, Yunus MM, Sopian K, Samduddin AH (2003) Energy potential from municipal solid waste in Malaysia. Renew Energy 29(4):559–567
20. Kloeden PE, Platen E (1999) Numerical solution of stochastic differential equations. Springer, Berlin
21. Laurikka H, Koljonen T (2006) Emissions trading and investment decisions in the power sector-a case study in Finland. Energy Policy 34:1063–1074
22. Luoranen M, Horttanainen M (2007) Feasibility of energy recovery from municipal solid waste in an integrated municipal energy supply and waste management system. Waste Manag Res 25:426–439
23. Luoranen M, Horttanainen M (2008) Co-generation based energy recovery from municipal solid waste integrated with the existing energy supply system. Waste Manag 28:30–38
24. Medina M (1997) The effect of income on municipal solid waste generation rates for countries of varying levels of economic development: a model. J Solid Waste Technol Manag 24(3):149–155
25. Møller J, Boldrin A, Christensen TH (2009) Anaerobic digestion and digestate use: accounting of greenhouse gases and global warming contribution. Waste Manag Res 29:813–824
26. Murphy JD, McKeogh E (2004) Analysis of energy production from municipal solid waste. Renew Energy 29:1043–1057
27. Øksendal B (2000) Stochastic deferential equations. Springer-Verlag, Berlin
28. Onu C (2000) Sustainable waste management in developing countries. In: Proceedings of the biennial congress of the institute of waste management of Southern Africa, WasteCon'00, Cape Town, South Africa, pp 367–378
29. Papageorgiou A, Barton JR, Karagiannidis A (2009) Assessment of the greenhouse effect impact of technologies used for energy recovery from municipal waste: a case for England. J Environ Manag 90:2999–3012
30. Parrot L, Sotamenou J, Dia BK (2009) Municipal solid waste management in Africa: strategies and livelihoods in Yaoundé, Cameroon. Waste Manag 29:986–995
31. Point Carbon (2009) Carbon market indicator. 12.08.2009, http://www.pointcarbon.com
32. Reimann D (2009) CEWEP energy report II (Status 2004–2007). Confederation of European waste to energy plants
33. Rentizelas A, Tolis A, Tatsiopoulos I (2009) Biomass district energy trigeneration systems: emissions reduction and financial impact. Water Soil Air Pollut J Focus 9(1–2):139–150
34. Rubin ES (2007) Learning rates and future cost of power plants with CO_2 capture. IEA international workshop on technology learning and deployment, Paris, France, 11 June 2007
35. Shreve R (2004) Stochastic calculus for finance II: continuous-time models. Springer-Verlag, Berlin

36. Tatsiopoulos I, Tolis A (2003) Economic aspects of the cotton-stalk biomass logistics and comparison of supply chain methods. Biomass Bioenergy 24:199–214
37. Trigeorgis L (1996) Real options. The MIT Press, Cambridge
38. Tsilemou K, Panagiotakopoulos D (2006) Approximate cost functions for solid waste treatment facilities. Waste Manag Res 24:310–322
39. Troschinetz A, Mihelcic J (2009) Sustainable recycling of municipal solid waste in developing countries. Waste Manag 29(2):915–923
40. Tuhkanen S, Pipatti R, Sipilä K, Mäkinen T (2000) The effect of new solid waste treatment systems of greenhouse gas emissions. 5th international conference on greenhouse gas control technologies (GHGT-5), Cairns, Australia
41. Unnikrishnan S, Singh A (2010) Energy recovery in solid waste management through CDM in India and other countries. Resour Conserv Recycl 54(10):630–640
42. Williams P (2005) Waste treatment and disposal, 2nd edn. Wiley, Chichester

Waste Management in Greece and Potential for Waste-to-Energy

Efstratios Kalogirou, Athanasios Bourtsalas, Manolis Klados and Nickolas J. Themelis

Abstract In Greece the daily production of Municipal Solid Waste (MSW) is estimated to be 15,000 tones, which means roughly 5.4 million tons per year, from which 77% is deposited in Landfills, 23% is recycled and composted. The European Union Legislation for Sanitary Landfills (1999/31/EC), imposes the decrease of biodegradable waste that are deposit to sanitary landfills; thus WtE methods of MSW is one of the best, in terms of affordability in a competitive world and environmental friendly, proposed solutions. Waste-to-Energy methods produce steam and/or electricity. Also, the weight of MSW is reduced up to 70–80% and the volume up to 90%, and finally the land area requirements are very small. Our proposal for the WtE technology implementation in Greece is the construction of MSW WtE plants in all major cities operating with an annual capacity of 200,000–400,000 tones. The required land area will be only 4–7 ha. The basic income of such plants is the gate fee, varying from 50 to 80 €/ton. The second income comes from selling of the produced electricity to the Public Power Corporation for 87.85 €/MWh (referring to the biodegradable fraction of MSW), according to the new Greek law for renewable energy sources (L. 3851/2010). Additional income comes from the recovered metals of the bottom ash. Furthermore, there is a considerable prospect for state subsidy of the whole investment, according to the Greek Development Law.

E. Kalogirou (✉) · A. Bourtsalas
WTERT Greece–SYNERGIA, A.I.T. 19 km. Peania Markopoulou Ave,
19002 Peania, Attica, Greece
e-mail: synergia@synergia.com.gr

M. Klados
INTRAKATSA, 19 km Peania Markopoulou Ave,
19002 Peania, Attica, Greece

N. J. Themelis
Earth Engineering Center, WTERT USA, Columbia University,
500 West 120th St., 10027 New York, NY, USA

A. Karagiannidis (ed.), *Waste to Energy*, Green Energy and Technology,
DOI: 10.1007/978-1-4471-2306-4_9,

1 Introduction

Economic development leads to higher consumption rate of goods and services and, thus, there is an increase with regard to the generation of solid wastes. As the rate of waste generation is higher than the rate of economic and population growth; several waste management methods are applied in order to minimize the environmental impact and make it an affordable solution in a competitive world. In the generally accepted waste management hierarchy, the first priority is waste reduction, followed by recycling and composting of clean biodegradable organic wastes as well (food and yard wastes). The EU promotes recycling over other waste treatment methods for recovering materials and energy, the latter either in the form of electricity/heat or production of waste derived fuels. In this way, physical resources are protected since paper, metals, glass, plastics that are recovered from the waste stream demand less resources and energy than the use of "virgin" materials. Also, the energy recovery provides electricity and heat to industrial, commercial and domestic consumers, and also minimizes the volume of wastes to be disposed. The goal of combining these approaches, as well as the additional option of composting, is to minimize the loss of resources to final inert landfill disposal.

Landfilling is the most common method for waste management in many EU Member States and in some cases this dependency exceeds 80%. The EU Landfill Directive of 1999 obliges Member States to progressively reduce the amount of organic waste going to landfill to 35% of the 1995 levels within 15 years aims to reduce such a loss of resources. This clear policy direction has put emphasis on waste management systems that increase and optimize the recovery of resources from waste—whether as materials or as energy. Accordingly, the member states are adopting Waste-to-Energy (WtE) and also mechanical–biological treatment (MBT) methods for the recovery of energy and materials from municipal solid wastes (MSW) and non—hazardous industrial wastes. WtE facilities can combust either as-received MSW (stoker or "mass burn" technology) or pre-processed "refuse-derived" fuels (RDF or SRF). The latter have higher calorific values and can be used both in dedicated WtE plants and as fuel substitutes in cement kilns and coal-fired power plants. In order to protect the environment from the emissions in energy recovery facilities, EU adopted the regulation on emission limits from waste incineration plants (Directive 2000/76/EC), while the regulation on renewable energy sources (RES) (Directive 2001/77/EC), includes the biogenic fraction of wastes [1–3].

Even though there are studies regarding the energy content of MSW and the possibilities for implementing WtE for power generation in Greece, none to our knowledge presents the energy potential for Greece in terms of power generation potential per region. This paper investigates this potential, focusing on electricity production, by examining the application of mass burn and RDF/SRF utilisation options for managing MSW in Greece, in the light of experience gained in E.U. and the U.S., and the National Plan for Waste Management of the Ministry of Environment.

Table 1 Chemical analysis of different waste streams

Waste stream	Content (%)							(MJ/kg)	
	Water	Carbon	Hydrogen	Oxygen	Nitrogen	Sulphur	Ash	HHV	LHV
Food waste	70	48	6.4	37.6	2.6	0.4	5	7.08	3.96
Paper and cartons	6	43.5	6	44	0.3	0.2	6	14.49	13.03
Plastic	2	60	7.2	22.8	0	0	10	26.11	24.48
Textile	10	55	6.6	31.2	4.6	0.2	2.5	20.17	18.47
Wood	20	49.5	6	31.2	4.6	0.2	2.5	15.96	14.16
Garden	60	47.8	6	38	3.4	0.3	4.5	7.94	5.16
Glass	2	0.5	0.1	0.4	0.1	0	98.9	0.23	0.16
Metals	3	4.5	0.6	4.3	0.1	0	90.5	1.55	1.35
Others	20.5	20.91	2.39	12.78	0.4	0.1	42.93	6.72	5.69

Waste can be a major source of renewable energy. The chemical analysis of different waste streams are presented in the Table 1.

2 Current Waste Management in Greece

Greece is a European country situated in the south-eastern part of it, with population, as of the census of 2001, 10.964.020 inhabitants. Its population is predicted to be 11,295,000 inhabitants in 2011. Administratively, Greece consists of thirteen peripheries, with total land area of 131,621 km^2. The climate of Greece is mostly Mediterranean type, which features mild, wet winters and hot, dry summers. Moreover, the MSW generation in Greece in 2001 was 4,529,585 tones, whereas it is forecasted that in 2011 the generation will be 5,981,290 tones and in 2025 it will be 7,625,648 tones. In 2010, the daily MSW production was around 15,000 tones, which correspond to 5.4 million tones of MSW on an annual basis. Focusing on the Attica region, where Athens, the capital of Greece, is placed, the MSW production reaches the quantity of 6,500 tones daily, which equals to 2.4 million tones of MSW per annum [4].

In general, the waste management in Greece depends especially on dumps and sanitary landfill sites. According to the European Union, uncontrolled disposal sites (dumps) are illegal and have to shut down. The deadline was extended for Greece, until the 1st January of 2011. Nevertheless not all of the dumps have been closed, but new ones have been created, as there is no sustainable waste management plan in Greece. The only exceptions are the five Mechanical and Biological Treatment plants located in Athens, Kalamata, Chania, Heraklio and Kefalonia. On the other hand the products of these plants, such as RDF and compost, have no responding market and as a result in the most cases they are disposed to landfill.

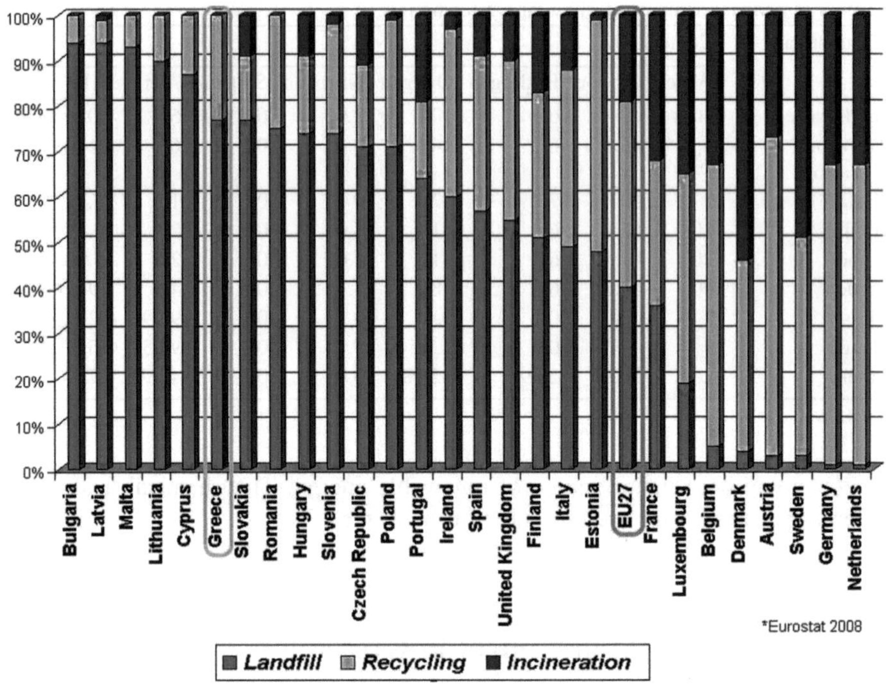

Fig. 1 Waste management statistics across Europe for 2008 (Eurostat)

All the above have driven Greece to the bottom of the European's sustainable waste management gradation. These facts are loudly verified by Eurostat, whose statistics for the year of 2008 for Greece give 77% landfill, 23% recycling and composting. In the Fig. 1 the statistics by Eurostat for the waste management across Europe on 2008 are presented.

The composition of MSW in Greece across the thirteen peripheries of the country is presented in the Fig. 2.

The waste management system is plagued by a number of problems, some of which include inadequate management, lack of technology and human resources, a shortage of transportation vehicles and insufficient funding. In 2010, there are 102 organizations working for the collection, transportation and disposal of residues for the 13 peripheries. Moreover, in Greece there are 25 Waste Transfer Stations, whereas the total WTS that are under construction are 107.

In 2010, the number of sanitary landfills in Greece was 77, for the convenience of 7,861,586 inhabitants, with annually dynamic of 3,031,570 tones. The total sanitary landfills that are under construction or under development are 146. Moreover, there are 3,036 uncontrolled disposal sites located throughout Greece, of which 316 are active, 429 are under reconstruction, and 2,291 are reconstructed. The only Periphery that has not any uncontrolled landfills is West Macedonia.

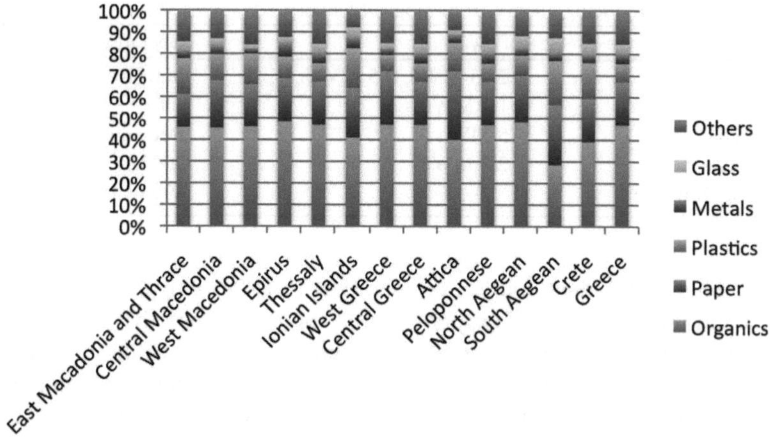

Fig. 2 Composition of Greek MSW

Table 2 Contribution of HE.R.R.Co S.A

Index	2006	2007	2008	2009
Population covered (million)	4.3	6.1	6.6	7.6
Contracted municipalities (#)	337	446	610	648
Sorting centers (#)	12	15	18	22
Recycling bins (#)	25,103	51,602	76,530	98,177
Collection vehicles (#)	95	140	236	327

As regards recycling, communal collection points have been set up in the main towns using 1,100 L containers for glass, paper, cardboard and plastic. These are then emptied by a Refuse Collection Vehicle, in a co-mingled form, and delivered to a materials recovery facility (MRF) for segregation. The system is poor at present as the infrastructure is not in place for regular collection from villages and rural areas. More government promotion is needed to encourage the wider population to recycle, e.g. posters, campaigns, education etc. The Hellenic Recovery Recycling Corporation (HE.R.R.Co S.A.) is responsible for the recycling and reuse of MSW in Greece. It was founded in December 2001 by industrial and commercial enterprises, which, either supply packaged products to the Greek market, or manufacture different packaging items. The Central Union of Municipalities and Communities in Greece (KEDKE) has a shareholding of 35% in the System's capital. In the Table 2 the contribution of HE.R.R.Co S.A. in recycling and reuse of materials is presented [5, 6].

In Greece there are five MBT facilities, which are presented below:

Table 3 Facilities producing energy from biogas recovery

Periphery	Power	Operation
Attica	23.5 MWe, 9.5 MWth	March-01
Thessaly	1.7 MWe	June-08
C. Macedonia	5 MWe	December-06
Crete–Chania	2.3 MWe	2005
Peloponnese–Kalamata		1998–2002. Restarts 2010

- Attica: This MBT plant started its operation in 2004 with a daily capacity of 1200 t and it operates 250 days per annum. The technology that is implemented is mechanical separation and aerobic biological treatment of the biodegradable fraction. The products are the secondary fuel RDF (470 tpd), the compost (120 tpd) and also 23 tpd of ferrous metals and 0.36 tpd of aluminium are recovered. As there is no market for the products of this plant, they are disposed to nearest landfill of Ano Liossia.
- Chania (Crete): This MBT plant started its operation in 2005 with an annual capacity of 70,000 t and it operates 260 days per annum. This plant recovers recyclables (9,000 tpa of paper, 5,200 tpa of plastic, 1,800 tpa of ferrous metals and 600 tpa of aluminium) and also producing 20,000 tpa compost. The 35% residuals are disposed to the nearest landfill.
- Kefalonia (Ionian Islands): This MBT plant started its operation in 2009 with an annual capacity of 25,000 t. The applied technology of this plant is bio-oxidation, through which the overall waste mass is reduced by 36% and the biodegradable fraction is diverted from landfilling by 60%. The product of this plant is used as daily cover material for the nearest landfill.
- Heraklion (Crete): This MBT plant started its operation in January of 2010 with an annual capacity of 75,000 t. The applied technology of this plant is bio-drying, through which the final product is reduced at least by 25% in terms of moisture. Also ferrous metals are recovered.
- Kalamata (Peloponnesus): The MBT plant of Kalamata is the first plant implementing this technology in Greece. The technology that is implemented is mechanical separation and aerobic biological treatment of the biodegradable fraction. It was constructed in 1997, having an operational capacity of 32,000 tpa, nevertheless there were operational difficulties and as a result the plant shut down. There are plans to restart the operation of the plant.

In Greece there are five facilities produce energy from biogas recovery. Concisely, these facilities are presented in the Table 3.

Facilities producing Energy from biogas recovery

In the Fig. 3 is presented the lower heating value (LHV) of Greek MSW

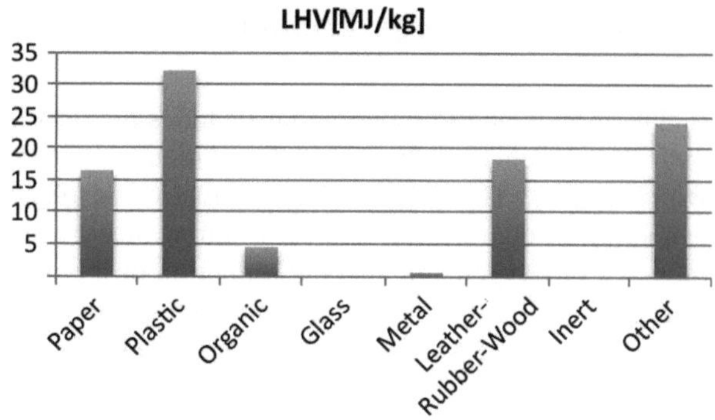

Fig. 3 LHV of Greek MSW (in MJ/kg)

3 Waste-to-Energy Technology

3.1 Introduction-Legislation

The European Union Legislation for Sanitary Landfills (1999/31/EC), imposes the decrease of biodegradable waste fraction which is land filled. Hence, thermal treatment methods of municipal solid waste along with recycling at the source and composting of pre-sorted fraction of waste is almost the only solution to such problems. Additional advantages are the volume and weight reduction of municipal solid waste and the energy production (with the possibility for heat for district heating, industrial heating purposes or for cooling, Waste to Energy Plants/ WtE). The Greek legislation for the Incineration of wastes is the Joint Ministry Decision JMD 22912/1117/2005 (in harmonization with the European Union directive for Incineration of Waste 2000/76/EC).

The European Union defined the optimum hierarchy for the waste treatment methods through the 2008/98 EU directive. According to this directive, reduction, reusing and recycling are the first stages, followed by efficient energy recovery methods, which have been upgraded in the hierarchy list [7].

Especially, the European Union, via the directives that it issues, indirectly promotes the application of thermal treatment methods (incineration, gasification and pyrolysis) along with recycling at the source and composting of pre-chosen organic fraction of waste, as an effective method of reducing the quantities of MSW that are deposited in sanitary landfills, while simultaneously producing energy in the form of heat and electricity by exploiting the heating value of MSW. With fossil fuel reserves decreasing dramatically, the utilisation of MSW as fuel looks even more promising [8].

3.2 Waste-to-Energy Technology

Incineration is an old process that includes the development of high temperatures, with the presence of flame for the oxidation of the MSW, (chemical compound with oxygen). The target of this complicated physicochemical operation is the evaporation, degradation and destruction of organic elements in MSW, with the presence of oxygen and also the reduction of the weight and the volume of MSW. The products of the incineration include gaseous compounds (for example CO_2, NO_x, acid gases, PAHs, eje.), which need to be further treated in the state of the art flue-gas cleaning system before the final emission in the atmosphere. The inert solid residues (bottom and fly ash) represent the 20–30% of the incinerator's initial feed (MSW quantity), and possibly contain some important inorganic pollutants, like heavy metals. After stabilization and solidification, the fly ash can disposed in sanitary landfill, while the bottom ash is relative inert and in general can be used for construction applications (roads, earthworks, in mines, etc).

Thermal treatment methods (incineration) with energy recovery represent the majority of waste to energy processes that are in operation across Europe. For the utilisation of produced heat and the recovery of energy, modern incinerators have special boilers for steam production. Then, the produced steam is used either straight for heating applications, or via a suitable steam turbine and generator for the production of electricity. In the following Fig. 4, a typical MSW incineration plant for energy production is presented. The most important process in such an incineration plant is the state of the art flue gas cleaning system, for the chemical cleaning of the produced in the WtE plant gaseous pollutants. The major systems are dry or wet scrubbers, electrostatic filters, fabric bag filters and cyclones, activated carbon filters, chemicals (like NH_3, CaO, Ca $(OH)_2$). The optimum selection of the flue-gas cleaning system is based in the composition of the gaseous pollutants under treatment (depending also on the composition of the feed of MSW) and the emission limits of the plant, according to the directive 2000/76/EC [9–11].

3.3 Emission Levels in Thermal Treatment

The best available antipollution techniques during the whole thermal treatment process of MSW, under the very strict emission limits of the 2000/76/EC directive, and similar in USA and other countries, lead to the environmental acceptance of the MSW incineration plants worldwide, and also promote these incineration methods as friendlier for the environment, even when compared with typical human activities such as using fireplaces or fireworks (see Table 4).

Even worse, the largest amounts of dioxin emissions are released to the open burning landfills, which is a really common phenomenon in Greece especially on the summer period. According to Dr. Nikolaos Mousiopoulos, Professor of the Aristotle University of Thessaloniki, the uncontrolled fire in Tagarades Landfill

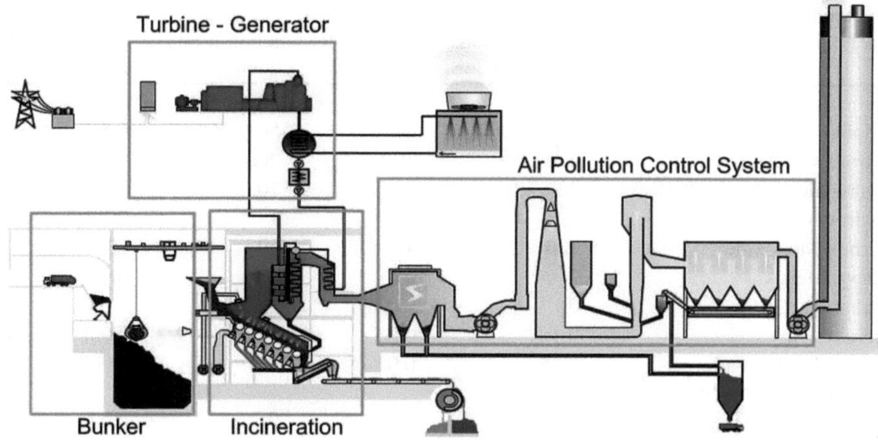

Fig. 4 Typical MSW incineration plant with electricity generation

Table 4 Dioxin emissions

Modern WtE plant	1	0.01 ng/m^3
Modern WtE plant for hazardous waste	1	0.01 ng/m^3
Non controlled incineration (i.e. Fireplaces)	1,000	10 ng/m^3
Fireworks	10,000	100 ng/m^3
Burning landfill	100,000	1,000 ng/m^3

(Thessaloniki) on 2006 summer generated 3 g of toxic dioxins per day. In contrast, the 88 WtE plants operating in USA, which burn more than thirty million tones of waste, release less than 10 g of dioxins at a period of a whole year, as the Emeritus Professor Nikolaos Themelis, director of the Earth Engineering Center of Columbia University, presented in Special Permanent Council for the Environmental Protection of the Greek Parliament on 26 July 2006 [5, 12].

Four years ago, the U.S.A. EPA announced that, MSW incineration plants produced 2800 MW electricity, with smaller emissions compared to any other power plant. MSW incineration plants are more than 800 in all over the world (around 120 WtE facilities have been constructed within the last 10 years). Although in Europe, in U.S.A. and in Japan, the MSW thermal treatment is the most popular technology for the MSW treatment, however, Greece, is almost the only country in the European Union that does not include WtE method in the national waste management plan. Pioneer countries in the application of waste to energy (MSW thermal treatment methods) are Switzerland, Sweden, Netherlands, Denmark, Germany and France, Belgium, Austria and Norway [13, 14].

In Europe, the contribution of dioxin produced in Municipal Solid Waste/WtE Plants is less than 0.7% of the total dioxin production. In Brescia of Italy there is a modern WtE Plant with a nominal capacity of 340,000 tones MSW annually (two

Table 5 Stack emissions from Brescia WtE Plant (All units are in mg/Nm^3)

The values correspond to dry air, normal conditions, 11% O_2	Plant authorization limits 1993	Design plant limits 1994	European union limits 2000	Actual operating data 2005
Particulate matter	10	3	10	0.4
Sulphur Dioxide	150	40	50	6.5
Nitrous Oxides (NO_x)	200	100	200	<80
Hydrochloric Acid (HCl)	30	20	10	3.5
Hydrofluoric Acid (HF)	1	1	1	0.1
Carbon Monoxide	100	40	50	15
Heavy Metals	2	0.5	0.5	0.01
Cadmium (Cd)	0.1	0.02	0.05	0.002
Merucy (Hg)	0.1	0,02	0.05	0.002
Polycyclic Aromatic Hydrocarbon (PAH)	0.05	0.01		0.00001
Dioxin (TCDD Teq)	0.1	0.1	0.1	0.002

lines for the combustion of household waste, went into operation in 1998) and 170,000 tones of biomass, producing 50 MW of electrical power and 100 MW of thermal power for district heating. This plant is only within 300 m distance from the first houses within the city of Brescia (located halfway between Milan and Padua). From the following Table 5 it is well understood that the gaseous emissions from the stack are much lower than the limit values of the European Union directive for Incineration of Waste (2000/76/EC) [2].

3.4 Contribution of MSW Thermal Treatment to Global Warming

The contribution of MSW Waste to energy plants for the reduction of gaseous pollutants in the atmosphere is very important. MSW incineration plants have the obligation to obey to the very strict gaseous emissions according to 2000/76/EC directive, comparing with the emissions of other industrial activities. Problems from the past, such as increased dioxin emissions have now been surpassed, due to the state of the art flue-gas cleaning systems. Of course, the production of various gaseous pollutants depends also in the physicochemical characteristics of the MSW under treatment. The production of the CH_4 and CO_2 in a sanitary landfill varies among cities and countries. According to the Greek literature, the estimated production of CH_4 in Greek sanitary landfills, ranges between 30 and 250 m^3/ton dry MSW, for U.S.A. is around 62 m^3/ton MSW. The estimated production of CO_2 is 1.32 ton/ton MSW in U.S.A. and 1.5 ton/ton MSW in Australia and Israel. However, it is well known that the CH_4 which is produced in the abovementioned significant quantities in sanitary landfills is a very potent greenhouse gas, having a global warming potential of 21 times that of CO_2. The production of CH_4 is avoided with the application of MSW thermal treatment methods [5, 12].

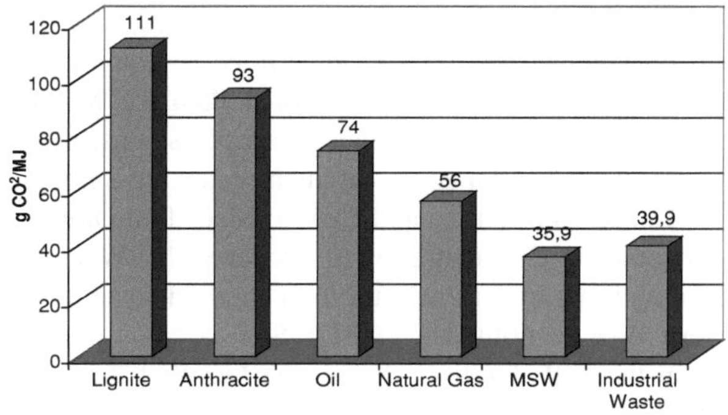

Fig. 5 CO_2 emissions by the combustion of different fuels [1]

In Fig. 5 below, it is well understood that the CO_2 emissions produced during the MSW combustion for energy production, are significantly less than the CO_2 emissions produced from conventional fossil fuels combustion such as lignite, petroleum, anthracite, natural gas. The produced CO_2 during the MSW incineration for energy production is counted as regenerated CO_2, a fact that contributes to the reduction of the production of «protogenic» CO_2 from fossil fuels, such as lignite. So, it contributes to the reduction of «new» inserted quantities of CO_2 to the atmosphere and finally to the Kyoto protocol targets [20]. Simultaneously the cost of mitigating 1 ton CO_2 with the use of MSW as renewable energy source in waste to energy plants, is significantly smaller (7–20 €) in comparison to the cost from other energy production plants which use other biomass forms (80 €), or solar power (photovoltaics, more than 1,000 €, [13]. In conclusion, the waste to energy plants have a significant contribution in the reduction of the atmospheric CO_2. The recovered energy produced from the MSW thermal treatment, reduces the emissions of gases which contribute to the Greenhouse phenomenon in two ways: (1) avoids the methane production and other greenhouse gases produced in sanitary landfills and (2) produces less CO_2 emissions compared to fossil fuels. Taking into consideration the above analysis, the MSW thermal treatment methods (waste to energy), are considered to be among the most efficient methods for solving the municipal solid waste management and treatment problem of Greece and other countries [13, 14].

4 Potential for Waste-to-Energy in Greece

4.1 Introduction

The National Plan of Greece for Waste Management for 2007–2013 (ex-Ministry of Environment, Planning and Public Works), foresees a number of Plants for RDF/SRF production in several areas of Greece, but does not mention the need for RDF-dedicated WtE facilities, because it is assumed that this material would be co-combusted in existing industrial plants. However, if this route does not materialize, as it did not for the MBTC of Ano Liossia, it would be necessary to build combustion plants fuelled by RDF. It must be noted that MSW Processing Plants have also been included in the National Plan for MSW for the Regions of Western Greece, Central Greece, Thessaly, Epirus, East Macedonia and Thrace; however, the number and capacities of such plants have not been defined to this date.

The cases of Attica and Central Macedonia are more complicated due to public acceptance. The quantities of produced MSW are very high and thus, based on international experience, the best method for energy recovery seems to be mass-burn WtE combined with RDF/SRF streams produced at communities far away from the WtE facility. Due to the fact that the major WtE power plants will be close to the consumers (Athens and Thessaloniki), the benefit is even higher due to lower electricity network losses. Combustion with energy recovery will also result in a 90% reduction in the volume of wastes to be landfilled, in case that beneficial uses of WtE ash are not developed. The resulting benefit will be quite high, both in terms of electricity supply and environmental quality in regions that are facing major problems in managing their wastes.

Waste-to-Energy methods produce steam and/or electricity. Also, the weight of MSW is reduced up to 70–80% and the volume up to 90%, and finally the land area requirements are very small. The WtE technology might be implemented in Greece by the construction of MSW WtE plants in all major cities operating with an annual capacity of 200,000–400,000 tones. The required land area will be only 4–7 ha. The basic income of such plants is the gate fee, varying from 50 to 80 €/ton. The second income comes from selling of the produced electricity to the Public Power Corporation for 87.85 €/MWh (referring to the biodegradable fraction of MSW), according to the new Greek law for renewable energy sources (L. 3851/2010). Additional income comes from the recovered metals of the bottom ash. Furthermore, there is a considerable prospect for state subsidy of the whole investment, according to the New Greek Development Law, rated at a 30–40% and depending on the selective site for the plant construction (30% equity and 40% loan). The basic actions required for the preparation of the construction of a MSW waste to energy plant are, in brief [4]:

- Locating the suitable site (land).
- Analytical Preliminary Environmental Study (approved by the Ministry of Environment, Energy and Climate Change).
- Final Environmental Study and all relative papers for the Public Power Corporation (Regulatory Authority for Energy) and the Ministry of Development (for installation, operation, electricity production eje).

4.2 Attica Case

In Attica region (capital of Greece) the daily production of Municipal Solid Waste (MSW) is estimated at 6,500 tones. This means around 2.4 million tones per year, from which 90% is deposited in one Sanitary Landfill, which is almost full. The

Table 6 Power production of the three scenarios for Attica region

Capacity (t/a)	400,000 MSW	700,000 MSW	700,000 MSW + 300,000 RDF
Lower heating value (MJ/kg)	9	9	10.8
Gross power (MW)	32.93	57.63	98.73
Net power (MW)	27.99	48.98	83.92
R1	0.6972	0.6972	0.6979
Net electrical energy (MWh/year)	223,929.08	391,875.90	671,387.41
Number of residents served	141,530	247,677	424,336

European Union Legislation for Sanitary Landfills (1999/31/EC), imposes the decrease of biodegradable waste which are deposit to sanitary landfills, so WtE methods of MSW is almost the only solution to such problems. The Greek legislation for the incineration of wastes is the Joint Ministry Decision 22912/1117/2005 (in harmonization with the European Union directive for Incineration of Waste 2000/76/EC).

For the Attica Region, the Greek Waste-to-Energy Research and Technology Council "SYNERGIA" has suggested the following tree scenarios, through which all the difficulties on the waste management in Attica will be overcome.

Two plants of 400,000 tpa, which is a medium capacity plant. Whether two plants of such capacity were constructed in opposite directions of the Attica Region, the residuals of recycling would be managed by the most environmental friendly method, producing electricity and increasing the lifetime of the landfills.

One plant of 700,000 tpa, which is a large capacity plant, according to the global reference. This kind of plant might be constructed at a site where already waste management facilities exist. Such a place in Attica might be at Ano Liossia Municipality, where the main landfill of Attica and a MBT plant exist.

One co-incineration plant of 700,000 tpa MSW and 300,000 tpa RDF, which is a large capacity plant and also implementing state-of-the-art techniques. This proposal is based on the previous one and also provides a treatment method for the produced RDF of Attica, which otherwise would be landfilled [6].

In the following Table 6 the power production of the three scenarios for Attica are presented.

4.3 Rhodes Case

At the very beginning of In February 2008, the Waste Management Company of Rhodes (DEKR), requested by Themelis Associates and the Earth Engineering Center of Columbia University to conduct a pre-feasibility study setting the following essential conditions to be met in selecting the technology to replace the existing sanitary landfill:

- The technology selected should solve the waste management problem of the island not only for a certain number of years but for generations to come.
- The technology should be proven for several years and be widely accepted in nations that are leading in the global effort for environmental protection.
- The proposed plant should be of capacity that can handle the MSW of all ten municipalities of the island of Rhodes as well as other commercial, light industrial and agricultural residues that cannot be recycled; plus the biosolids produced by the wastewater treatment plants in Rhodes.
- The proposed technology should be environmentally superior to a new Landfill.
- The proposed plant should be economically viable and not impose a very high gate fee, per ton of MSW processed, on the citizens of the island. In particular, it should conserve land by reducing the volume of residuals to be landfilled.

On the basis of the specified conditions by DEKR, the best suited thermal treatment technology for the first WtE facility in Greece is controlled combustion of as-received MSW on a moving grate and recovery of the energy in the combustion gases by means of a boiler and a steam turbine. Furthermore, the municipal waste management company of Rhodes (DEKR) plans to construct an integrated waste-management Environmental Park that will include composting of source-separated organics, recycling of source-separated recyclables (mostly paper, metals and some marketable grades of plastics), combustion of post-recycling MSW, and sanitary landfilling of the WtE ash that is not used beneficially [3].

The proposed WtE plant may consist initially of one line (grate, furnace, boiler, Air Pollution Control system) of annual capacity of 80,000 tones (10 t/h). However, the building size can provide for later expansion to two lines of 160,000 tones capacity. The two-line WtE facility (160,000 tones/year) will generate an estimated 96,000 MWh of net electricity (600 kWh/ton) for the grid. There will also be available another 80,000 MWh of thermal energy which may be utilized by an adjacent industrial operation that can make use of low pressure steam, such as a paper recycling plant. The capital cost of the two-line operation was estimated at €98 million and of the single-line plant €63 million. A preliminary estimate based on the assumptions that [15, 16]:

- The EU grant will amount to 30% of the capital cost of the two-line plant;
- The average price received for the electricity (about 50% of which is biomass energy and therefore renewable) will be €70 per MWh (according to the new Renewable Energy Law L.3851/2010); and the gate fee for the MSW will be €80 per ton of MSW, showed that the projected WtE would be economically viable, during the 25 year period that the capital investment is paid off, and an economic boon to the community thereafter.

The plant will most likely be financed as a Public Private Partnership. A special problem in Rhodes and other popular tourist destinations is that during the summer months the generation of MSW nearly doubles. To overcome this problem, the Rhodes authorities are considering the importation of MSW from other islands in the area, to complement the feedstock to the WtE plant; or the storage and use of industrial wastes from Rhodes and other places.

The high capital cost of the plant is explained by the fact that in addition to eliminate the need for landfilling on the island, it will utilize the heat of combustion to generate an appreciable amount of electricity for the island [17]. Also, it will include state of the art gas cleaning equipment consisting of dry scrubber (HCl and SO_2 removal), activated carbon non-catalytic selective reduction (for NO_x), activated carbon injection (for volatile metals and dioxins/furans), and fabric filter baghouse (for particulate matter). As a result of the highly sophisticated Air Pollution Control system, the projected emissions will be at lower levels than the E.U. stringent standards that are applied in nations like Denmark, Germany, France, the Netherlands, Sweden, and Switzerland. For example, the projected total emissions of dioxins and furans from the combustion of 160,000 tones of MSW annually will be 0.05 g TEQ. Efforts are also under way to attract industrial or commercial users who can use the low-pressure steam, which remains after generating electricity in the steam turbine, to heat or cool facilities that may be built within a few kilometers of the Eco-Park site [18].

The preliminary technical and environmental studies have been completed and the projected environmental impacts of this installation have been submitted to the Ministry of the Environment, Energy and Climate Change (YPECA). The chosen site for this project is on the northern part of the island and is adjacent to the existing regulation landfill that is now serving most of the population of Rhodes but is scheduled to fill up in less than three years. According to the present plan of Rhodes, in addition to a state-of-the-art waste-to-energy plant which, at the beginning, will process an estimated 300 tones of MSW per day, the Environmental Park that will be created at that site will include an aerobic composting plant for the production of soil conditioning compost, a Center for Recovery of Recyclable Materials (KDAY of Rhodes)., and a new monofill cell that will be used for disposal of the WtE ash that cannot be used beneficially on the island, as has been done in the island of Bermuda and many other nations [19].

5 Conclusion

The waste management plan in Greece has to be changed rapidly in order to be conformed to the European directives. Many efforts should be made in order to inform and persuade the society and the policy makers of Greece that modern waste to energy technology is the demanded step after recycling and composting at the source, in order to be severed by the landfill sites and the illegal dumps.

The research conducted on the existent MSW management system in Greece led to the conclusion that it has several assets and numerous liabilities. Currently, the MSW generated in Greece are mainly transferred either directly or indirectly through Waste Transfer Stations (WTSs), to sanitary landfills; also, some are disposed at illegal Uncontrolled Waste Disposal Sites (UWDSs). To alleviate this situation, the construction of Integrated Waste Management Facilities (IWMFs) has been planned, but not yet implemented. The above reasons render the study for an alternative SWM system obligatory.

In the search for long-term solutions to the existing problem, the advantages and disadvantages of the SWM system currently practiced were taken into consideration in order to develop an effective MSW management plan, which will greatly improve the quality of life in the Region of Attica and Rhodes.

Therefore, a preliminary assessment of WtE as a possible solution to the MSW issue in the Regions of Attica and Rhodes was carried out in this study. This alternative was chosen, because of its demonstrated environmental and economic viability throughout Europe and other nations. It is a well proven means of environmentally sound treatment of solid wastes that also generates renewable electricity and heat. Controlled combustion of as received MSW on moving grates allied with stringent Air Pollution Control (APC) technologies can consistently and reliably process not only untreated MSW, but also post recycling/composting waste residues in an environmentally safe fashion with minimal impact on the environment. Additionally, the volume of waste to be landfilled is reduced by 90%, resulting in alleviation of traffic congestion and the reduction of air pollution caused by trucks. Finally, the electrical and thermal energy produced by the processing of waste (replacing fossil fuels) is a major source of profit and also can be used for the operation and for cooling/heating of the WtE plant and/or neighboring facilities. For all these reasons, WtE is considered to be a long-term solution to the waste problem situated in Greece.

To sum up, the integration of WtE in Greece's Regional Plan for SWM will lead not only to compliance of the Region with the EU targets (Directive 2008/98) towards Sustainable Development, but also to the final solution of the MSW problem of the Region with the simultaneously production of renewable energy (reducing GHGs comparing with fossil fuels, Directive 20–20–20 and the relative new Greek Law 3851/2010) [20].

References

1. Bilitewski B (2006) State of the art and new developments of waste-to-energy technologies. Proceedings Venice 2006: biomass and waste to energy symposium, November 29–December 1, Venice, Italy
2. Bonomo A (2003) WTE Advances: the experience of Brescia, Keynotepresentation at the 11th North American waste-to-energy conference, Tampa FL
3. Psomopoulos CS, Themelis NJ (2009) Potential for energy generation in Greece by combustion of as received or pre-processed (RDF/SRF) municipal solid wastes. 2nd international conference on environmental management, engineering, planning and economics (CEMEPE) and SECOTOX conference, 21–26 June 2009
4. Kalogirou E (2009) Waste management in Greece and potential for waste-to-energy. ISWA Beacon conference–strategic waste management planning in SEE, Middle East Mediterranean Region, Novisad, 10 Dec 2009

5. Kalogirou E (2001) Mathematical models for the determination of different physicochemical quantities related with air pollution, using the reversed-flow gas chromatography technique. Ph.D. dissertation, N.T.U.A., Athens
6. Kalogirou E (2009) Waste to energy solution for municipal solid waste. 2nd international conference on environmental management, engineering, planning and economics (CEMEPE) and SECOTOX conference, 21–26 June 2009
7. Karagiannidis A, Bilitewski B, Tchobanoglous G, Themelis NJ, Wittmaier M, Tsasarelis Th (2008) Waste-to-energy on thermal treatment and energetic utilization of solid wastes. Nova Science Publishers Inc, New York
8. Karagiannidis E, Kalogirou K, Psomopoulos N, Themelis (2010) Waste-to-energy in United States of America–an analysis on the current situation. Technic annals, Technical Chamber of Greece, March–April 2010 pp 97–116
9. Lalas D, Gidarakos E (2007) Evaluation of general effects and cost of solid waste management. Local Government Institute, Athens
10. Lee SH, Themelis NJ, Castaldi MJ (2007) High-temperature corrosion in waste-to-energy boilers. J Therm Spray Technol 16:104–111
11. Millrath K, Themelis NJ (2003) Current trends in the waste to energy industry. Proceedings ASME international congress, Washington, D.C
12. Papageorgiou A, Karagiannidis A, Barton R, Kalogirou E (2009) Municipal solid waste management scenarios for Attica and their greenhouse gas emission impact. Waste Manag Res 27(9):928–937
13. Stengler E (2006) Developments and perspectives for energy recovery from waste in Europe. Proceedings Venice 2006: biomass and waste to energy symposium, November 29–December 1, Venice, Italy
14. Themelis N (2003) An overview of the global waste-to-energy industry. Waste Management World 2003–2004 Review pp 40–47
15. Themelis N (2007) Thermal treatment review, global growth of traditional and novel thermal treatment technologies. Waste Management World pp 37–45
16. Themelis NJ, Bourka A, Ypsilantis G (2009) Energy and materials recovery from municipal solid wastes at the island of Rhodes. 2nd international conference on environmental management, engineering, planning and economics (CEMEPE) and SECOTOX conference, 21–26 June 2009
17. Themelis N, Koroneos C (2004) Assessing waste-to-energy and lanfilling. Technic Chronics pp 1–2
18. Themelis NJ (2008) Developments in thermal treatment technologies. Proceedings NAWTEC 16, Paper 16–1927, Philadelphia
19. Tim Byrne (2009) Waste the greek and cypriot way. Waste Management World pp 54–58
20. Vehlow J (2006) State of the art of incineration technologies. Proceedings Venice 2006: biomass and waste to energy symposium, November 29–December 1, Venice, Italy

Incineration of Municipal Solid Waste in the Baltic States: Influencing Factors and Perspectives

Harri Moora, Viktoria Voronova and Rasa Uselyte

Abstract The three Baltic States are in the stage of changing their municipal waste management systems since they have to comply with the principles and targets of the European Union waste policy and directives. Over the past years, thermal treatment of municipal waste has been discussed more intensely in these countries as one of the waste management option that could help to reach the legal targets in a relatively short time. In general, the Baltic States have similar socio-economic characteristics, waste and energy sector developments and geographical conditions that form similar frameworks for the development of a waste management infrastructure, including possible waste-to-energy options. However, as experience from recent studies and projects shows, there are several local and regional factors that could significantly influence the economic success of large scale waste incineration. The paper attempts to identify and discuss these main influencing factors and perspectives for MSW incineration in the Baltic States. The main focus is on conventional mass-burn incineration. The specific issues in terms of technical, economic and environmental aspects are presented in the form of an illustrative case study based on the design and performance data of the first waste-to-energy facility in Estonia.

H. Moora (✉)
Stockholm Environment Institute Tallinn Centre,
SEI-Tallinn Lai 34, 10133 Tallinn, Estonia
e-mail: harri.moora@seit.ee

V. Voronova
Institute of Environmental Engineering, Tallinn University of Technology,
Ehitajate tee 5, 19086
Tallinn, Estonia
e-mail: viktoria.voronova@ttu.ee

R. Uselyte
Ekokonsultacijos UAB, J.Galvydzio str.3, 08236 Vilnius, Lithuania
e-mail: rasa@ekokonsultacijos.lt

A. Karagiannidis (ed.), *Waste to Energy*, Green Energy and Technology,
DOI: 10.1007/978-1-4471-2306-4_10,

1 Introduction

The three Baltic States—Estonia, Latvia and Lithuania—have recently experienced rapid economic growth, resulting in a significant increase of municipal solid waste (MSW) quantities, while their waste management systems still require much effort to be adjusted to the European state-of-the-art. A wide variety of technological options, increasingly diverse waste fractions, environmental restrictions and European Union (EU)-wide recovery targets require the decision makers to well consider the steps to be made. The solutions to municipal waste management should not only be environmentally sustainable but also cost-efficient and socially accepted. Therefore, waste management has become one of the key issues in governments, in the waste management sector as well as among the general public in all new EU Member States including the three Baltic States.

In spite of its lowest priority in the European waste management hierarchy, landfilling has been the predominant method for municipal waste management in all the Baltic States. The fact that landfilling is the worst option for MSW treatment is generally accepted. However, the choice of the most optimal waste management solution has been under heavy discussion.

Over the past years, thermal treatment of MSW has been discussed more intensely in the Baltic States. There exist several plans to build waste to energy (WtE) facilities in the Baltic States. At least one mass burn incineration is under construction in Estonia and the construction of the first incineration facility in Lithuania will start soon. Several other projects are in the preparatory phase.

Even though the 27 EU Member States are directly governed by the same overall legislation, including that on waste management, disposal or incineration, the importance of incineration differs widely from one EU member state to another [24, 25]. This is because the issue of waste incineration is complex and the success of this MSW treatment option depends largely on the framework conditions characteristic for a specific country or region [23, 26].

Energy recovery by waste incineration has a double function as a waste treatment method and a supplier of electricity and/or heat, thereby linking the systems of energy and waste management. Both systems are undergoing great changes in the Baltic States. There are also several other influencing factors (e.g. waste generation and content, waste and energy sector/market developments, environmental impacts and public opinion) that have to be studied carefully before starting to develop any plans for MSW incineration [26]. The Baltic States waste sector/market is relatively small in size. Therefore the already existing plans to build large-scale WtE facilities have resulted in many discussions and debates among other waste management actors. As the experiences from other European countries have shown, waste incineration could have a significant impact on the existing waste management system [1, 13, 14, 22–24].

This paper attempts to identify and discuss the main influencing factors and perspectives for MSW incineration in the Baltic States. The main focus is on conventional mass-burn incineration as it is probably the most suitable large-scale

WtE technology for the management of mixed municipal waste under the conditions in the Baltic States.

The specific issues in terms of technical, economic and environmental aspects related to large scale mass-burn incineration of MSW are presented in the form of an illustrative case study based on the design and performance data of the first WtE facility in Estonia. The environmental impacts of the new WtE facility were assessed by using the life cycle assessment (LCA) software tool WAMPS [17].

The information contained in this paper was derived from a series of earlier research projects, the aim of which was to analyse the environmental impacts and economic costs of planned WtE projects as well as evaluating possible alternative scenarios for MSW management in the Baltic States [1, 16, 17, 28, 31, 32].

The discussion on influencing factors of MSW incineration in this paper focuses mainly on the Estonian context. This is because of availability of data and a more established waste management system. However, most of the examples and discussions are also relevant for the other Baltic States which have a similar socio-economic structure and waste management and energy sector development as Estonia.

2 Plans for MSW Incineration

There is no experience in large scale MSW incineration in the Baltic States. However, for some years now, cement factories in the Baltic States have used refuse derived fuel (RDF) imported from other EU Member States and on a smaller scale produced in the first local mechanical–biological treatment (MBT) facilities.

All three Baltic countries have developed national waste management plans that foresee a place for possible municipal waste incineration. However, concrete projects to build waste incinerators are in very different stages of development in these countries (Table 1).

The construction of the first mass-burn incineration facility in the Baltic States started in 2010 and the new waste incineration unit of the Iru power plant close to the Estonian capital Tallinn is expected to begin generating electricity and heat from MSW in 2013. This WtE facility with annual capacity of 220,000 tons is supposed to incinerate MSW from all Estonia. Furthermore, plans for a second waste incineration plant in Tartu (central part of Estonia) with a 100,000 t/a capacity are under discussion. The city of Tartu initiated discussions about this plant because the regional landfill was closed in 2009. However, the project has been postponed due to uncertainties related to financing and the waste market.

There have been discussions for many years to build a waste incinerator in Riga, the capital of Latvia. However, currently there are no concrete plans for municipal waste incineration projects in Latvia.

There exist two projects to build mass-burn waste incineration plants in Lithuania. The Klaipeda region municipal waste management plan for 2010–2019 foresees that the mixed municipal waste collected from the Klaipeda waste

Table 1 Plans for WtE facilities in the Baltic States

Country	Number of projects/plants	Capacity (t/a)	Status
Estonia	Iru WtE unit	220,000	Under construction
	Tartu WtE plant	100,000	Planned
Latvia	–	–	–
Lithuania	Klaipeda WtE plant	245,000	Construction will start in 2011
	Vilnius WtE plant	250,000	Planned

management region (7 municipalities) will be incinerated. The construction of Klaipeda WtE plant at the Lypkiai local boiler house area will start in 2011 and it is expected to be completed by 2013. The total capacity of this plant is 245,000 t/a including up to 130,000 tonnes of MSW, 75,000 of biofuel and 50,000 of industrial waste. Klaipeda WtE plant is a co-operation project between the Finnish energy company Fortum and the local energy company Klaipėdos energija, mainly controlled by the Klaipeda city municipality.

A second WtE plant is planned to be constructed in Vilnius, the capital of Lithuania. A private company Regioninė komunalinių atliekų deginimo gamykla (Regional Municipal Waste Incineration Plant) had an intention to build a WtE plant in Vilnius (next to the current combined heat and power plant CHP-3 in Vilnius). The local people strictly opposed to this project. As a result of the Environmental Impact Assessment (EIA) process, the Vilnius Region Environmental Protection Department did not allow to build the plant, because the involvement of Vilnius municipality, especially in relation to the engagement of the public, was considered not sufficient. After this experience, the municipalities of Vilnius waste management region (8 municipalities) have decided to start a new a tendering process for the construction and operation of a waste incineration plant.

3 Factors Influencing MSW Incineration

Experiences in other EU Member States show that the role of waste incineration differs widely from one EU member state to another. This is because waste incineration is a highly complex waste treatment option, which involves large investments and that depends largely on the framework conditions characteristic for a specific country or region. In the EU both waste management and energy production are subject to extensive regulations. The legislation aims to set a suitable policy framework at the EU level, with specific targets, while leaving the choice of pathways and technological development to the individual players in the Member States (authorities and private sector).

In addition to the legal framework that is based on the same general requirements the three Baltic States have similar socio-economic characteristics, waste and energy sector developments and geographical conditions that form similar framework for the development of waste management infrastructure including possible WtE options. However, as the experiences from feasibility studies and

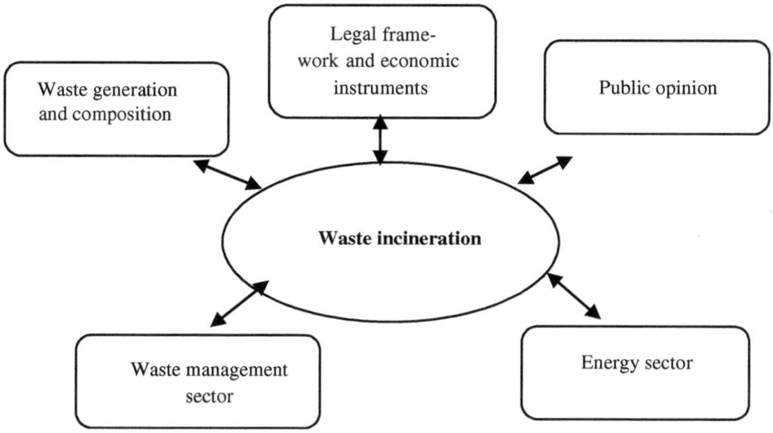

Fig. 1 Main influencing factors for municipal waste incineration

first WtE projects in these countries indicate, there are several local and regional factors that could significantly influence the success of waste incineration. Also the possible future trends and developments of these influencing factors have to be carefully studied. Given a service life of over 30 years, a waste incineration technology must also be able to function efficiently under the changing conditions in the future. The main framework factors that could influence the economic success of the new WtE facilities in the Baltic States are presented in Fig. 1.

3.1 Legal Framework and Economic Instruments

National waste policy and legislation in the Baltic States, as in all other EU-27 Member States, is governed by the EU policy and legislation.

The EU legislation on waste management is based on the Waste Framework Directive 2008/98/EC [8], which among others provides a definition of waste and sets out a general ranking of waste management methods, the so-called waste hierarchy. According to waste hierarchy waste generation should be prevented or reduced, and what is generated should be recovered by means of reuse, recycling and other recovery operations, thus reducing disposal/landfilling. A strong driver for improving the energy performance of waste-to-energy facilities is the Waste Framework Directive's new provision that allows high efficiency installations to benefit from a status of "recovery" rather than "disposal".

Recognising that not all waste can be prevented or recycled, the EU has also adopted directives on waste incineration and landfilling: Waste Incineration Directive 2000/76/EC [7] and Landfill Directive 1999/31/EC [6].

The Landfill Directive is arguably one of the most influential documents of the portfolio of the EU waste management regulations with direct influence on the development of waste recovery (including WtE) options [2, 10, 21, 29]. It sets

Table 2 Landfill taxes and bans (2010)

	Average gate fee for landfilling euro/tonne	Landfill tax in euro/ tonne	Ban of landfill of unsorted MSW
Estonia	45	12	Yes
Latvia	24.2	4.27	–
Lithuania	17.5	Planned in 2013, 22 euro/tonne	–

progressive targets for the reduction of the biodegradable fraction of MSW going to landfills to 75% of their 1995 baseline levels by 2006, 50% by 2009 and 35% by 2016. The Baltic States as other new Member States that rely heavily on landfilling, have made use of the allowance to postpone these targets by 4 years. Therefore they need to meet the respective diversion targets by 2010, 2013 and 2020.

The diversion of the biodegradable fraction of MSW places major challenges on all new EU Member States. Taking into account the current situation in the MSW management, it can be expected that the biodegradable waste diversion targets (especially the targets for 2013 and 2020) will be very challenging for the Baltic States. Consequently, there is an urgent need for action. Municipal waste incineration is one of the most realistic options that could help to reach these targets.

Many EU Member States with high waste recovery rates have facilitated the EU waste policy implementation by taxes on waste landfilling and landfill bans. Estonia introduced a pollution charge for municipal waste disposal (landfill tax) already in 1990. Until 2005, the rate was very low at EUR 0.10–0.20 per tonne. In 2006 it rose to EUR 7.8 per tonne and increases every year, reaching EUR 29.84 in 2015. Due to the landfill tax the landfilling fee has increased considerably over recent years. Estonia has also introduced a ban on the landfilling of untreated waste (including mixed municipal waste) (see Table 2).

Latvia has also introduced a landfill tax, applied as a natural resource tax. Lithuania intends to introduce a landfill tax in 2013. A ban on the landfilling of untreated waste has not yet been implemented in Latvia and Lithuania, due to lack of an alternative waste treatment capacity.

Based on the experience of Estonia, the legal requirements together with economic instruments such us landfill tax have resulted in favourable conditions for the development of new recovery facilities including waste incineration.

3.2 Municipal Waste Generation and Composition

The economic success of waste incineration depends directly on the available waste amount and waste composition. Investments in waste incineration presume a steady fixed stream of waste to ensure financial viability. The waste supply should be fairly stable in the whole life span of a WtE facility (up to 30 years).

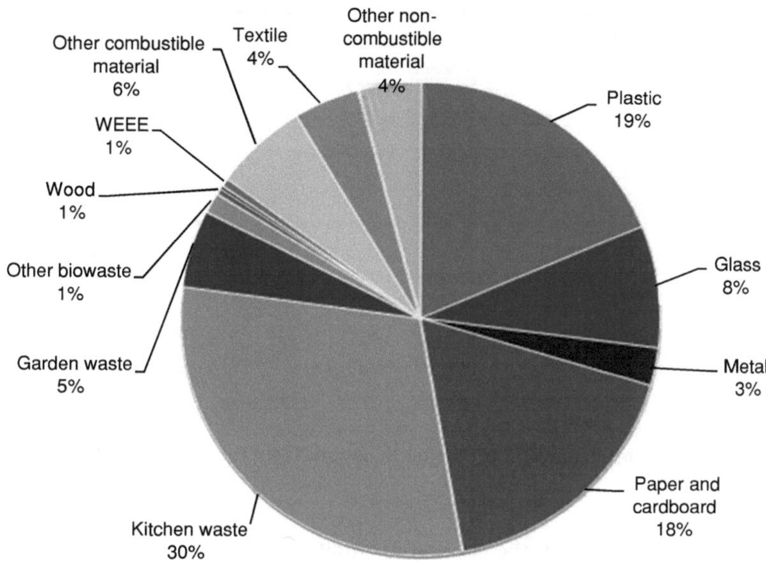

Fig. 2 Mixed municipal waste (landfilled) composition in Estonia (2008)

The energy content of waste, the so-called calorific value, depends on the composition of the waste and preferred to be as high as possible.

However, waste composition may change in time because of either additional recycling or changes in the socio-economic situation in the collection area. Both changes can significantly alter the amount of waste and its calorific value. Therefore, data on waste generation and composition as well as forecasting these waste trends are essential for the planning and development of a waste incineration project.

The availability and quality of data on MSW generation and composition in the Baltic States have been quite poor. To specify and validate the mixed municipal waste composition data in Estonia, a country-wide sorting analysis of mixed municipal waste was carried out in 2008 [18].

The results of the sorting study indicate that even at a relatively high rate of recycling the landfilled mixed municipal waste has a relatively high calorific value (10.5 MJ/kg) due to the high share of combustible materials such as plastic and paper (see Fig. 2).

The quantity of MSW in the Baltic States has rapidly risen as a result of economic growth and increasing consumption. In Estonia approximately 540,000 tonnes of municipal waste (400 kg per person) were generated in 2008. The respective figures in Latvia and Lithuania 1.2 million tonnes (407 kg per person). Earlier forecasts show that MSW will continue to grow rapidly; generation of waste was projected to increase by approximately 50% from 2005 to 2020 [12].

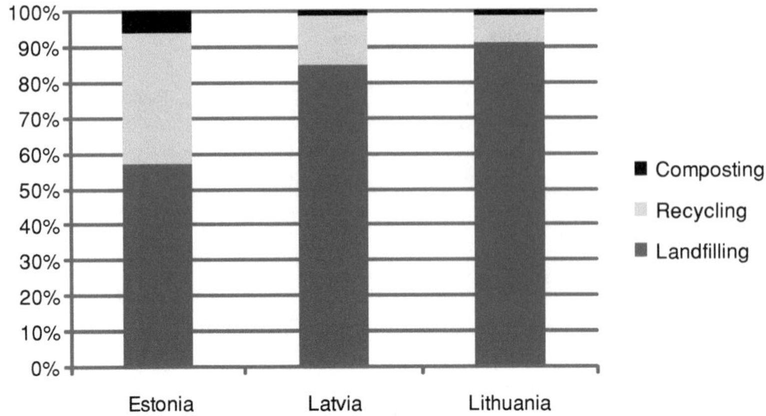

Fig. 3 MSW treatment methods in the Baltic States at 2008 [10]

However, fluctuations in the economic situation could lead to changes in waste generation. This is well illustrated by the impact of the unexpectedly serious global economic decline that has significantly influenced the municipal waste generation rate and made earlier waste generation forecasts questionable. Recent indicators show that MSW generation in Estonia dropped during 2008 and 2009 in correlation with the Gross Domestic Product (GDP) by almost 25%. It can be anticipated that municipal waste generation in Estonia and in the other Baltic States will start to grow along with the recovery of economy. However, it is very difficult to predict the growth rate and time line. Since the number of population is expected to remain roughly the same in all Baltic States, a possible economic development will be the key driving force for changes in waste volumes in the next decade.

Since earlier waste generation forecasts cannot be used anymore, the smaller amount of available mixed municipal waste may significantly influence the number, capacity and financial costs of the possible waste incineration projects in all three Baltic States.

3.3 Waste Management Sector Developments

The success of waste incineration depends largely on the development of the waste management sector including the commercial competition with other waste management options. Here the interest and ability of public sector waste management authorities to control and regulate the local waste management market plays important role.

Landfilling has traditionally represented the easiest and the cheapest option for MSW management. This is the reason why landfilling is still dominating MSW treatment option in the three Baltic States (see Fig. 3). In 2008, the lowest share of municipal waste landfilled was in Estonia (57%) and the highest in Lithuania (91%). Most of the new landfills in the Baltic States are owned by the local municipalities or Regional Waste Management Centres (RWMCs) established by municipalities. All new regional landfills that are established have received financial support in terms of EU funding.

Since the municipal waste management system must comply with the principles and targets of the European waste policy and directives, the role of landfilling will decrease significantly in coming years. Estonia has already achieved a relatively high share of MSW recycling. The main driver for the development of a separate waste collection and recycling systems have been the recovery targets of the EU Directive on Packaging and Packaging Waste 94/62/EC [5] and increasing landfilling gate fees driven by the growing landfill tax [11].

As the Estonian experiences show, the material recycling has been also influenced by the recent fluctuations of market prices of virgin and recycled materials. The high share of impurities in the source separated waste is another limiting factor of an efficient recycling system. The same aspects influence the recycling of collected organic waste. The experiences of composting of organic waste from households show that the quality of compost is low [17]. Another limitation for environmentally beneficial use of compost is the very low market demand for such product. In Estonia, today most of the municipal biodegradable waste based compost is used as a filling material and for landscaping the landfills.

Mechanical–biological treatment as an alternative treatment option for mixed municipal waste has gained recently a lot of attention in the Baltic States. Since there are many problems in running separate collection systems of different municipal waste streams, the MBT is seen as a relatively easy and low cost alternative for treatment of mixed municipal waste. Many private waste management companies as well as municipalities have started to invest into new MBT facilities. However, as the experiences of the first simple and low-cost MBT facilities in Estonia show, the quality of the produced RDF is relatively low. The remaining residual fraction contains significant amount of heavy metals and other harmful or disturbing substances and therefore this should be landfilled.

MBT facilities and conventional mass burn WtE plants compete for the same mixed municipal waste that is available in the region. In Estonia MBT facilities planned or under construction alone already exceed the available municipal waste amount. In Lithuania and Latvia investments to MBT are also a subject of State support that disturbs a fair competition on the waste management market.

In this new situation the public authorities play crucial role. In all three Baltic States the local municipalities are responsible in planning waste management system and organising the waste collection and treatment. Compared to Latvia and Lithuania majority of the municipalities in Estonia are small and therefore they are not able to manage the waste treatment tasks that have been imposed on them. The increased liberalisation and free competition pressure from the government has lead to the situation where the waste management market in Estonia is in a high content controlled by the private sector. Municipalities have very limited ability to direct the waste to certain waste treatment facilities. This has occasionally caused legal problems regarding the ownership of the waste. The extremely liberal waste management market has lead to intense commercial competition between different waste recovery facilities and companies. This also could influence the economy of the possible waste incineration. The WtE facility under the development in Estonia have not managed to sign any waste delivery contracts with municipalities.

In Latvia and Lithuania the municipalities have formed regional waste management centres that plan and develop the waste management on the regional level. They also participate directly in the MSW management by delegating certain functions of waste management, e.g. planning of new waste management infrastructure, operation of regional landfills, civic amenity sites, green waste composting sites, organising tenders for municipal waste collection, collection of fees from waste holders etc. [32]. Therefore, they have stronger waste flow control and general influence over the waste management sector.

3.4 Energy Sector Developments

Large scale WtE facilities, such as municipal waste incineration in combined heat and power plants, could generate significant amounts of energy and are therefore important players in the local energy markets—especially in small countries. It is thus important to establish whether an incineration facility can be integrated into the local legal framework and infrastructure of energy sector. The potential for heat utilisation, defined by the availability and accessibility to the district heating network, is also a very important aspect.

The Baltic energy markets are all undergoing a transition towards new sources of energy and significant changes will take place in the coming years [19]. The biodegradable fraction of MSW is a part of biomass definition, thus it counts as a renewable energy source. The use of MSW for energy production can contribute to achieving the 20% renewable energy goal and the 20% reduction in CO_2 emissions agreed upon at the EU level. Heat and electricity from waste are replacing the energy generated by conventional power plants which still predominantly use fossil fuels.

The Baltic States have all committed themselves to the EU energy goals of 20–20–20 by 2020 (goals which commit EU countries to reduce their energy consumption and emissions of greenhouse gases by 20% and ensure that 20% of energy consumption is covered by renewable energy by 2020) [3]. Twenty percent is the European average but individual goals have been set for each country in order to reach this total average of 20%. By 2020, the percentages of renewable energy in final energy consumption for the Baltic countries should be 25% for Estonia (currently 18%), 23% for Lithuania (currently 15%), and 42% for Latvia (currently 31.4%) [15]. All three countries have introduced policy instruments and support schemes to promote renewable energy generation.

In Estonia, the development within the renewable energy sector is mainly focused on wind energy and biomass/waste based CHP. As a support scheme, producers of renewable energy can sell electricity to a state owned energy company at a fixed high feed-in tariff. There is also support available for CHP and heat production to switch to renewable fuels.

Like the other Baltic countries, Latvia is also obliged to increase the proportion of renewable energy, which could be achieved by installing more biomass/waste based energy production. Several projects are outlined for the modernization of the district heating segment and for making changes in the composition of the fuel utilised there. The emphasis is on the implementation of renewable energy based technologies, especially biomass/waste type solutions.

The Lithuanian energy sector is already in the reorganization stage in order to cope with the shutdown of the Ignalina NPP. The government has launched several initiatives such as the liberalization of the energy market, increasing energy efficiency, broader use of renewable energy sources and promotion of small energy producers. Promotion of renewable energy is one of the main goals of the energy policy in Lithuania. Particular focus is laid on the generation of thermal energy, and 48 million Euros have been allocated for the construction and renovation of bio-fuel/waste boilers and CHP power plants until 2013. In addition, the government has approved a special policy to buy green electricity. The price is by 50–60% higher for biomass energy compared to the current average level of traditional resources, and these prices are state guaranteed until 2020. It is obligatory to buy the supplied green energy and there is a possibility to receive a 40% tax advantage for connecting to the electricity grid.

To have high energy efficiency, waste incineration in a large-scale CHP should take place in large district heating networks where the WtE facilities can function as base load heat providers with both diurinal and seasonal variations.

As in many other Eastern European countries, district heating has a relatively large share in the Baltic States. Most major cities have big district heating networks dating back to the Soviet time. For example in Estonia the share of district heating in heat consumption is approximately 70% [17]. Most of the district heating systems are, however, small and heat is supplied from relatively small-scale boilers (in Estonia 80% of the boilers are less than 1 MW). The largest district heating networks in the Baltic States are located in larger cities (e.g. Tallinn, Riga, Vilnius, Klaipeda, Kaunas), where the share of district heating is close to 90%. The predominant fuels used in district heating systems vary. However, most of the large scale power plants still use fossil fuels (oil, natural gas).

Energy policy and the increasing costs of fossil fuels have given an impetus to local heat producers to transfer from their sometimes rather old and inefficient production technology to the modern CHP technology fuelled by renewable fuels including waste.

3.5 Public Perception

Changes in waste management arrangements in local areas are gaining more attention in media. As a result of greater publicity and higher awareness, many people and organisations are opposing the new waste management facilities. In areas of no public awareness of waste incineration plants, there is usually resentment and distrust towards the environmental and technical performance of such a facility. The experiences in the Baltic States show that the distrust to decision makers and developers may lead to a situation where new waste management facilities of any type are rarely welcomed by the residents close to where the facility is to be located. Public opinion on waste management issues is widely varied and can often be at the extreme ends of the scale. The public opposition to the development of a waste incinerator in Vilnius is a good example of this. Due to the failure to discuss with the public, the project was turned down.

Waste incinerators are still generally perceived as great pollutant sources. Public perception may not be so strong when the WtE facilities are planned to be built in the territory of already existing energy facilities. However, if the same plant has caused environmental problems earlier, the opposition could even stronger.

Therefore, it is important to communicate the waste incineration technology, as well as local and global environmental/health impacts, in a trustworthy and detailed manner. Here the development of the Iru WtE unit could be taken as a positive example. The public, especially the people living in the neighbourhood, was involved in an early stage by using public information meetings and hearings. In addition to EIA, several other studies were ordered from independent institutions to provide comprehensive and independent information about the impacts of the planned WtE unit. This all ensured that the opposition to the Iru WtE was low among the public.

4 The Case of Iru Waste-to-Energy Unit in Estonia

In 2006, the Estonian state owned Energy Company Eesti Energia AS started preparations for the construction of a waste-to-energy unit located in the outskirts of Tallinn, the capital of Estonia.

The main aim was to diversify the production portfolio of Eesti Energia as the company currently produces most of its energy from fossil fuels with rather high CO_2 emissions. In addition there was a need to replace and add to the rather old and oversized energy production units that are not flexible enough to meet the needs of the changing heat and electricity market.

The WtE unit is an extension of the existing Iru CHP plant which has generated electricity and heat in the same location for over 30 years. This CHP plant has been one of the main suppliers of district heat for Tallinn and the city of Maardu. It currently uses natural gas as its main fuel and oil as a reserve fuel.

The new unit will annually burn up to 220,000 tonnes of the mixed municipal waste currently deposited in landfills. This will replace nearly 70 million m^3 of natural gas currently used each year. The facility will use municipal waste mainly from the Tallinn region (the city of Tallinn and the surrounding municipalities).

Table 3 Main parameters of the Iru WtE unit

Thermal treatment unit	Mass burn—MARTIN grates
Number of units	1 × 2.7 tonnes per hour
Capacity	220,000 t/a
Energy efficiency (R_1)	1.28
Heat	50 MW
Electricity	17 MW

More than 50% of the total municipal solid waste in Estonia is generated in this region. However, since the territory of Estonia is small (45,226 km^2), waste can also be collected from other parts of the country.

The construction of the waste incineration unit of the CHP plant started in 2010 and the unit will start to generate electricity and heat from waste in 2013. As the first plant for the thermal utilisation of municipal waste in the Baltic States, the Iru WtE facility can be considered as a pilot project for similar WtE projects in the region.

4.1 Incineration Technology

Modern mass burn waste incineration technology was chosen for the WtE unit as the most reliable, economically feasible and proven technology for commercial use. This type of incineration technology is flexible enough to burn different waste streams without pre-treatment and for producing heat and electricity at a price that is acceptable and competitive in the market. The mass burn technology is the most common commercially used waste incineration solution in the neighbouring Scandinavian countries and there are over 400 similar plants around Europe [9, 27]. A WtE unit that runs on technology similar to that of Iru will soon be completed in the Finnish capital Helsinki.

The available mixed municipal waste with the average calorific value of 10.5 MJ/kg and ranging from 8 to 15 MJ/kg will be incinerated on a modern air-cooled moving grate. At a waste throughput of 27.5 tons per hour (220,000 t/a) it converts about 82% of the energy in the waste into electricity and heat. The thermal energy capacity of the WtE plant will be 50 MW, while the electricity generating capacity is planned to be 17.3 MW. This should complement and partly replace the existing capacities of the Iru CHP plant.

According to the new Waste Framework Directive energy efficiency calculation formula R_1 the energy efficiency of the Iru WtE unit is as high as 1.28. As such the WtE unit complies with the energy efficiency criteria of the directive and can be considered as a R_1 recovery operation of waste (Table 3).

The main components of the energy block are as follows:

- Waste receiving hall and storage bunker
- Automatic feeding and mixing system (crane and waste feed hopper)

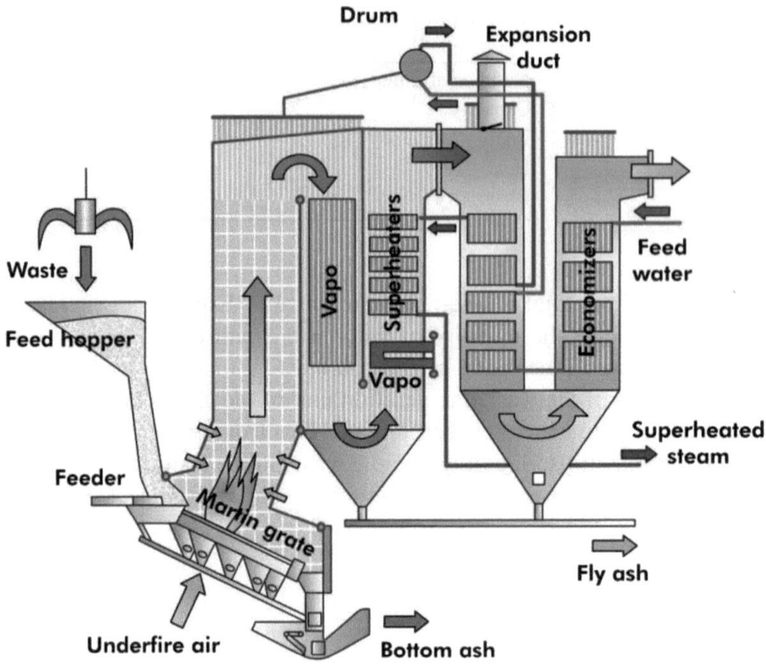

Fig. 4 The main components of the energy unit of the Iru WtE facility [35]

- Combustion unit (MARTIN type reverse-acting grate)
- Vertical heat recovery boiler (CNIM)
- Steam turbine and generator
- Pollution control system including flu gas treatment plant (CNIM/LAB semidry system) (Fig. 4).

The MARTIN reverse-acting grate is the key component of the combustion system. It consists of several parallel runs inclined at an angle of 26°. Each grate run has its own drive and feeding device. Grate bars made from a wear and temperature-resistant chromium-steel alloy are assembled to form grate steps. Alternating fixed and moving grate steps make up a grate run.

The reverse-acting movement ensures that the grate surface is always covered by a protective layer of waste or ash. Thermal wear due to heat irradiation from the furnace does not occur, and consequently grate bar life times are long. Water cooling is not needed, even with very high waste heating values.

The combustion air is divided into underfire air, which passes the grate surface, and overfire air, which is injected into the furnace above the grate. Each grate run is divided into several under grate air zones. The underfire air is distributed as needed locally—it is not required to cool the grate bars.

There is a defined pressure drop as the underfire air passes through the grate surface, providing a uniform distribution within each zone. The air gaps between the grate bars are constantly kept clean by means of a relative movement of the grate bars with regard to their adjacent bars at the end of each stroke.

The heat released from the combustion of waste is recovered in a water tube boiler which forms an integrated unit with the grate. The boiler is of the vertical type, top-supported, and includes one steam drum and five vertical passes. The superheated steam produced by the boiler feeds a turbo-generator set of the back pressure type. Three steam bleeds are provided to ensure both the feeding of the district heating network and the feeding of internal consumers (air pre-heater, deaerator, etc.)

Many of the fittings of the new unit will be installed in the existing plant. The unit will use the 202.4 m chimney of the Iru plant. The turbine and the generator will be installed in the existing building and the existing office buildings and other auxiliary rooms will be used as well.

The pollution control system of this incinerator includes a flue gas cleaning process, a wastewater treatment unit, an odour and noise control system and an ash management system.

Waste management will be conducted in sealed rooms to prevent the spread of offensive odours beyond the unit. Trucks will drive into the building and dump their loads directly into a deep bunker. The air needed for combustion of the waste will be drawn from the waste unloading room and the waste bunker. This will ensure that there will constantly be a low air pressure in these rooms and that an inward draught will be created when the door is opened.

The flue gas cleaning process comprises an active carbon and semi-dry lime scrubbing process followed by particle removal in a fabric filter followed by a two-stage wet scrubbing process. The waste scrubbing process will remove a vast majority of HF, HCl, SO_2 and Hg left from the semidry stage. In order to avoid wastewater from the flue gas cleaning process, the small amount of wastewater from the wet process is evaporated in the boiler.

Reduction of dioxin takes place by adding activated carbon to the flue gas prior to the fabric filter, where dioxin and activated carbon are collected together with fly ash and FGT-residues.

Reduction of NO_x from the combustion process will take place in a selective non-catalytic reduction (SNCR) process by injecting ammonia water (NH_4OH) into the first pass of the boiler, thus securing compliance with the Waste Incineration Directive 2000/76/EEC.

Since the facility will utilise a semi-dry flue gas cleaning system, it is designed with zero wastewater discharge. This is accomplished via reuse of wastewater produced by the facility.

Separate systems will be implemented for the drainage, treatment and discharge of rainwater, including roof water, so that it does not mix with the potentially or actually contaminated wastewater streams. Surplus rainwater which cannot be stored on site, will be discharged to the public sewer.

The facility will utilise water from the Pirita River and to a lesser extent ground water. Under normal operating conditions, the water consumption is approximately 6.5 m^3/hour.

The main solid waste streams generated by the WtE unit are bottom ash, fly ash and flue gas treatment residue.

Bottom ash constitutes the largest percentage of solid waste resulting from the combustion process. After burnout of the waste at the end of the grate, the combustion bottom ash falls down the bottom ash chute into the water bath of the wet ash extractor. Bottom ash consists of inert materials from the combustion process such as glass, metal, earth and other fractions. The bottom ash is magnetically screened to recover ferrous metal. Separated metallic scrap will be sent to recycling. Ash is stored in a separate bottom ash bunker with sealed surfaces. The bottom ash bunker offers a temporary storage capacity of approximately 900 tonnes. This is equivalent to the amount of bottom ash produced over a period of approximately 4 days. It is planned to dispose of bottom ash at a landfill. Also possibilities to use it as a construction aggregate substitute are studied.

The flue gas treatment residue containing fly ash, calcium-based salts, lime and activated carbon (or coke) is collected in the hopper(s) of fabric filters. The flue gas treatment residue is transported pneumatically to two fully enclosed silos/steel tanks. It is planned to send the flue gas treatment residue either to a special hazardous waste landfill or to Germany for filling up old coal mines.

4.2 Economic Aspects

The economic cost of a WtE facility is influenced by local circumstances related to the size and design of the factory, legal aspects, labour cost, the cost of consumables, potential for heat utilisation, market price for energy, etc.

The total investment costs of the Iru WtE unit are approximately 98 million euros. Although the technology supplier carries out all the engineering, procurement and construction works (EPC contract) providing a fully-equipped facility ready for operation ("turn of the key"), the capital costs are relatively low—445 euros per tonne of installed capacity. Based on the experience of other similar WtE facilities the capital costs are generally 600–900 per tonne of installed capacity [17, 30]. The low investment costs have been attained mainly by integrating the new energy unit tightly with the existing power plant. Capital costs lower than that have been achieved in Europe only by very experienced energy companies who already own or operate several plants and have the in-house competence for construction and management of waste incineration plants.

The capital costs can be split into different components. In Table 4 the total capital costs are split into four main components. For each main component, the percentage of the total capital costs related to the specific component is shown.

It is difficult to estimate the operational costs of the WtE unit since they depend on several variable cost items such as cost for residue disposal, maintenance, salaries, etc. It could be expected that the operational costs will be about 50–70 euros per tonne of waste. The European average operational costs per tonne fall

Table 4 General distribution of the total capital costs of Iru WtE unit

Component	Percentage of capital costs (%)
Thermal processing equipment and flue gas treatment (grate, boiler, etc.)	60
Energy production equipment and electrification (steam turbine, generator transformers, etc.)	15
Civil works	10–15
Miscellaneous	10–15

Table 5 General distribution of the operational costs of Iru WtE unit

Component	Percentage of operational costs (%)
Labour and consumables	25–35
Maintenance	20–30
Residues (management and disposal)	40–50

into the same range (45–70 euros per tonne of installed capacity) [17, 30]. The operational costs can be split into different components as indicated in Table 5.

Sale of energy is a significant element in the economy of waste incineration. The potential energy production and income from energy sale depend heavily on waste composition (calorific value), the potential for heat utilisation and market price for energy (heat and energy).

The average market price for electricity (Nord Pool) was about 45 euros/MWh in 2010 and the competitive heat price was about 30–35 euros/MWh. Additionally, there is a subsidy for electricity produced in CHP (32 euros/MWh).

The estimated income from energy sale covers up to 80% of the total costs. In Europe, the average is about 40% [26].

Net treatment costs can be calculated based on the estimates of costs and potential income from sale of energy. Using the costs presented above a rough estimation of the net costs of waste incineration shows that the gate fee for MSW treatment at the Iru WtE unit is approximately the same as the current average landfill gate fee in Estonia (45 euros/tonne). In the case of higher energy prices in the future and other favourable conditions the gate fee for MSW incineration could be even lower (ca 20%). However, if one or more of the critical preconditions fail (especially waste supply, calorific value of waste or energy prices), the actual net treatment costs may be severely influenced.

4.3 Environmental Impacts

A number of environmental impacts are linked to the incineration of waste. In addition to site- specific impacts that are studied usually in the frame of EIA, indirect impacts/emissions should be taken into account when assessing the net impacts of WtE facilities. As incineration of municipal waste should fit into the overall waste management system of the region, it should be compared with alternative waste management options. Life cycle based environmental assessment methodologies can help to identify an overall, optimal environmental solution for managing MSW, without risking that the decision (e.g. to build an incineration plant) will result in a more negative overall impact [4, 33, 34].

Since the Iru WtE unit will be built as an extension to the already existing plant, the local site-specific environmental impacts will be relatively small. The summary of an EIA report indicates that the potential increase in direct air emissions is 0.01–1%. However, the indirect atmospheric emissions will notably fall. This is because municipal waste will replace the current fossil fuels such as natural gas and oil. Waste incineration with energy production will partly offset the emissions that occur when energy (both electricity and heat) is produced from fossil fuels. This is especially important concerning the climate change impact in terms of greenhouse gas emissions.

Large scale municipal waste incineration has to be discussed within the context of the overall waste management strategy. As part of the environmental impact assessment an additional Life Cycle Assessment study was carried out [20]. The aim of this study was to evaluate on how MSW incineration in the Iru WtE unit will influence the life-cycle based environmental impacts of the Estonian municipal waste management system. The incineration-based scenario was compared with an alternative scenario where legal waste management targets in Estonia are achieved by intensive material and biological recycling of municipal waste. For the environmental impact assessment the waste management situation in 2000 was taken as a starting point or a base scenario. The LCA model for waste management planning WAMPS was applied for assessing the environmental impacts of the studied waste management scenarios [17]. In this paper the results of the LCA study are expressed in terms of a climate change impact (GHG emissions) as the most important global environmental impact category of waste management.

Both studied scenarios are in compliance with the legal requirements and recycling targets of the relevant EU directives. It was assumed that waste composition remains the same during the studied period (2000–2020). It was also assumed that after the economic crisis the amount of MSW will continue to increase. For both scenarios it is also assumed that all landfills in Estonia will be equipped with a landfill gas collection system by 2010 at the latest and the landfill gas recovery rate will increase by up to 50% by 2020. Before 2010 the collected gas is flared and after 2010 it is used for electricity and heat production, which is substituting oil shale based electricity and natural gas based heat used for district heating. The energy produced in waste incineration will replace the electricity produced from oil shale that has a very high climate change impact in terms of CO_2 emissions, and heat from natural gas.

Base scenario (scenario 0)—In 2000, waste management in Estonia primarily involved landfilling of MSW (92% of the total MSW). There was no landfill gas collection in landfills at that time. Only a small amount of packaging waste

Table 6 Municipal Solid Waste management scenarios

Scenario	Material recycling (%)	Biological recycling (composting) (%)	Incineration (%)	Rest waste (landfilling) (%)
2000 Base scenario	4	4	0	92
2020 Scenario 1	27	15	45	13
2020 Scenario 2	27	37	0	36

(mainly PET-bottles and cardboard) was collected separately and sent to recycling. There was no centralised collection system for biodegradable waste. Approximately 17,000 tonnes of biodegradable waste (mainly garden waste) were composted by the households (4% of the total MSW). It is assumed that the share of home composting will remain the same till 2020.

Material recycling with intensive incineration (scenario 1)—This scenario is a projection for 2020, where the dominant option of MSW management in Estonia is incineration. 45% of total MSW generated in Estonia is incinerated in the Iru WtE unit. A large amount of the generated heat could be utilised since Tallinn has large dwelling areas with district heating system. In this scenario increased amounts of recyclable materials (mainly packaging, paper, cardboard and metals) are collected separately and recycled to meet the recycling targets of the EU Packaging Directive. About 30% of the waste material is expected to be recycled. As incineration is already contributing to the reduction of biodegradable waste, the share of biological recycling is not expected to exceed 15% of the total MSW. Centrally collected kitchen waste is composted by using the static composting method with forced aeration. Collected garden waste is composted in open windrows. Intensive material recycling and incineration leads to a relatively small amount of rest waste, which is landfilled (13% of the total MSW).

Material recycling with biological recycling by composting (scenario 2)—This scenario is a projection for 2020, where the legal targets are achieved by material and biological recycling. Also in this scenario material recycling is expected to amount to up to 30% of the total MSW. The Landfill Directive requirement to divert biodegradable waste away from landfilling is met by increasing composting to 37% of the total MSW. An increased amount of wet biodegradable waste is composted by using the centralised reactor-composting method (without gas collection and energy recovery). It is assumed that the remaining waste will be deposited in a landfill (Table 6).

The results of the scenario analysis regarding net GHG emissions are shown in Fig. 5. The diagram shows net GHG emissions from the waste management system minus saved emissions in the background system. When the emissions from the studied waste management scenario or waste management practice are lower than the saved emissions in the background system then net result is negative.

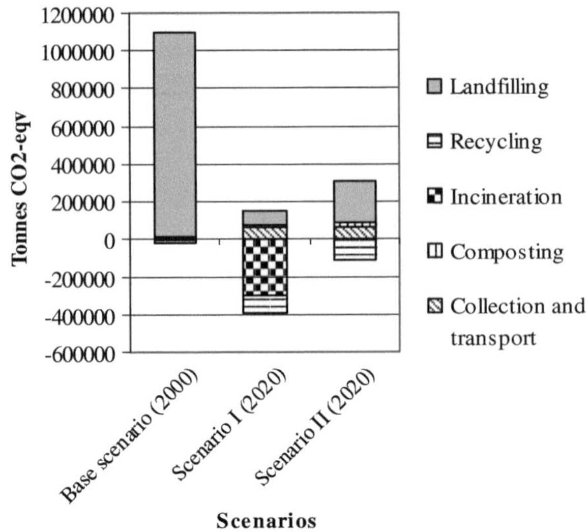

Fig. 5 Emissions of net GHG from the studied waste management practices and scenarios, 2000–2020 (CO_2-equivalents, tonnes)

When comparing the two scenarios we can see that the incineration scenario (scenario 1) has a higher climate protection potential than the alternative scenario (scenario 2). In the incineration scenario where high rates of recycling and incineration with energy recovery are attained, the net emissions of CO_2-equivalents are even negative. The reason for the negative net GHG emissions is a relatively low amount of waste sent to landfills as well as a high share of material recycling (avoided primary production of materials) and the recovered energy in Iru WtE unit (avoided emissions as a result of replacing heat and electricity produced from natural gas and oil shale in the background system). In Estonia electricity produced from waste replaces oil shale based electricity which has a high climate change impact in terms of CO_2 emissions.

Incineration gives approx. 75% and recycling almost 25% of the total avoided emissions. In scenario 2 GHG savings are attained mainly due to material recycling and the avoided emissions from landfilling. As in this scenario composting without energy recovery is applied, the net GHG emissions are higher than in the incineration scenario.

Direct emissions from landfills continue to be a major source of GHG emissions till 2020 despite of the fact that the landfilling rate will significantly decrease and a relatively high share of landfill gas is recovered in both studied scenarios. GHG emissions from waste collection and transport will increase until 2020 due to increased recycling. In scenario 2 a higher collection rate of biodegradable waste causes slightly more emissions of CO_2-equivalents. However, collection and transport of waste accounts for a relatively small amount of the estimated net GHG emissions in both future scenarios.

5 Conclusions

The three Baltic States are in the stage of changing their municipal waste management systems since they have to comply with the principles and targets of the European waste policy and directives. Over the past years, thermal treatment of MSW has been discussed more intensely in these countries as one of the waste management options that could help to reach the legal targets in a relatively short time.

In general, the Baltic States have similar socio-economic characteristics, waste and energy sector developments and geographical conditions that form similar frameworks for the development of a waste management infrastructure, including possible WtE options. However, as experience from feasibility studies and the first WtE projects shows, there are several local and regional factors that could significantly influence the economic success of waste incineration. Since the service life of a waste incinerator is usually over 30 years, the possible future trends and developments of these influencing factors have to be carefully studied before a decision for waste incineration is made.

The legal framework on waste management in the three Baltic States is based on the same general requirements. However, it is important how the policy implementation is facilitated. Estonian experience shows that a relatively high landfill tax together with a ban for landfilling of unsorted MSW have resulted in favourable conditions for the development of new recovery facilities, including waste incineration. Calculations of the economic costs of the new Iru WtE unit in Estonia indicate that incineration of waste has already today a competitive advantage in terms of a lower gate fee compared to landfilling and other new mixed waste recovery options such us MBT.

Investments in waste incineration presume a steady fixed stream of waste with high calorific value, to ensure financial viability. The results of recent sorting studies show that mixed municipal waste in the Baltic States has a relatively high calorific value due to the high share of combustible materials. However, municipal waste composition as well as the amount may change in time because of either additional recycling or changes in the socio-economic situation in the region. This is well illustrated by the impact of the recent economic crisis that significantly reduced the mixed municipal waste generation rate and made earlier waste generation forecasts questionable. This has lead to the situation that there might be not enough mixed municipal waste for all the planned waste recovery facilities. For example, the planned Iru WtE unit will treat most of the mixed municipal waste that today is landfilled (approximately half of the total municipal waste generated per year). As such it will significantly influence the economic performance of all other municipal waste management options. However, conventional mass-burn incineration plants have a competitor in the form of MBT facilities, because they compete for the same mixed municipal waste that is available in the region.

In this new situation with a more liberal waste market and the ever increasing commercial competition between different recovery facilities, the ability and willingness of public authorities to regulate and control the waste sector plays a crucial role. In Estonia, local municipalities have less control over the waste market than in Latvia and Lithuania where municipalities have formed regional waste management centres that participate directly in MSW treatment services and infrastructure projects.

Due to the EU energy policy goals the Baltic energy markets are undergoing a transition towards new sources of energy. Municipal waste contains a large amount of biological and renewable materials, and is therefore a promising source of renewable energy. As a consequence, WtE option is becoming more interesting as a potential contributor to energy security and diversification and matches the growing demand for renewable energy. All three Baltic States have introduced policy instruments and support schemes to promote renewable energy generation. Another favouring factor is the high potential for heat utilisation of WtE facilities due to rather cold climate and existing big district heating networks in larger cities. As the studied Iru WtE unit show, the relatively high value of district heat and support schemes for RES make the average energy revenues much higher in the Baltic States than in many other European countries.

Public perception is an important factor that should be taken into account when developing WtE facilities. The experience in the Baltic States shows that distrust of decision makers and developers may lead to a situation where new waste management facilities of any type are not welcomed by the residents close to where the facility is to be located. Waste incinerators are still generally perceived as great polluters and it is very difficult to convince the public that health and environmental risks are under control. Therefore, it is important to well plan and conduct the public involvement process and to communicate the technology related, as well as the environmental and health impacts, in a trustworthy and detailed manner.

From the decision-makers perspective, when developing sustainable waste management plans at the national or regional level, it is also important to take into account the full life-cycle and the related environmental and economic benefits/trade-offs associated with alternative waste management options for achieving the targets. As the results of the LCA study of Iru WtE unit show, waste incineration with energy recovery can partly offset the emissions that occurred when energy was produced from fossil fuels. This is an important aspect in the context of climate change, since the oil shale combustion technology currently used in Estonia to generate electricity has a very high climate change impact. This means that the municipal waste management scenario where a high share of recyclable waste fractions are sent to material recycling and the maximum amount of rest waste is incinerated with energy recovery should be preferred from the environmental impact point of view.

In general, it can be concluded that thermal treatment of MSW in large scale WtE facilities has a relatively good outlook in the Baltic States. Although the initial investments are relatively high, the favourable conditions in the energy sector allow the WtE facilities to treat municipal waste at a relatively low cost. Therefore, it can be expected that WtE provides an environmentally and economically efficient way to meet the stringent EU waste management targets. However, large scale municipal waste incineration has to be discussed within the context of an overall waste management strategy, rather than as a single option.

References

1. Autret E, Berthier F, Luszezanec A, Nicolas F (2007) Incineration of municipal and assimilated wastes in France: assessment of latest energy and material recovery performances. J Hazard Mater B139:569–574
2. Bulkeley H, Askins K (2009) Waste interfaces: biodegradable waste municipal policy and everyday practice. Geog J 75(4):251–260
3. CEC (2008) Communication from the commission to the European parliament, the council, the European economic and social committee and committee of the regions. COM (2008) 30 final
4. Damgaard A, Riber C, Fruergaard T, Hulgaard T, Christensen TH (2010) Life-cycle-assessment of the historical development of air pollution control and energy recovery in waste incineration. Waste Manage 30:1244–1250
5. EC (1994) European Parliament and Council Directive 1994/62/EC of 20 December 1994 on packaging and packaging waste. Ofiicial Journal, L 365
6. EC (1999) Council Directive 1999/31/EC of 26 April 1999 on the landfill of waste. Ofiicial Journal, L 182
7. EC (2000) Directive 2000/76/EC of the European Parliamnet and of the Council of 4 December 2000 on the incineration of waste. Official Journal, L 332/91
8. EC (2008) Directive 2008/98/EC of the European Parliament and the Council of 19 November 2008 on waste and repealing certain Directives. Official Journal, L 312/3
9. Ecoprog (2010) Waste to energy. The Worldwide Market for Waste Incineration Plants 2010/2011. http://www.ecoprog.com/en/pdf/studies/studie_waste_incineration.pdf. Accessed 20 Jan 2011
10. EEA (2007) Greenhouse gas emission trends and projections in Europe 2007. European Environment Agency, EEA Report No 5, Copenhagen
11. EEA (2009) Diverting waste from landfill—effectiveness of waste policies in the European Union. European Environment Agency, EEA Report No7, Copenhagen
12. ETC/RWM (2007) Environmental outlooks: municipal wastes. Working paper 2007/1, Copenhagen. http://waste.eionet.europa.eu/publications. Accessed 20 Jan 2011
13. Gohlke O (2009) Efficiency of energy recovery from municipal solid waste and the resultant effect on the greenhouse gas balance. Waste Manage Res 27:894–906
14. Grosso M, Motta A, Rigamonti L (2010) Efficiency of energy recovery from waste incineration, in the light of the new waste framework directive. Waste Manage 30:1238–1243
15. Ministry of Foreign Affairs of Denmark The Trade council Baltics (2008) Market opportunities, Sector analyses. Energy and Environment. http://tradecouncil.baltics.um.dk/en/menu/MarketOpportunities/SectorAnalyses/EnergyAndEnvironment/?printmode=True. Accessed 20 Jan 2011
16. Moora H (2007) Installation of waste incineration unit in Iru power plant—Life cycle based environmental and economic assessment. SEI-Tallinn. (In Estonian)
17. Moora H (2009) Life cycle assessment as a decision support tool for system optimisation—the case of waste management in Estonia. PhD thesis, Tallinn University of Technology, Tallinn, Estonia
18. Moora H, Jürmann P (2008) The analysis of the quantity and composition of mixed municipal solid waste in Estonia—Municipal solid waste composition study. SEI-Tallinn (In Estonian)
19. Moora H, Lahtvee V (2009) Electricity scenarios for the Baltic States and marginal energy technology in life cycle assessments—a case study of energy production from municipal waste incineration. Oil shale 26(3):331–346

20. Moora H, Voronova V, Reihan A (2009) The impact of municipal solid waste management on greenhouse gas emissions in Estonia. Interdisciplinary Aspects of Climate Change. Frankfurt am Main: Peter Lang Publishers House, pp 311–325
21. Morrissey A, Phillips P (2007) Biodegradable municipal waste (BMW) management strategy in Ireland: a comparison with some key issues in the BMW strategy being adopted in England. Resour Conserv Recycl 49(4):353–371
22. Münster M, Meibom P (2010) Long-term affected energy production of waste to energy technologies identified by use of energy system analysis. Waste Manage 30:2510–2519
23. Papageorgiou A, Perkoulidis G, Karagiannidis A, Kalogirou S (2010) Integrated assessment of a new waste to energy facility in central Greece in the context of regional perspectives. Waste Manage 30:1395–1406
24. Ragosnig AM, Warthe C (2008) Energy efficiency in waste to energy and its relevance with regard to climate control. Waste Manage Res 26:70–77
25. Ramboll (2006) Waste to Energy in Denmark. PE Offset prints. http://viewer.zmags.com/showmag.php?mid=wsdps. Accessed 20 Jan 2011
26. Rand T, Haukohl J, Marxen U (2000) Municipal Solid Waste Incineration—Requirments for a Successful Project. World Bank Technical Paper No. 462
27. Stantec (2010) Waste to energy. A technical Review of Municipal Solid Waste Thermal Treatment Practices. Final Report No 1231–10166
28. Sundqvist JO (1999) Life cycle assessments and solid waste—Guidelines for solid waste treatment and disposal in LCA. Stockholm, AFR—Report 279
29. Taseli B (2007) The impact of the European landfill directive on waste management strategy and current legislation in Turkey's specially protected areas. Resour Conserv Recycl 52(1):119–135
30. Tsilemou K, Panagiotakopoulos D (2006) Approximate cost functions for solid waste treatment facilities. Waste Manage Res 24:310–322
31. Uselytė R, Moora H (2010) Strategic Environmental Impact Assessment (SEIA) of Klaipeda Region Municipal Waste Management Plan. Ekokonsultacijos, Vilnius, January 2010. Report commissioned by Klaipeda Regional Waste Management Centre
32. Uselytė R, Silvestravičiūtė I (2009) Status of Waste Management in Lithuania. Ekokonsultacijos, Vilnius, November 2009. Report commissioned by the Lithuanian Ministry of Environment
33. Winkler J (2005) Comparative evaluation of life cycle assessment models for solid waste management. Int J Life Cycle Assess 9(6):156–167
34. Winkler J, Bilitewski B (2007) Comparative evaluation of life cycle assessment models for solid waste management. Int J Life Cycle Assess 27:1021–1031
35. CNIM Environment (2011) Technological overview. http://www.energia.ee. Accessed 20 Jan 2011

Waste-to-Energy in Eastern and South Eastern Europe

Saša Malek

Abstract Waste-to-Energy (WtE) incineration is a crucial part of modern waste management, providing safe waste disposal together with electricity and heat production. Future projections show that especially new European Union (EU) member states, who are trying to catch up with the economic growth, can expect further growth in waste amounts in the coming decades. At the same time, these countries are mostly landfilling their wastes, contributing to the worst environmental impacts and greenhouse gas emissions, while the EU is encouraging safe waste disposal and the diverting of waste from landfills. Many are therefore already struggling to live up to the rising environmental standards either because they are bound to the EU legislation or because of the increased environmental awareness of the citizens and of international standards in general. The capacities for Waste-to-Energy (WtE) treatment seem to be already saturated in some of the old member states. This has had a profound effect on the WtE market, which was continuously growing since 1995 but now seems to be losing its expansion pace. All eyes have therefore turned to countries with undercapacities in waste incineration, such as Great Britain, Italy and Spain on one hand, and Eastern and South Eastern European countries [some of them being EU member states already, some not (yet)] on the other. But the path to a second large expansion still has to face some challenges. One of them is overcoming the Not in My Back Yard (NIMBY) effect—a few countries of the former Eastern bloc will need some time to forget the bad memories of past environmental abuse due to the harsh industrialization. They will also need to inform themselves better of new innovative technologies. Many of them also suffer from limited financial means, less stable economies and lack the basic data to establish future plans in the field of waste. Nonetheless, the

S. Malek (✉)
Institute of Resource and Energy Technology, Technische Universität München,
Petersgasse 18, 94315, Straubing, Germany
e-mail: sasa.malek@wzw.tum.de

A. Karagiannidis (ed.), *Waste to Energy*, Green Energy and Technology,
DOI: 10.1007/978-1-4471-2306-4_11,

future role of WtE in Europe, may it be in the East or West, is certain. WtE completes a fully matured waste management system, enabling the waste producer to choose the best possible treatment option. With WtE, waste becomes an important source of renewable energy, mitigating the climate change effects and saving earth's valuable resources and raw materials. The second WtE expansion will therefore be a part of the new era with resource independent and climate neutral societies, leaving the fossil-based economy in the past.

1 Introduction

In spring 2010 several environmental groups decided to issue a public warning against serious and potentially dangerous waste disposal problems in Eastern Europe. The warning was issued in the light of the recently published Eurostat figures, revealing Eastern Europe as a region with the lowest recycling rate in Europe. Environmentalists have therefore decided to point out the most urgent waste management (WM) problems that occur there, such as poor legislation, insufficient infrastructure, lack of environmental awareness outside cities and a shortage of political will to tackle waste management problems with illegal waste dumping, which together with non-existent recycling poses a threat to human health and the environment [86].

Meanwhile, the European Union (EU) is increasingly encouraging environmentally safe waste disposal and the diverting of waste from landfills. More recently life-cycle thinking has been introduced as a guiding principle of resource management by which the strategic goal of moving towards more sustainable patterns of consumption and production has been set with a view to decoupling resource use and waste generation. Nevertheless, waste quantities in Europe are still on the rise [16]. At the same time many of the new member states are mostly over 80% dependant on the worst possible waste management practice—landfilling. Many of the EU-12 are therefore already struggling to live up to the environmental standards of the EU and are confused about finding the right waste management options for their specific regional needs. A comparison across Europe shows that countries with progressive waste management systems in place have both a high proportion of waste incineration as well as high rates of material recycling [92]. Out of the countries that have a high share of landfilling, only France has an incineration treatment share that exceeds 30%. Nonetheless, low shares of material recovery and biological treatment in France are the reason why more than 20% still is landfilled [3]. During the last 10–20 years, several research groups as well as consultants have been analysing the environmental impacts of waste incineration in comparison to other waste treatment options. The review of 38 case studies shows the following [17, 69, 117]:

- Landfilling is the main treatment option in Europe and clearly the worst environmental option, according to the studies and the EU Framework Directive on Waste (2008/98/EC).
- Material recycling, waste incineration and biological treatment are complementary options that need to be expanded in order to replace landfilling.
- Material recycling generally leads to lower environmental impacts than incineration. However, this is valid only if the material recycling is based on well source-separated and clean material fractions. In addition, incineration could lead to lower environmental impacts regarding the global warming potential, photo-oxidants and toxicity when it comes to some paper products. The same is valid for eutrophication and toxicity in the case of plastics.
- Incineration leads to a lower environmental impact than landfilling.
- To reach the best environmental results for material recycling and the biological treatment of organic combustible material, waste incineration is necessary for treating residues arising during pre-treatment and processing at the material recycling facilities and the biological treatment plants.
- Considering the environmental impacts on climate change, many studies show that Waste-to-Energy (WtE) poses a greater benefit than mechanical biological treatment (MBT), while there are only small differences between mono- and co-incineration of refuse derived fuel (RDF). The crucial part is the energy efficiency of the thermal treatment plants. For the non-thermal treatment plants, the amount and quality of the recyclable material output, the energy demand and the emissions are relevant.
- Due to different local conditions and opportunities for development, the distribution of waste treated by material recycling, waste incineration and biological treatment must be allowed to vary.
- Regional differences will lead to different distributions optimal for different regions in Europe.

Implementation of WtE is therefore a sign of a mature WM system that offers best possible waste treatment options taking into account the entire WM hierarchy. Understanding the significant environmental benefits of such WM systems, it is to expect that many European countries with landfill-based WM, are to implement WtE in the near future. This is especially valid for Eastern and Central European countries that are new members of the EU but used to be part of the so called Eastern bloc of communistic countries (such as Estonia, Latvia, Lithuania, Poland, Czech Republic, Slovakia, Hungary, Slovenia, Romania and Bulgaria) and for the South Eastern European countries that used to be part of the Eastern bloc of communistic countries but are either official EU candidates today (such as Croatia, Macedonia and Montenegro) or officially recognised potential EU candidates (such as Albania, Bosnia and Herzegovina and Serbia). Progress towards WtE due to environmental benefits can be seen also in other Eastern European countries that are not part of the EU ascension process (such as Belarus and Ukraine), which indicates that the future development of the WtE market will focus in the East and South-East of Europe.

2 The State of Waste-to-Energy in Europe

Waste-to-energy combustion is an important technology for municipal solid waste (MSW) management that refers to modern treatment practices for waste that cannot be recycled in other ways. It is therefore offering great opportunities for the reduction of volume of waste that would otherwise have to be landfilled, as well as for generating heat and power. It has been used widely in Europe and Japan without any adverse health effects [87]. EU member states can be categorised into three waste management groups, clustered according to their strategies for diverting MSW away from landfill and their relative shares of landfilling, material recovery and incineration. Since only eight of the EU-27 countries maintain high levels (each more than 25%) of both material recovery and incineration and most of the new member states belong to the third waste management group with worst WM practices (less than 25% of both—material recovery and waste incineration), landfilling is still the predominant option in the EU [18] (Fig. 1). But this might change soon. Broadly speaking, waste management has improved in almost all EU Member States in the recent years, as more waste is being recycled and less landfilled. According to Eurostat's statistics in 1995, 62% of MSW was landfilled. In 2009 this has fallen to 38% and is continuing to fall. EU policy instruments, such as the Landfill Directive 1999/31/EC (ban on landfilling of untreated waste) and the Packaging Directive 1994/62/EC (separated waste collection) have been so far mostly effective in new member states in terms of decreasing the waste being landfilled [17]. In addition, the combination of the Renewable Energy Directive 2009/28/EC, that sets targets for electricity supply from renewable sources, and the new Waste Framework Directive 2008/98/EC, that introduces a new waste management hierarchy, according to which waste-to-energy incineration is considered a recovery operation rather than disposal (providing certain efficiency circumstances), could increase WtE.

Currently, there is a huge gap in the WtE development in Europe, resulting in several EU countries already experiencing a hold in the expansion of the WtE due to a possible national overcapacity, while the majority of the EU countries still have under-capacities or no waste incineration at all. The result is a significant decrease in the overall investments into the WtE technology in Europe in the last few years, since the investment into waste incineration in the countries with most developed WM seems to be slowing down.

A study made by the Swiss consultancy Vaccani, Zweig and Associates showed that orders for incineration plants (including upgrades of existing plants) in Europe fell from 29 in 2007 to 14 in 2008 and finally to only 9 in 2009 [31]. Between 1995 and 2001 the first period of continuous growth of the WtE market in Europe took place, when orders accumulated up to a total of 74,629 tons per day (tpd), which corresponds with the annual average growth of about 10,661 tpd. WtE orders grew also in the second period 2000–2010, where they achieved an annual average of 11,740 tpd, reaching its peak in 2005 with an annual average of 16,389 tpd in 2005. The ordered capacities in the two following years still achieved above

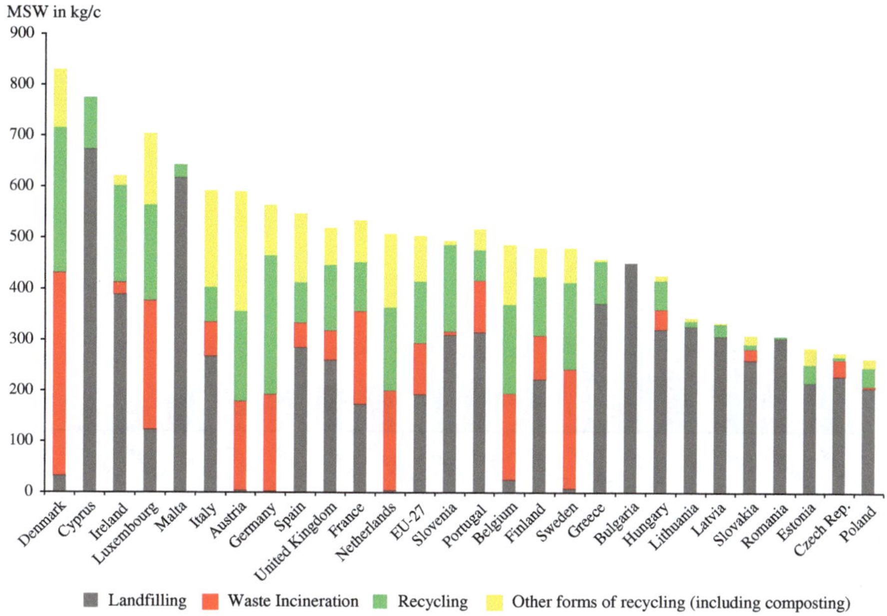

Fig. 1 MSW treatment (in kg per capita) in EU-27 in 2009 [30]

average annual order volumes (15,884 tpd in 2006 and 15,084 tpd in 2007). Finally, the WtE expansion phase was followed by a major decline in orders; when average annual WtE orders fell to 8,627 tpd in 2008 and to 8,756 tpd in 2009 [32].

Another report made by the Swedish Waste Management Association [3] considers the expansion of waste incineration between the years 1997 and 2005. In this period, the total amount of MSW treated by incineration increased from 36 to 49 million tons. The largest increase took place in Germany (4 million tons) and Italy (2 million tons). Considering the amount incinerated per capita, Denmark and Switzerland have been the leading countries during the entire period. During 1997–2005 Portugal and the Czech Republic (re)opened their first incineration plants. A substantial expansion of waste incineration capacity, in relation to the number of inhabitants, has also taken place in Denmark, Portugal and Austria.

Taking into consideration an average recycling rate of 42% for household and commercial waste in Western Europe (which includes countries mentioned in Fig. 2) or a minimum 300 kg per capita rate for residual waste that remains as not suitable for recycling, [95] estimated that about 134 million tons out of a total of 230.8 million tons of household and commercial waste generated in Western Europe would currently be suitable for waste incineration. However, six Western European countries have already reached overcapacities in WtE. These are: to a greater extent Luxembourg, Sweden, Denmark and the Netherlands. Additional countries with overcapacities or with the possibility to reach overcapacities soon are Germany,

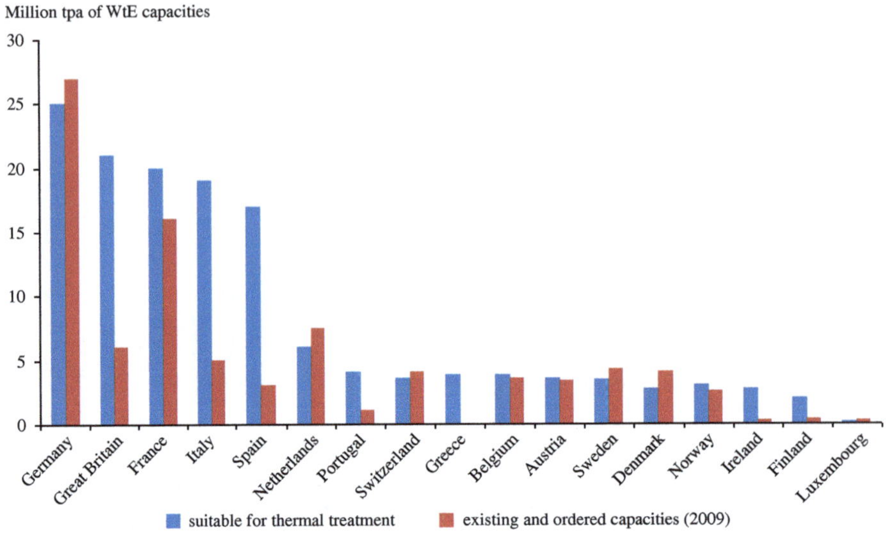

Fig. 2 There are currently about 500 waste incineration and RDF plants in Western Europe [95]

Switzerland, Belgium and Austria. Figure 2 shows which countries already have more existing and ordered capacities than waste suitable for thermal treatment.

On the other hand, Western Europe as a whole still has about 50 million tons of waste suitable for incineration that is not being incinerated yet, so its' WtE capacities are used somewhat over 60%, which is especially valid for countries such as Great Britain, Spain and Italy [95]. This has been helping the WtE market to recover again as of 2010, where the WtE orders grew again up to an annual average of 10,264 tpd. Amongst Western European countries that are not experiencing a saturation of the WtE market yet, most prospective seem to be Great Britain, where many developers (operators) are present [34]. The latest list of Private finance initiative funded projects [12] by the United Kingdom's Government Department for Environment, Food and Rural Affairs (DEFRA) contains 32 projects of which 22 include WtE and RDF plants, while the UK Without Incineration Network is citing around 80 potential locations for WtE [91]. In 2010, Great Britain was the third year in a row the biggest and most dynamic market of WtE in Europe, followed by Italy, whose WtE orders in 2010 represented the biggest share or 23% of the entire ordered volume in Europe that year (2,360 tpd of 10,264 tpd), making the country the principal market for WtE plants in Europe in 2010 [32]. An expansion of WtE is also expected in Eastern Europe, which includes countries listed in Fig. 3. The trend has been set by many announced plans for WtE orders in the coming years.

Assuming a recycling rate of 38% for the 67.1 million tons of the generated household and commercial waste in Eastern Europe (including Turkey) or a minimum 200 kg/capita rate for residual waste that remains as not suitable for recycling, a total of 42 million tons of waste should be available for waste

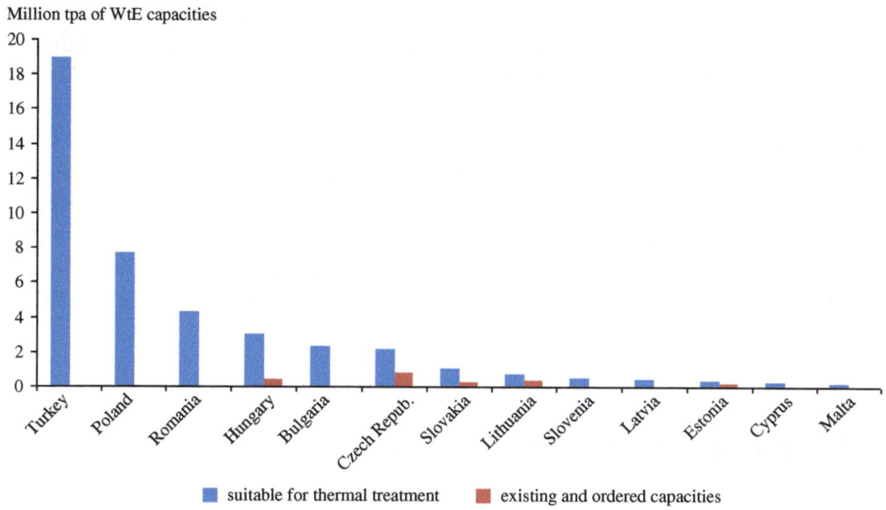

Fig. 3 Less than ten waste incineration plants are currently operating in Eastern and Central Europe [95]

incineration in Eastern Europe. Since the region almost knows no waste incineration, it could represent a big market potential of the next decade. The currently existent and ordered future WtE incineration capacities sum up to 2.1 million tons per year, which represents about 5.5% of the total possible WtE capacity [95].

According to the estimations of the Waste-to-Energy Research and Technology Council Germany (WtERT Germany), more than 50 WtE and RDF projects are in progress and/or planned in the entire Eastern Europe, taking into consideration 13 countries (Poland, Hungary, Czech Republic, Slovakia, Slovenia, Romania, Bulgaria, Estonia, Lithuania, Ukraine, Croatia, Serbia and Bosnia and Herzegovina). However, with the exception of the Polish projects and the projects in Lithuania and Estonia, the realisation of planned capacities in other countries is still rather uncertain. Furthermore, some European countries have a policy which limits incineration. In order to avoid the further expansion of waste incineration, the Czech government issued a list of priority funded projects in 2009, according to which waste incineration was not to be financially subsidized [47].

However, the Czech ministry later changed its position on energy recovery from waste, enabling the funding of WtE projects from public sources through the Operational Programme on the Environment [97]. According to an announcement in 2007, France has decided to limit incineration together with landfilling, even though a large share of the country's waste is treated with incineration at the present. Also, at a policy level Scotland has stated that at most 25% of the waste shall be treated by incineration. The present Irish government is also against incineration and considers implementing an incineration tax with the purpose to obstruct a large scale development of incineration and promote composting and small MBT-facilities as an alternative to landfilling. According to their waste plans, Spain and Portugal want to focus on composting in preference to incineration since compost is demanded for soil improvement [3]. It is therefore fair to conclude that, in spite of overall insufficient WtE capacities in Europe, the development of the market is still facing many challenges. Largest investors in WtE so far are dealing with overcapacities on one hand, while the states with undercapacities either struggle with the public acceptance of the technology, the lack of a WM strategy or the lack of investment potential on the other, making it difficult to predict the direction of the growth of WtE as well as its speed. In addition, the predictions about the expansion of WtE are mostly based on the recent national waste production trends. Taking into consideration all possible circumstances that can affect waste generation in the future, such predictions can be full of uncertainties.

3 Waste Generation and Prediction

Each country's waste generation profile varies according to numerous facts, such as economic growth, population density and consumer behaviour, while waste management choices depend on the existing waste management facilities, infrastructure and governance structures [18]. Achieving a more or less accurate prognosis is therefore a difficult task. The EU is on the policy level committed to reducing waste generation, but in reality it is not succeeding. Even though the waste generation has fallen from being around 2.8 billion tons in 2004 to 2.6 billion tons in 2008, there are differences among different waste streams. Recent projections show that electrical and electronic equipment waste (WEEE) seems to be one of the fastest growing waste streams. Also waste generation from construction and demolition has increased, as has packaging waste and hazardous waste. Increasing are also sewage sludge generation and marine litter [17]. The generation of MSW has slightly fallen since 2003 (from 514 to 512 kg/c), but there are substantial differences in waste generation between countries, up to a factor of 39 between the member states, largely due to different industrial and socio-economic structures. Similarly, MSW generation per person varies by a factor 2.6 between countries, amounting to 512 kg per person in 2009 in EU-27. It has increased between 2003 and 2009 in 18 out of 27 EU countries. However, the growth of MSW generation in EU-27 has been slower than that of gross domestic product (GDP), thus achieving a relative decoupling for this waste stream. The growths in waste volumes were driven mainly by household consumption and by an increasing number of households [16]. Only a few countries have made significant progress with stabilising or reducing the amount of waste generated. Those showing a decrease or stabilisation of MSW generation per citizen over the last ten years are Germany, Bulgaria, Czech Republic, Lithuania, Poland, Slovenia and Slovakia, and it is not clear how much of the effect was due to different economic patterns and how much due to underdeveloped waste collection and recording systems [39].

According to the European Topic Centre on Resource and Waste Management (ECT/RWM) MSW generation in the EU is projected to increase by around 25%

by 2020 in comparison to 2005. However, large differences still exist between countries. In the EU-15, the generation of MSW is projected to increase by approx. 22% in 2020 compared to 2005 and by 33% in 2030, by an annual growth rate of 1.4% till 2020 and 1.2% till 2030. Except for a few countries, the projected increase in waste generation is between 22% and 43% [85]. Waste generation in EU-12 is projected to grow faster than in the EU-15. Furthermore, the MSW in the EU-12 is supposed to increase by around 50% in 2020 compared to 2005 and by around 92% in 2030, considering annual growth rates of 3% from 2005 to 2030. Nevertheless, the variations between countries are significant: Slovenia is supposed to have a comparable growth to the EU-15, i.e. from 4.5% (until 2020) to 7.1% (until 2030). Latvia should also experience a considerably lower growth than the rest of EU-12, while Bulgaria is expected to decrease its MSW generation, which corresponds to its national trends since 1998. On the other hand Poland, the Czech Republic, Hungary, Malta and Slovakia all have a projected growth of more than 100% in 2030 compared to 2005 (Table 1) [85].

When comparing the predictions of the ECT/RWM for 2030 with the real waste generation data in EU in 2009 provided by Eurostat (Fig. 4) it becomes evident that the role of the six most populated countries in EU (Germany, France, United Kingdom (UK), Italy, Spain and Poland—70% of EU's population) in contributing the biggest share of MSW (69%) should not change too much. In 2030 their share should rise a bit (to 72%) and UK and Italy should become the second and the third largest waste producer in the EU, overcoming France. Analysing all EU-27 countries individually, the biggest jump in rankings amongst the EU-27 with regards to MSW generation is projected to happen to the Czech Republic. The latter is expected to switch its position from being the 18th biggest MSW producer in 2005 to being the 12th biggest producer in 2030 (6 places), while Hungary is expected to raise its MSW production from place 13 to 9 (4 places) in the EU-27. The biggest fall in rankings is expected to happen to Austria (from place 9 to 15, 6 places). Figure 5 includes the countries that are supposed to change their ranking position with regard to MSW production in 2030 according to ECT/RWM predictions.

However, a comparison of the ECT/RWM waste generation projections for 2010 and the actual MSW generation in 2009 (which represents the closest data to 2010 that is currently available) reveals how vague such projections can be (Table 2).

Generally, the predictions for 2010 are much higher than the results for 2009: they predict an 8.7% raise in MSW in 2010 (in comparison to 2005), while the actual results for 2009 show an overall increase in MSW generation for less than 1% in comparison to 2005. For the EU-12 the indicated possible "mistake" in the estimations seems to be similar: MSW increased for 0.8% in 2009, while according to the projections it was expected to rise for 9.6% more.

Even though we could assume a much less accurate projection for the EU-12, due to the lack of historical statistical data, the inconsistency between the national statistical methods used in the past and in some cases severe regional changes in the national territories of the countries that have emerged in the 1990, this does not seem to be the case. The projections appear to be rather equally overly-pessimistic for the EU-12 and EU-15. The reasons for that are complex and could probably be

Table 1 Projected growth in MSW generation in the EU-27 in 2020 and 2030 in comparison to 2005 [85]

Country	Projected MSW growth in 2020 (%)	Projected MSW growth in 2030 (%)
Germany	15.2	24.4
United Kingdom	27.1	42.9
France	22.7	33.4
Italy	29	42.9
Spain	27	33.7
Poland	66	131.1
Netherlands	3.7	10.1
Romania	56.4	93.3
Austria	12.1	10.5
Belgium	15.1	21.6
Greece	33.1	42.2
Portugal	31.4	58.0
Hungary	62.1	108.4
Sweden	22.3	32.2
Bulgaria	−15.4	−24.3
Denmark	16.4	22.3
Ireland	30.1	38.6
Czech Republic	63.5	108.8
Finland	16.5	24.8
Slovakia	54.3	107.1
Lithuania	31.8	53.6
Slovenia	4.5	7.1
Latvia	18.7	25.2
Estonia	43.7	70.7
Cyprus	45.7	69.6
Luxembourg	72.4	118.9
Malta	63.7	109.5
EU-27	26.48	41.87
EU-15	22.32	33.12
EU-12	50.59	92.49

traced back to various factors. Among them are a possible raise of awareness about waste in the EU, a more and more coherent waste policy across the region, the slow but on-going decoupling of environmental pressures from the economic growth [85] and last but not least the financial crisis that has hit Europe in summer 2007 and has left it in a recession [14].

Noticeably, the predicted MSW growth of Bulgaria is negative, even though its GDP annual growth rate has been between 2005 and 2009 the third highest in the EU. In addition, according to the Eurostat data on MSW generation Estonia has experienced very varying trends in MSW generation since 1998, a condition which has "stabilised" in a negative growth trend since 2005 (with the exception of 2007). Still, the predictions assume a positive MSW growth of about 14% in 2010. On the other hand Hungary's MSW quantities were rising constantly since 2001

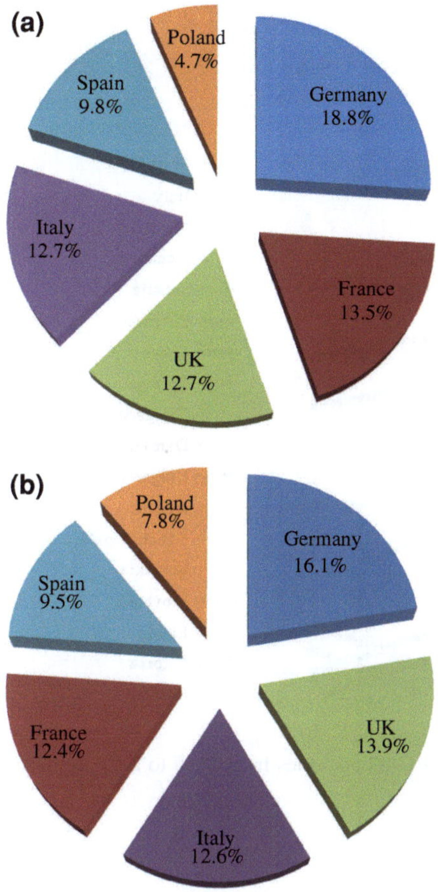

MSW generation in EU in 2009 (in %)

	%
Germany	18.8%
France	13.5%
UK	12.7%
Italy	12.7%
Spain	9.8%
Poland	4.7%
EU-27	100%

MSW generation in EU in 2030 (in %)

Germany	16.1%
UK	13.9%
Italy	12.6%
France	12.4%
Spain	9.5%
Poland	7.8%
EU-27	100%

Fig. 4 MSW production shares of the 6 most populated countries in the EU in 2009 (*upper*) and in 2030 (*lower*) (according to Eurostat data and the ECT/RWM prognoses), in percentage [85]

(with the exception of 2004), but this trend turned around in 2007, when its GDP annual growth fell back.

The predicted "big jumps" in MSW production in countries with no waste incineration, even if they will prove themselves to be correct, do not necessarily mean a naturally prospective market for waste incineration development. They can mean, however, that there will be more pressure for the country to install better waste management in the future in order to improve its environmental conditions. Nonetheless, in a society focused on climate change mitigation, WtE has proved to have significant potential as a renewable source, which might represent a high motivation for the future implementation of this technology.

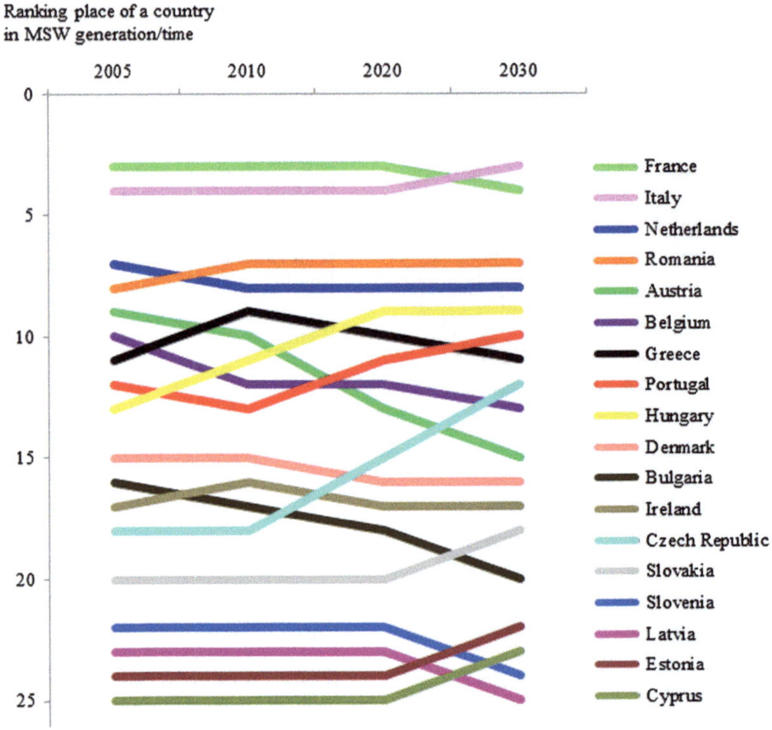

Fig. 5 Predicted changes in rankings of some of EU-27 countries from 2005 to 2030 according to their MSW generation [85]

4 The Role of WtE in Renewable Energy Production and CO_2

As seen previously, Western European countries with WtE overcapacities are not expected to order a lot of new plants in the shortcoming future, but some of them might nevertheless significantly increase their renewable energy contribution from WtE until 2020. According to an analysis made by the Confederation of European Waste-to-Energy Plants (CEWEP) [52] about renewable energy contribution from various waste processing methods (excluding agricultural and industrial food waste and grown biomass) for the years 2006/2007, waste incineration with energy recovery (excluding RDF and co-incineration) provided by far the largest quantity of renewable energy in Europe. The EU-27 renewable energy target to be achieved by 2020 is about 2,700 TWh, thus the gap to be filled (between 2005 and the 2020 binding target) amounts to approximately 1,500 TWh. CEWEP estimates that about 95 TWh of this gap could be provided by energy from waste, using a combination of all possible waste treatment methods (direct incineration with

Table 2 Projected growth in MSW generation in the EU-27 (including Norway and Switzerland) in 2005 and in 2010, compared to actual MSW growth in 2009 [30, 85]

Country	MSW generated in 2005 (in 1000 t)	MSW generated in 2009 (in 1000 t)	Actual growth of MSW in 2009 (compared to 2005) in (%)	Projected growth for 2010 (compared to 2005) in (%)	Difference in (%)
EU-27	253,837	256,017	0.9	8.7	7.8
EU-15	216,417	218,306	0.9	8.4	7.5
EU-12	37,419	37,712	0.8	10.4	9.6
Germany	46,555	48,101	3.3	2.5	-0.8
UK	35,121	32,507	−7.4	8.5	15.9
France	33,366	34,504	3.4	10.6	7.2
Italy	31,664	32,500	2.6	10.4	7.8
Spain	25,683	25,090	−2.3	17.0	19.3
Poland	12,169	12,053	−1.0	7.9	8.9
Netherlands	10,178	10,107	−0.7	−6.0	−5.3
Romania	8,173	8,507	4.1	19.4	15.3
Austria	5,084	4,941	−2.8	7.0	9.8
Belgium	5,024	5,277	5	6.6	1.6
Switzerland	4,940	5,460	10.5	10.3	−0.2
Greece	4,853	5,154	6.2	15.7	9.5
Portugal	4,694	5,185	10.5	7.5	−3.0
Hungary	4,646	4,312	−7.2	15.9	23.1
Sweden	4,347	4,486	3.2	10.3	7.1
Denmark	3,990	4,590	15.0	9.0	−6.0
Bulgaria	3,680	3,561	−3.2	−5.9	−2.7
Ireland	3,041	2,953	−2.9	18.2	21.1
Czech Rep.	2,954	3,310	12.1	13.7	1.6
Finland	2,506	2,562	2.2	6.5	4.3
Norway	1,968	2,269	15.3	8.5	−6.8
Slovakia	1,558	1,745	12.0	4.4	−7.6
Lithuania	1,287	1,206	−6.3	7.4	13.7
Slovenia	845	913	8.0	−1.4	−9.4
Latvia	716	753	5.2	6.8	1.6
Estonia	587	464	−21.0	14.1	35.1
Cyprus	553	620	12.1	13.0	0.9
Luxembourg	313	349	11.5	13.7	2.2
Malta	251	268	6.8	22.0	15.2

energy recovery, landfill gas utilization, RDF co-incineration in both cement kilns and power plants, anaerobic digestion and waste wood incineration). From that WtE incineration could potentially contribute around 60 TWh (4%). Adding this to the amount that WtE was already supplying in 2006 (38 TWh) the total contribution to renewable energy from WtE could potentially reach 98 TWh, which is enough to supply 22.9 million inhabitants with renewable electricity and 12.1 million inhabitants with renewable heat [9].

Converting this into Joules, it would mean that in 2006 WtE supplied 136 PJ and by 2020 this amount can grow to 240 PJ for the whole EU, including Norway and Switzerland. In a more ambitious scenario, where landfill is totally replaced by a combination of recycling and WtE, the total maximum potential contribution by WtE across the EU-27 is 257 PJ of renewable energy, which represents an increase versus the 2006 level for about 1.9 [52]. In fact, WtE plants could provide twice this amount as not all of the energy produced from waste is considered to be renewable [9]. According to the European Directive on Renewable Energy Sources the biodegradable fraction of municipal and industrial waste is considered as biomass, thus a renewable energy source. Currently, across Europe most states recognize energy from waste incineration as renewable energy. The percentage of the energy from WtE that is classified as renewable varies—being the highest in Denmark (80%) and in general 50% in other countries, which is why on average the latter is a good estimate [9]. The projections are that by 2020 the relative contribution of WtE to the renewable energy volume of individual countries will decrease, despite the fact that the WtE renewable energy quantity in absolute terms will increase by 82% for the EU-27. This is due to the fact that the total amount of renewable energy across EU must grow much faster so that all countries will meet the ambitious targets set by the EU Commission per country. The current contribution share from WtE to the total energy consumption of a country is still modest. However, looking at the WtE share of the total renewable energy produced by a country during 2006, it becomes clear that for a number of countries the contribution of WtE to their renewable energy production is very significant: the Netherlands (14.3%), Belgium (13.3%), Denmark (12.5%), and Germany (7.5%). These are predominantly countries that do not have a strong share of hydropower. The countries which are supposed to have in 2020 the highest percentage contribution of WtE to their renewable energy production, assuming they will meet the binding target, are Denmark (6.3–9%), Sweden (4.7–6.7%), the Netherlands (4.4–6.3%), Czech Republic (3.3–4.7%) and Germany (3.0–4.3%) [52].

As seen in Fig. 6, the amount of renewable energy produced from WtE in the EU-27 summed up in 2006 to 124 PJ and is predicted to amount to 226 PJ in 2020. Furthermore, among the EU-12 countries Poland, Czech Republic and Romania are expected to contribute the biggest renewable energy share in 2020 (together 16,6 PJ). On the other hand, Great Britain, Germany, Sweden and the Netherlands are supposed to significantly increase their contribution of renewable energy from WtE as well. Combining this with the assumption that these countries (with the exception of Great Britain) already have an overcapacity for waste incineration and are therefore not expected to order many new WtE capacities in the near future, this could be potentially explained through the fact, that they might most significantly improve their WtE-facilities efficiency for energy recovery. Secondly, the (predicted) rise of the waste quantities suitable for incineration would enable the countries with WtE overcapacities to finally reach their WtE capacities, resulting in a higher energy output from waste. This trend has been proven in Germany, where WtE plant orders have recently dropped most dramatically in

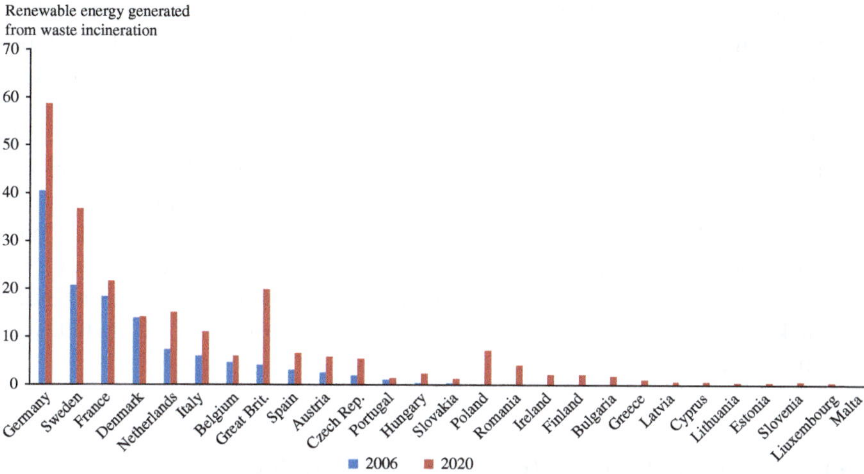

Fig. 6 Renewable energy generation of WtE plants in EU-27 in 2006 and 2020 in PJ [52]

Europe [between 2007 and 2010 the ordered capacities have dropped from 3,928 tpd (2007), to 2,044 tpd (2008), 761 tpd (2009) and finally to only 605 tpd in 2010], [31, 32] but the country still managed to produce more energy from waste between 2005 and 2009. This was achieved through a rise in the efficiency of the plants in the first place, for which a more efficient and thus higher electricity production was most important. Germany managed to increase its absolute electricity production from waste by 39% (from 5.51 million MWh in 2005 to 7.67 million MWh in 2009). In addition, the amount of electricity that actually entered the grid rose for 45% (from 3.95 million MWh in 2005 to 5.72 million MWh in 2009). The increase of heat production played a less important role, due to the fact that the electricity production is less bound to the location of the plant, while heat production depends on season and is restricted to sites with the need for regular heat utilisation. Thus, heat recovery from waste increased for 7% (from 13.2 million MWh in 2005 to 14.2 million MWh in 2009). Moreover, the amount of waste incinerated in Germany rose for about 3 million tons (from 15.9 million tons in 2005 to 19.07 million tons in 2009) [33].

Improving the energy efficiency of existing less performing installations and building of new efficient WtE plants in Europe to treat currently landfilled waste could potentially reduce greenhouse gas (GHG) emissions by 2020 by approx. 45 million tons of CO_2–equivalents. This represents the annual CO_2 emission of 22 million cars. WtE offers therefore a significant potential to contribute to the reduction of CO_2 emissions [115]. Next to energy recovery, WtE reduces GHG also through metal recovery from bottom ashes [43].

The EU does not have a common Kyoto target. However, the member states have agreed to make a 20% reduction in their GHG emissions compared to 1990 by 2020. This would mean that the EU-27 would have to reduce its greenhouse gas emissions by a total of 1,112 million tons CO_2–equivalents compared with 1990.

If the quantity of 511 million tons CO_2–equivalents already saved in 2007 is deducted from this, the remaining reduction required by 2020 is a further 600 million tons CO_2–equivalents. Of this, the waste sector in the EU-27 could contribute between 19% (with a saving of 112 million tons CO_2–equivalents per year) and 32% (with a saving of 192 million tons CO_2–equivalents per year) [11]. Providing enough renewable energy to supply several dozen millions of European citizens and preventing the CO_2 emissions of several dozen million cars will clearly benefit the European environment, making WtE as one of the most important renewable energy sources of the future. Some of the Eastern European countries are therefore already in the process of abandoning the landfilling based WM, making plans for more sustainable treatment of waste, which includes the production of energy.

5 The Status of WtE in Individual Countries of South Eastern and Eastern Europe

Municipal WM in Central and Eastern Europe is to a large extent dependant on landfills with low technical standards and plagued by illegal dumping. In the recent decades the amounts of household waste to be disposed increased rapidly in all countries independent of their different development level. However, the new waste amounts burden the environment, putting the countries under pressure to change their waste management ways. The collection of waste is mostly well organized, but significant problems exist in the separate waste collection. Consequently, the mixed waste is regularly collected and landfilled in more or less suitably equipped landfills in the immediate vicinity of the settlements. Waste collection fees are usually difficult to obtain and if so, they mostly remain on the level of symbolic financial contributions. The available funding is therefore often not sufficient to build a modern waste collection, recycling and recovery infrastructure. At the same time waste deposition on landfills is often inexpensive and the polluter pays principle not installed. In addition, informal systems of garbage collectors exist, which concentrate mainly on economically interesting waste fractions, such as metals, glass, paper and packaging. The local governments also lack technical and methodological knowledge for waste handling as well as waste related data needed for prognostic assessments [116].

Even worse is the situation in South-Eastern Europe (SEE), where the waste management development is driven by the EU accession process. The waste collection, transportation, treatment, disposal, and the available WM infrastructure are insufficient, do not meet the EU standards and contribute consistently to the pollution of air, water and land resources. The waste producers are mostly unfamiliar with the techniques and benefits of waste prevention and are often not having to take the financial responsibility according to the polluter pays principle. That is why the region is in an urgent need for national initiatives for the minimization of waste on the municipal and the industrial level. Resources, including waste, are not used efficiently, while the national strategic plans as well as the laws mostly prove to be ineffective. There is no clear strategic basis for determining the investment priorities in the waste sector, nor any performance requirements or specifications defined, making it difficult to obtain the required waste management standards. Furthermore, there are no reliable waste data. Therefore, investments into the waste management infrastructure in the SEE region are crucial to reduce environmental burdens [4].

5.1 The Baltic States

Among the Baltic States (Lithuania, Latvia and Estonia) Lithuania and Latvia seem to have similar problems in the WM. Both were forced to close the EU-incompliant former Soviet dumps and both struggle with the implementation of separate waste collection. Until 2006 Latvia reduced its dumpsites from 558 to 99 and 270 dumps have been recultivated. Eight landfills compliant with the EU legislation were already operating in 2008 (Ziemeļvidzeme, Ventspils, Liepāja, Pierīga, Zemgale, Maliena, Dienvidlatgale and Austrumlatgale) [49].

Until the end of 2011 Lithuania had to close 800 dumps, from which only 35 were larger than 5 ha, and all of them were unequipped and poorly designed. For the closure of old dumpsites and the opening of new ones 155 million € from the state budget and the EU were used in the period 2000–2006. In the second stage (2007–2013) of the country's WM renewal 130 million € were assigned and already distributed (in 2010) to close the remaining dumps and to construct new capacities in the 10 planned regional WM centres (Alytus, Šiauliai, Tauragė, Klaipėda, Vilnius, Kaunas, Panevėžys, Marijampolė, Telšiai and Utena). These manage 11 regional non-hazardous waste landfills and should solve the problem of the recycling of biodegradable waste as well as enable the recovery of secondary raw materials and the production of energy from waste. So far Lithuania has built 13 waste composting facilities, 68 bulk waste acceptance facilities and 347 container sites for secondary raw materials [15, 19, 76, 99]. In the future, most likely 10 sorting lines (one in each WM centre) and 40 new composting plants should be built [44]. However, the implementation of the new waste management is plagued by high costs and troubles that arose when defining the locations of new landfills and allocating the needed finances. The closed former Soviet dumps, usually located near cities and towns, were replaced by new landfills located even further away from the citizens, taking the waste management collectors even up to 2 h to reach them and adding up to their time-loss and costs. Consequently, high fees motivated citizens and small business to avoid paying by illegal waste dumping. In addition, money for WM ran out early also due to inadequate timing and planning of the financial means [99]. Separate waste collection is developing very slowly also in Latvia, where it is not available to residents outside regional centres. In addition, the recent economic crisis caused a significant fall in prices and in the demand for the use of secondary raw materials [49].

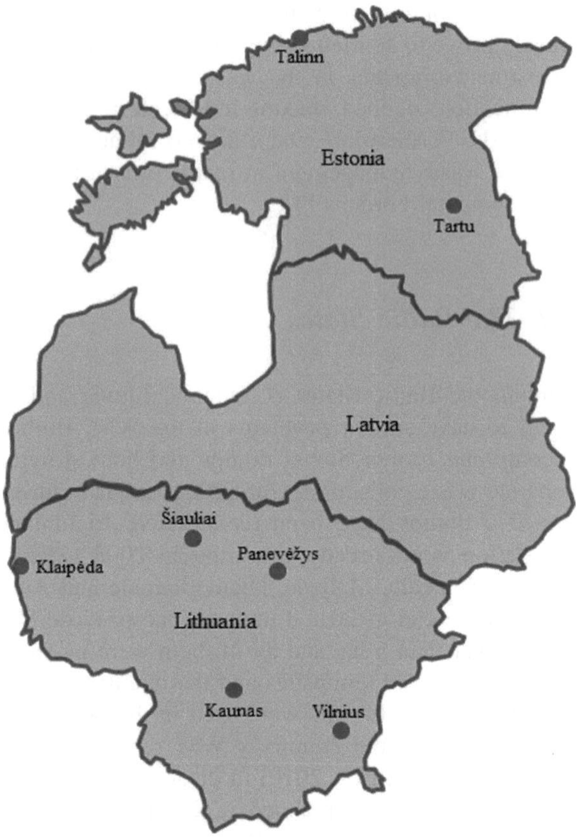

Fig. 7 Possible locations for WtE plants in the Baltic region [34, 44, 98]

Until 2009 eight projects were implemented for the construction of waste stations and composting sites in largest cities (Riga, Jelgava, Rēzekne) and small rural districts (Malta, Dagda, Aloja) [75]. A national waste management plan with regional WM plans will be prepared to cover the period 2013–2020. There are more than 230 million € available for the Latvian WM from the European funds in the period 2007–2013, with additional funding to be provided from the state and the local governments [54]. When it comes to waste incineration, Lithuania is planning the implementation of at least 3 WtE plants for the thermal treatment of 360–420 kilo tons of waste. The Lithuanian ministry of the environment expects that about 40% of the total annual MSW produced will be suitable for WtE plants. The Plan of National Energy Strategy 2007–2012 foresees the possible WtE locations at Vilnius (20 MWel, 50 MWth from 250,000 t/a MSW), Kaunas (15 MWel, 50 MWth), Klaipėda (25 MWel, 50 MWth), Šiauliai (25 MWel, 50 MWth) and Panevėžys (25 MWel, 50 MWth) [8] (Fig. 7). A feasibility study for the WtE plant in Kaunas was done in 2008, but no construction permits have been issued yet. Šiauliai and Panevėžys are preparing the feasibility studies on WtE. The first Lithuanian WtE plant in Klaipėda has already received the permission for construction. It is designed as a public private partnership (PPP) project (Fortum Heat Lietuva) and represents also the first WtE plant in the Baltics. The over 140 million € worth plant should start operating in 2013 and is supposed to combust 272,000 tons of waste annually to produce 85 MW of electricity and heat [34, 44, 98]. In addition, by the end of 2010 the European Investment Bank announced that it is considering providing funds for the construction of the WtE plant in Klaipėda [100]. However, the planned Vilnius plant is still facing harsh public opposition, even though official findings by independent experts were presented to the public in December 2010 by the Environmental Protection Department of the Vilnius region, stating that planned regional WtE plant would be within the allowed limits [101].

Estonia's biggest waste problem is related to the oil-shale industry. MSW forms only about 3% of the total amount of waste. Estonia too has decreased the number of landfills from 170 to 30 in 2007. In 2009 all with the EU legislation non-compliant landfills were closed, leaving in operation only 11 landfills from which 6 are for hazardous waste [20, 93]. Estonia already has a smaller incinerator for hazardous waste (capacity 2,000 t/a) at Tartu, which was recently renovated, and two facilities for co-incineration (cement and ceramic factories), which mainly incinerate liquid hazardous waste. However, the plans made to establish two MSW incineration plants near Tallinn and Tartu (Fig. 7), and an RDF incineration-line at the Kunda Nordic Cement plant, are already in progress and should satisfy the waste incineration capacities in the country [118]. The WtE plant in Iru, on the outskirts of Tallinn, will be constructed by the French Constructions Industrielles de la Méditerranée (CNIM). The plant should cost about 97.5 million € and will combust up to 220,000 tons of waste annually to produce 50 MW of heat and 17 MW of electricity. It should start operating in 2012 [88]. Furthermore, the Kunda Nordic Cement plant began with waste-fuel incineration in 2009 [102] while in February 2010 Veolia and the Tallinn Landfill started to manufacture about 20,000 tons of waste fuel from 40,000 tons of waste [103] to sell it to the Kunda Nordic Cement plant and to a Latvian cement factory. In addition, Ragn Sells launched the construction of a waste fuel plant at the Suur-Sõjamäe (Tallinn) landfill, which should cost 14.7 million € and process about 120,000 tons of waste annually. The plant should be completed by autumn 2011 [104].

5.2 *Eastern and Central European Countries*

The segment about Eastern and Central European countries includes EU member states, such as Slovenia, Hungary, Slovakia, the Czech Republic and Poland and the non-EU country Ukraine.

Most waste produced in Ukraine (88%) comes from the mining industry, followed by the rest of the industry (10%), while the MSW accounts for only 2%. Consequently, most waste is generated in the industrial areas in the east (Donetsk and Dnipropetrovsk), the Central Ukraine (Kiev and Shytomyr) and in the touristic

Odessa. Waste treatment is based on landfilling. Ukraine has one of the largest landfill sites in the world, with about 7,773 landfills (in 2005) stretching over 160,000 ha, of which only about 1000 landfills are legal [45].

From a total of 4 existing incinerators in Ukraine (Kiev, Dnipropetrovsk, Kharkiv, Sewastopol) that were established during the Soviet Union era, only 2 are in operation (Kiev—175,000 tons per year, and Dnipropetrovsk). Both plants treat only about 2.5% of the household waste generated by 1.3 million inhabitants, however, they do not fulfill the European environmental standards [6, 45, 73]. The reconstruction of the waste incineration plant in Kiev, which would improve the flue gas cleaning, enable the production of power (11 MW) and heat (270 Gcal) and should cost about 21 million €, was postponed [48]. To encourage the construction of new WtE facilities, the Ukrainian government signed in summer of 2009 a contract with EcoEnergy Scandinavia on the construction of 14 incineration plants in Ukrainian cities, into which Sweden wants to invest more than 20 billion €. The start of the construction of two WtE plants was planned before the end of 2009, but so far, only the Donetsk incinerator (470 t/a) was announced [105] (Fig. 8). In addition, the Romanian company Mentor Group Holdings proposed the delivery of six turnkey plants for sorting and the thermal treatment of waste for the Ukrainian Mykolaiv region [5].

In contrast to Ukraine, the new EU member states already display a more established WM with serious WtE planning. Poland is the first amongst them trying to implement a major WtE project consisting of 11 WtE plants of a total 2.41 million tons of annual capacity, from which 8 will cost at least 1.2 billion € and will be funded by the EU (estimated 883 million €). The main reason behind the project, which will change the entire waste management organisation in Poland, was the pressure to fulfil the targets set by the EU on one hand, and to catch the last possible application deadline for the EU-funds money for such projects in the period until 2015 (30th of June 2010). To fulfil the requirements of the targets set by the EU legislation, Poland needs to reduce its landfilling of biowaste for 75% [35, 71, 106].

The EU has already started to fine Poland for poor waste disposal 40,000 € per day in 2010. In case Poland would not succeed to improve its WM until the end of 2010, it would have to pay fines of 1.8 million €, reaching up to 250,000 € of fines daily in 2013, while being restricted to obtain EU funding [46]. This would mean a great loss, since according to the Operational Program "Infrastructure and the Environment", about 1.4 billion € of funds are allocated for the Polish WM [71, 106]. From the 12 proposed WtE plants in Poland (Szczecin, Koszalin, Gdansk, Olsztyn, Białystok, Warsaw, Bydgoszcz–Toruń, Poznań, Łódź, Upper Silesian Metropolitan cities, Krakow), only nine applied for EU funds [46, 71], leaving the planned Warsaw project to a PPP scheme, while Olsztyn was withdrawn [107] (Fig. 9).

In order to be able to implement waste incineration, waste delivery to plants must be assured, which is why Poland also decided to change the main Waste Management act in order to shift the responsibility for waste from the hands of the private sector back to the municipality [83]. According to Prof. Tadeusz Pająk, 12

Fig. 8 Possible locations for WtE plants in Ukraine [48, 105]

incinerators won't be enough for Poland that generates around 13 million tons of waste annually, which is why until 2020 Poland will have to ensure the thermal treatment for additional 1.4 million tons of waste [35].

Unlike other Eastern European countries, waste incineration has a long tradition in the Czech Republic. In 1905 the first waste incineration plant that was already then producing electricity from waste was built in the Austro-Hungarian Empire in today's Brno in the Czech Republic [82]. Today's Czech Republic has a relatively unbalanced distribution of waste incineration in the country: in 2001 there were three MSW incineration plants in operation, 67 incinerators of hazardous waste (operation was stopped in six of them) and in addition, waste was also incinerated in 2002 in four cement factories. At present time there are still three MSW incinerators operating: SKO Praha, Malešice (310,000 t/a), SAKO Brno (210,000 t/a) and SKO Liberec (96,000 t/a) [40, 74]. The Praha and Liberec MSW WtE plants have been constructed in 1998 and 1999, respectively, while the SAKO Brno incineration plant was totally renewed from 2002 to 2010, when it re-started its operation. The renewal's total cost 90 million € was financed 60% by the EU [10, 72]. Czech Republic has a total budget for WM improvements until 2013 of 913.5 million €, of which 776.5 million € will be provided by the EU, according to the Czech Operational Program "Environment" [26]. Different sites for the placement of new MSW WtE plants were discussed (Fig. 10), such as Žatec, Kroměříž, Jihlava, Polička or Pardubice, Plzeň, Prague, Opatovice, Mydlovary, Ostrava and Mladá

Fig. 9 Possible locations for WtE plants in Poland [71]

Boleslav. Currently, new MSW WtE plants are planned for the Plzeň region [2, 78]. Municipalities Pardubice and Hradec Králové still plan to build a WtE plant together (20,000 t/a), which should cost about 68 million € [7]. The planned Plzeň incinerator will be placed next to the local landfill and should combust from 60,000 to 100,000 t/a of waste. The project will apply for EU subsidies in the first half of 2011; its construction should start in 2012, enabling it to start operating in 2016. The planned Karviná incinerator, a 200 million € project that would be placed in the Moravian-Silesian region at the Czech most eastern border to Poland, provokes thoughts about the heavy industrial pollution of the region and raises concerns amongst Polish communities. The plant should burn 200,000 t/a at first and should gradually increase its capacity to 350,000 t/a. Another plant is proposed in the Vysočina region. It would burn about 100,000 t/a and should start operating in 2018 [36, 109–111].

Recent problems in the various fields of the WM sector of Slovakia have brought the country to the attention of the European Commission (EC). Consequently, Slovakia is now facing possible charges in front of the European Court because of the insufficient transposition of the WEEE Directive (due to deficits concerning the collection of WEEE and mobile phones) [36], and because of the

Fig. 10 Possible locations for WtE plants in the Czech Republic and Slovakia [2, 79, 108, 109, 111, 112]

insufficient transposition of the Directive on end-of-life vehicles (due to failing in the definitions of hazardous materials and the obligations of the vehicle manufacturers) [37].

To make it worse, Slovakia is already being sued by the EC for various landfilling problems, such as for the failure to comply with the EU rules ensuring that the landfill Považský Chlmec near Žilina will not significantly harm the environment [67] and for failing to adopt a national strategy for the reduction of biodegradable waste going to landfills [68].

The Slovakian Operational Programme "Environment" reserves about 570.6 million € for waste management, of which 485 million € will be contributed by the EU [29]. In 2007, five incineration plants for industrial waste were in operation, six incineration plants for medical waste, four plants for co-incineration and one facility for the incineration of carcass fat. MSW is incinerated at two large scale incinerators in Bratislava (capacity 140,000 t/a) and Košice (80,000 t/a). The Bratislava incinerator has been completely reconstructed in 2003 to become compliant with EU emission limits. The flue gas cleaning of the Košice plant was modernised in 2004. Until 2013, about 20 million € will be invested for the improvement of the efficiency of the plant, which releases more than half of its steam into the air. New investments include a plan for a WtE plant in Žilina (80,000 t/a capacity) and a waste processing facility for RDF, which will be located at the landfill Trnava, about 50 km from Bratislava. The costs of the plant are estimated to be 2.3 million € [112] (Fig. 10).

Waste is for Hungary of particular importance, since the country is not rich in raw materials and energy resources. The landfilling of waste is declining and so far about 328 landfills have been recultivated. Until 2013/14 85 modern landfills should be established. Separated waste collection is available to 55% of the population. The amount of MSW increased until 2006, despite the decreasing trend of total waste generation, which can be explained by the change in consumption

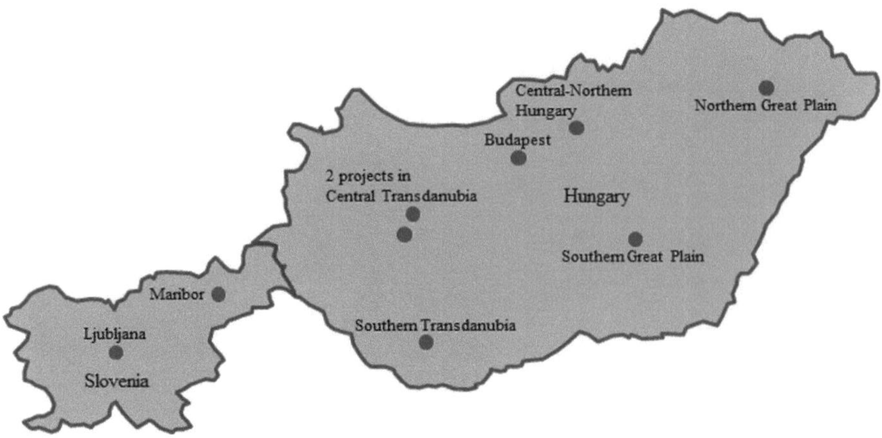

Fig. 11 Possible locations for WtE plants in Slovenia and Hungary [21, 42, 50, 70, 80]

structure and the extension of statistical records. [1, 21] Following the expiration of the National Waste Management Plan (NWMP) 2003–2008 Hungary is developing a new one for the period 2009–2014, which will follow the already adopted new National Environmental Program 2009–2014 [15, 21].

The old NWMP planned and realized the renewal of the only MSW incineration plant in Budapest (capacity: 420,000 t/a) between 2002 and 2005 to improve its energy efficiency. In addition, the inspection of hazardous waste incinerators in 2005 was finalized, leading to the closing or modernisation of those not meeting the requirements, by which the emissions from waste incineration decreased for 30%. Furthermore, the co-incineration of waste with energy recovery exists at some cement kilns—at the Holcim plant Nyergesújfalu-Tát (75,000 t/a), Beremend and Hejöcsaba (each 10,000 t/a), and at power plants—Mátra (200,000 t/a) and Vértes (20,000 t/a). There are plans for the establishment of co-incineration in power plants Bakonyi and Pécs. According to some estimations waste incineration will succeed the landfilling of waste until 2015. The old WM plan supported the construction of six additional waste incineration plants. Planned incinerators are the Budapest project (2020, 300,000 t/a, Central Hungary region), 2 projects in the Észak-Dunátúl (Central Transdanubia region)—the Közép-Duna project (2010, 125,000 t/a) and the Észak-Balaton project (2015, 125,000 t/a), the Észak-Magyarország project (2010, 200,000 t/a, Central-Northern Hungary), the Észak-Kelet project (2015, 150,000 t/a, Northern Great Plain), the Dél-Dunántúl project (2015, 100,000 t/a, Southern Transdanubia) and the Del-Alföld project (2020, 220,000 t/a, Southern Great Plain) [21, 42, 70, 78] (Fig. 11).

The proposed total capacity of the WtE plants would treat 1.22 million tons of waste, which represents about 28% of the entire MSW produced in 2009. However, the acceptance of waste incineration in the country is extremely low, making the permitting of new waste incinerators impossible so far. After strong protests the proposed 150,000 t/a incinerator in the Hungarian town Hajdúböszörmény, which was on the list of 50 controversial projects highlighted by the CEE Bankwatch Network and Friends of the Earth Europe, was cancelled in 2008 [22, 41]. Another drawback is the high dependence on EU subsidies as the state has little money for financially demanding investments, such as for large waste incineration plants. According to the Operational Program "Environment and Energy" for Hungary, about 2.6 billion € are meant for the priority Health and clean settlements (together with waste management) of which the EU will contribute 2.2 billion € [1].

Waste management in the smallest country in the region—Slovenia is organized on a national, regional and local level, which includes the development of about 15 regional waste management centers that will offer waste treatment in terms of the five-step waste management hierarchy. So far only one WM center (Celje) has been completed to offer the complete hierarchy-based waste treatment. Furthermore, Slovenia is still facing several environmental problems related to waste. Exceeding the permitted CO_2 emissions by 4.6%, the country failed to meet the requirements of its Kyoto obligations in 2008, which has led to fines of 80 million € over four years. Furthermore, in the same year the Slovenian Court of Audit, which carried out an audit of the appropriateness of the treatment of separately collected waste from 2005 to 2008, concluded that the Slovenian environmental ministry failed to establish an effective system for the treatment of the separately collected waste, endangering the 2013 WM goals as well as seriously breaching the duty of good business. Consequently, the Slovenian environmental minister had to resign [50]. Today only one RDF plant exists in Celje, which burns the light waste fraction and sewage sludge to produce about 13 MWth and 2,1 MWel [50]. Two more WtE plants are planned—one RDF plant in Ljubljana and another one in Maribor, for which the planning is still in the initial phase [50]. Meanwhile. Slovenia stopped co-incinerating waste at the cement plant Lafarge due to several years of severe protests of the citizens [51].

5.3 South East Europe

The designation of South East Europe includes countries of the Balkan Peninsula (the mentioned Croatia, Bosnia and Herzegovina and Serbia) together with Romania and Bulgaria. These differ greatly in size as well as in their position with regards to the EU.

Bulgaria and Romania are both the youngest members of the EU and display worst WM practices in the union. Officially Bulgaria landfills all of its waste, but according to the claims of the Bulgarian environmental minister in the media, the landfilling rate has lowered to 90% in 2010 [56]. Romania landfills 99% of its waste.

Troubles in the Bulgarian waste management sector lead to the European Commission (EC) launching an infringement procedure against Bulgaria in

Fig. 12 Possible locations for WtE plants in the Romania and Bulgaria [53, 55]

October 2007. The trial in the European Court of Justice is expected to last at least two years and could cost the state millions of euros of fines [57]. The procedure is based mostly on Sofia's failure to improve its WM infrastructure in accordance with the EU legislation demands, signified by a lack of a system and of installations for the recovery and disposal of the city's household waste, the lack or inadequacy of temporary storage sites and the lack of adequate pre-treatment of the waste. The waste problems in Sofia started in 2005, when residents living close to the city's landfill Suhodol started staging rallies, demanding to close the dumpsite and blockading the delivery to it, which accumulated in a political crisis. The dumpsite was reopened end of December 2007 in order to prevent a health crisis [113].

Since Bulgaria has almost none financial means on its own, it relies mostly on the EU funds. The Bulgarian Operational Program "Environment" 2007–2013 allocates 366.7 million € for the improvement and the development of the waste treatment infrastructure, of which the EU will contribute 311.7 million € [27]. Similarly, economically one of the weakest EU countries with the lowest GDP per capita in the EU-27 (46 per capita in Purchasing Power Standards (PPS)), Romania is facing limited possibilities for the sorting and collection of separated waste fractions because of a small number of economic agents who would be willing to recycle these sorted materials [84]. Romania's biggest problem too is the lack of WM infrastructure, which is contributing to difficulties in the legislation implementation, resulting in rural areas having almost no organised waste management and a lack of interest for investments into this sector [25].

To address these problems, Bulgaria adopted the National Waste Management Program 2009–2013, which includes the Action Plan and sets strategic priorities in WM [15]. The Romanian ministry for the Environment announced in March 2011 that by 2015 Romania will invest 1.2 billion € to improve its waste management situation [81]. In 2009 the preparation of 30 projects for integrated waste management under the National Waste Management Plan and Regional Waste Management Plans started, for which funds of 1,168 billion € (934 million € of EU grants) have been allocated in the Operational Program "Environment" 2007–2013 [28]. In Bulgaria, the most important improvement measures planned are the construction of the Sofia's waste treatment plant and the opening of the most modern WEEE recycling plant in Eastern Europe (Fig. 12). The Sofia MBT plant will be built by the Bulgarian-German consortium "Stanilov—Heilit". It should start operating in 2012 and will cost around 106 million €, from which about 50 million € for the main waste depot and the compost equipment should be provided by the EU funds. The authorities are also contemplating the option for RDF, which could supposedly help with the lifting of the EC court procedures [58–65, 89]. The WEEE plant in Novi Iskar near Sofia opened in June 2010. Its costs were 20.5 million €. It was developed and built by the German Adelmann Umwelt and while is being operated by Nadin Jsc. The plant has two recycling lines—one for refrigerators and one for monitors/tv-sets with the ability to recycle other electronics as well [38]. Furthermore, to improve the planning of the WM, end of January 2011 the government urged Bulgarian municipalities to start making research on the future needed waste treatment facilities [66]. The country also plans to unveil a strategy for construction waste in early 2011 [90].

Romania on the other hand has already fully implemented separate MSW collection in urban areas. Its' further main WM objectives are to reduce the volume of unsorted waste from the estimated 3.75 million tons to 2.2 million tons annually, to improve recycling of packaging waste by 2013, to close (by 2017) 238 existing MSW landfills which do not comply with EU regulations, to construct 65 new landfills and to invest in containers for separated collection, garbage trucks, sorting facilities, composting and recycling and the construction of new warehouses [53, 84, 94]. End of 2010 Adama Technologies announced their plan to establish a landfill in Bucharest that would incorporate a landfill-gas utilization system for the production of electricity. The landfill would receive 750,000 tons of waste annually of which Adama estimates to earn a minimum of 22,000 € a day, while reaching a profit of 181,000 € a day in four years' time [96]. The investment potential in the landfilling sector in Romania for the next 20 years accumulates up to 5–10 million €, with estimated annual returns of 2 million € [80].

So far, co-incineration of waste in cement kilns exists in Romania [93]. The new Romanian Waste Management Strategy 2009–2015 considers that energy recovery from MSW should be made possible after the year 2009 at the level of 17% of the total MSW generation, which represents 1.5 million tons annually. WtE facilities should be constructed and operated for minimum 150,000 t/a (equal to a population of 300,000). Feasibility studies for the locations Timişoara, Bucharest and Braşov have been made. The plans for possible WtE facilities

Fig. 13 Possible locations for WtE plants in Croatia, Bosnia and Herzegovina and Serbia [23, 24, 77, 114]

emphasize the WtE capacities of 150,000 t/a with the production of 5.5 MWel and 27 MWth each in Timişoara and Braşov, a total capacity of 750,000 t/a for Bucharest with the first module of 150,000 t/a with the production of 6 MWel and 28 MWth and two additional plants in the North–East of 300,000 t/a capacity, each with the production of 11 MWel and 54 MWth [53] (Fig. 12).

The situation in other non-EU South-Eastern countries is similar, but is expected to improve gradually because of the EU accession process. Croatia plans to invest a total of 3.37 billion € in the modernization of its waste management in the period 2005–2025. About 300 million € will be provided for the planned Zagreb WtE plant [77]. However, the Zagreb waste incineration plant proposal has been facing a heavy opposition from various environmental groups, which is why the project has not been progressing much [13] (Fig. 13). Bosnia and Herzegovina and Serbia experience major problems in the waste management infrastructure, but are willing to support potential WtE projects. According to the Serbian government, Serbia has the potential to cover 3–5% of their total energy consumption from waste. In addition, Serbia has the capacity to install in total seven WtE plants (Belgrade, Novi Sad, Niš, Šumadija region, southern Serbia), suggests the British consultancy EC Harris. The Counties Kraljevo and Novi Pazar are building with the support of the German Medsorga already a WtE plant, which is expected to cost up to 60 million € to produce 12–15 MW of energy [114] (Fig. 13). In Bosnia, the group Prevent announced the construction of a 3.5 million € worth incineration plant for the treatment of the leather industrial waste [23], whereas Sarajevo is in the middle of a feasibility study for an incinerator [24] (Fig. 13).

6 Conclusion

European countries, whether they lie in the east or west, differ greatly in their generation and their approach to waste treatment. To a certain extent the European Union has been encouraging better environmental approaches in waste management among its member states and those Eastern and South Eastern European countries who are in the process of becoming EU members. However, other non EU candidates from the former Eastern bloc are striving to improve their waste management as well, which seems to be happening in concordance with the global raise of environmental awareness. Waste-to-Energy can help a nation achieve not only the most suitable and environmentally friendly waste management possible, but also offer help in the crossover to a non-fossil society. Predictions say that by 2020 only energy from waste incineration could supply at least almost 23 million Europeans with renewable electricity and heat (with a total of 98 TWh). Furthermore, the expansion of WtE could help Europe annually save CO_2 emissions of 22 million cars, not to mention the saving of resources, such as minerals and metals, which is achieved through urban mining and the recycling of bottom ashes. So far only Poland and Lithuania are making concrete progress in the implementation of WtE amongst the new member states. Poland needs to tackle its growing waste amounts and the growing expenses due to EU fines because of not complying with rules. It is trying to do so by implementation of at least 11 WtE plants and an overall change of the waste management system. In 2013 Lithuania will deliver the first WtE plant in the Baltics. According to several public announcements and literature sources, it is to expect that more than 50 WtE plants are to be constructed in the former Eastern bloc, of which most of them are in the early stages of planning and exposed to many uncertainties. It is important to understand that the underlying conditions in the countries are very different. For example, due to severe abuse of the environment by the industry during the communistic times in the Eastern European countries, this region is showing a raising environmental awareness, resulting in some cases (the Czech Republic, Hungary) in severe public opposition against waste incineration. Other problematic aspects of this region might be the lack of electricity/heat transfer infrastructure, the lack of financial stability, which becomes even more important in the light of an economic crisis, missing of sufficient financial means and the inability to use the EU-funds money for such investments due to lack of experience, organisation and a high corruption rate. Whether WtE will reach its second expansion peak in the East and South East of Europe and which waste management technologies a country will really adopt depends on several factors—financial, historical, political, and cultural. The biggest challenge is therefore to assess the social/historical, environmental/geographical and economical aspects of an individual country or a group of countries that potentially share some specifics in the mentioned fields.

References

1. Anders-Clever E (2010) Ungarns Abfallwirtschaft wartet auf politische Weichenstellungen. Sekundär-Rohstoffe. October, pp 20–21
2. Arnika (2005) Arnika: construction of ten new incinerators was discussed in the Czech Republic in the last year. (Online); (cited 2011 May 5). Available from: http://english.arnika.org/press-releases/item/73-construction-of-ten-new-incinerators-was-discussed-in-the-czech-republic-in-the-last-year
3. Avfall Sverige (2008) Energy from waste–an international perspective. Malmö, Avfall Sverige, Report No.: U2009:05
4. Baltzar E, Petrovska A, Sevic D, Uzunov D, Varbova V, Zhechkov R (2009) Speeding up investments in the waste sector–a manual for waste utilities in South Eastern Europe. Regional Environmental Center for Central and Eastern Europe, Szentendre
5. ВКурсе vkurse.ua: Mentor Group Holdings предложила построить в Николаевской обл. 6 мусороперерабатывающих заводов. (Online); 15.10.2010 (cited 2011 May 5). Available from: http://vkurse.ua/business/postroit-v-nikolaevskoy-obl-6-musoropererabatyvayushchikh-zavodov.html
6. Brink P, Bassi S, Vergnon M (2008) Benefits of environment in ENP countries. Methodology test case: Ukraine. Working document for DGENV of the European communities. Institute for European Environmental Policy
7. CEE Bankwatch Network Bankwatch.org: Opatovice municipal waste incinerator, Czech Republic. (Online); (cited 2011 May 5). Available from: http://www.bankwatch.org/project.shtml?w=330665&s=1680840
8. CEWEP (2010) CEWEP country report 2010 Lithuania. CEWEP country report. Brussels, Belgium: Confederation of European Waste-to-Energy plants, Report No.: 2010
9. CEWEP (2009) Waste-to-Energy–a cost effective and reliable sustainable energy source. Position paper on sustainable energy from waste. Brussels: Confederation of European Waste-to-Energy plants
10. CNIM (2010) The Brno Waste to Energy plant. Presentation at the first WtERT annual meeting Europe
11. Dehoust G, Schüler D, Vogt R, Giegrich J (2010) Climate protection potential in the waste management sector–Examples: municipal waste and waste wood. (UBA-FB). Dessau-Roßlau, Germany: Umweltbundesamt, Report No.: 001347/E
12. Department for Environment, Food and Rural Affairs (DEFRA). Local authority funding—PFI projects. (Online); 2011 (cited 2011 April 10). Available from: http://www.defra.gov.uk/environment/economy/waste/widp/pfi-projects
13. Dnevnik AV hr: 'Nisam sretan zbog odluke o spalionici'. (Online); 17.12.2008 (cited 2011 May 5). Available from: http://dnevnik.hr/vijesti/hrvatska/nisam-sretan-zbog-odluke-o-spalionici.html
14. EC-a (2009) Economic Crisis in Europe: causes, consequences and responses. European economy. Luxembourg: European commission–directorate-general for economic and financial affairs; Report No.: 7/2009
15. EC-b (2009) Environment Policy Review–Part 3. Comission Staff Working Document. Brussels, Belgium: European Commission; 2010. Report No.: SEC(2010) 975 final
16. EEA-a (2010) The European environment–state and outlook 2010: synthesis. Copenhagen: European Environment Agency
17. EEA- b (2009) Diverting waste from landfill–effectiveness of waste management policies in the European union. EEA Report. Copenhagen: European Environment Agency; Report No.: 7/2009
18. EEA-c (2007) The road from landfilling to recycling: common destination, different routes. Brochure. Copenhagen: European Environment Agency

19. EEA-d European Environment Agency. EEA.europa.eu: The European environment- state and outlook 2010–Waste (Lithuania). (Online); 2010 (cited 2011 March 14). Available from: http://www.eea.europa.eu/soer/countries/lt/soertopic_view?topic=waste
20. EEA-e European Environment Agency. EEA.europa.eu: The European environment- state and outlook 2010–Waste (Estonia). (Online); (cited 2011 May 2). Available from: http://www.eea.europa.eu/soer/countries/ee/soertopic_view?topic=waste
21. EEA-f European Environment Agency. EEA.europa.eu: The European environment state and outlook 2010–Waste (Hungary). (Online); (cited 2011 March 15). Available from: http://www.eea.europa.eu/soer/countries/hu/soertopic_view?topic=waste
22. Eger A, Kropacek I (2008) CEE bankwatch.org: waste incinerator plans up in smoke in Hungary, campaigners call for EU funds for recycling. (Online); 11.03.2008 (cited 2011 May 5). Available from: http://www.bankwatch.org/newsroom/release.shtml?x=2081251
23. eKapija (a) ekapija.ba: prevent BiH planira izgradnju postrojenja za termički tretman organskog otpada u Visokom. (Online); 11.03.2011 (cited 2011 May 5). Available from: http://www.ekapija.ba/website/bih/page/408439
24. eKapija (b) ekapija.ba: Spalionica otpada u Sarajevu: Uskoro javni poziv za izradu studije izvodljivosti. (Online); 12.10.2010 (cited 2011 May 5). Available from: http://www.ekapija.ba/website/bih/page/355263
25. Enache E (2010) A SWOT analysis on the waste management problem in Romania in 2010. Theor Appl Econ 17(3):101–108
26. European Commission–Regional Policy (Inforegio). Inforegio: Czech Republic–operational programme 'Environment'. (Online); 2007 (cited 2011 May 5). Available from: http://ec.europa.eu/regional_policy/country/prordn/details_new.cfm?gv_PAY=CZ&gv_reg=ALL&gv_PGM=1029&LAN=7&gv_per=2&gv_defL=7
27. European Commission–Regional Policy (Inforegio). Inforegio: Bulgaria–operational programme 'Environment'. (Online); 2007 (cited 2011 May 2). Available from: http://ec.europa.eu/regional_policy/country/prordn/details_new.cfm?gv_PAY=BG&gv_reg=ALL&gv_PGM=1007&LAN=7&gv_per=2&gv_defL=7
28. European Commission–Regional Policy (Inforegio). Inforegio: Romania–operational programme 'Environment'. (Online); 2007 (cited 2011 May 5). Available from: http://ec.europa.eu/regional_policy/country/prordn/details_new.cfm?gv_PAY=RO&gv_reg=ALL&gv_PGM=1038&LAN=7&gv_per=2&gv_defL=7
29. European Commission–Regional Policy (Inforegio). Inforegio: Slovakia–operational programme 'Environment'. (Online); 2007 (cited 2011 May 5). Available from: http://ec.europa.eu/regional_policy/country/prordn/details_new.cfm?gv_PAY=SK&gv_reg=ALL&gv_PGM=1236&LAN=7&gv_per=2&gv_defL=7
30. Eurostat (2009) Eurostat.ec.europa.eu: municipal waste generated and treated in (Online); 2010 (cited 2011 May 5). Available from: http://epp.eurostat.ec.europa.eu/
31. EUWID-a Aufträge für Verbrennungskapazitäten sinken 2009 in Europa um ein Viertel. Europäischer Wirtschaftsdienst–Recycling und Entsorgung. 2010 April;(16): p 13
32. EUWID-b Aufträge für Verbrennungskapazitäten in Europa wieder deutlich gestiegen. EUWID Recycling und Entsorgung. 2011 February;(8): p 11
33. EUWID-c MVA erzeugen zwischen 2005 und 2009 mehr Energie und verbrennen mehr Abfälle. Europäischer Wirtschaftsdienst–Recycling und Entsorgung. 2011 April;(13): p 7
34. EUWID-d Fisia Babcock liefert Kessel für erste MVA im Baltikum. Europäischer Wirtschaftsdienst–Recycling and Entsorgung. 2010 March;(9)
35. EUWID-e Polnische Abfallwirtschaft braucht bis 2020 weitere Behandlungsanlagen. Europäischer Wirtschaftsdienst–Recycling & Entsorgung. 2010 September;(38)
36. EUWID-f EU-Kommission fordert Umsetzung von EuGH-Urteil zum Abfallrecht. Europäischer Wirtschaftsdienst–Recycling and Entsorgung. 2010 November;(44)
37. EUWID-g Zweite Mahnung für Slowakei wegen Altautorichtlinie. Europäischer Wirtschaftsdienst–Recycling and Entsorgung. 2011 February;(6)
38. EUWID-h Bulgariens erste Anlage zum E-schrottrecycling in Betrieb. Europäischer Wirtschaftsdienst–Recycling and Entsorgung. 2010 July;(28)

39. Gallop P (2008) Fuelling the fire–European investment bank financing for the incineration industry. Central and Eastern Europe Bankwatch Network, Praha
40. Government of the Czech Republic Waste management plan of the Czech Republic. Government regulation. Prague, Czech Republic: Ministry of the Environment of the Czech Republic; 2003. Report No.: No. 197/2003 Coll
41. Herczeg M Evaluation of waste policies related to the landfill directive–Hungary. ETC./RWM working paper. Copenhagen, Denmark: European Topic Centre on Resource and Waste Management;. Report No.: 7/2008
42. Illés Z Hungarian waste management policy. (2009). A presentation held in Ankara, Turkey form 24th to 25th of August
43. ISWA (2009) Waste and climate change. White Paper. Vienna, Austria: International Solid Waste Association
44. JSC (2010) "Naftos grupė". WtE development in Lithuania. Poster session summary for the 5th CEWEP Congress on Waste-to-Energy. Antwerp, Belgium: Confederation of European Waste-to-Energy Plants
45. Kisliakova N, Pavliska O, Mtibelashvili T (2006) Umweltpolitik, -strategien und -programme der Länder Russland, Ukraine, Georgien. Wien, Austria: Wirtschaftskammer Österreich (WKÖ), Österreichische Gesellschaft für Umwelt und Technik (ÖGUT); Report No.: 2006
46. Kuhn M (2010) Am Scheidepunkt. Recycling Magazin. October pp 30–33
47. Kuhn M (2009) Stiefkind Verbrennung—Tschechien will MVA finanziell nicht fördern. Recycling Magazin. August pp 28–30
48. Kyiv Post. KyivPost.com: Kyiv postpones construction of new waste incinerator. (Online); 05.11.2010 (cited 2011 May 5). Available from: http://www.kyivpost.com/news/city/detail/88854
49. Latvian Ministry of Environment. ENVIRONMENTAL POLICY STRATEGY 2009-2015. Riga, Latvia: Ministry of Environment Latvia; 2009. Report No.: No. 517 of 31 July 2009
50. Malek S (2010) Waste incineration in Slovenia: a life-cycle assessment of environmental impacts of potential waste-to-energy options. Master Thesis
51. Malovrh P. Delo.si: Streljanje na Lafarge Cement, sosežig prepovedan. (Online); 09.03.2011 (cited 2011 May 5). Available from: http://www.delo.si/novice/slovenija/streljanje-na-lafarge-cement-sosezig-prepovedan.html
52. Manders JLC (2008) The renewable energy contribution of "Waste to Energy" across Europe. study. Confederation of European Waste-to-Energy Plants, Brussels
53. Marcu V, Atudorei A (2010) WtE development in Romania. Poster session summary for the 5th CEWEP Congress on Waste-to-Energy. Antwerp, Confederation of European Waste-to-Energy Plants, Belgium
54. Ministry of Foreign Affairs of Denmark- The Trade Council, Baltics. (2010) DDK 1.2 billion for improving waste management in Latvia. (Online); 11.10.2010 (cited 2011 May 2). Available from: http://tradecouncil.baltics.um.dk/en/menu/MarketOpportunities/LatestNews/DDK12BillionForImprovingWasteManagementInLatvia.htm
55. Novinite–Sofia news agency (a). Novinite: Sofia City Hall Announces Winning Waste Plant Bidder. (Online); 16.06.2010 (cited 2011 May 5). Available from: http://www.novinite.com/view_news.php?id=117205
56. Novinite Sofia News Agency (b). Novinite.com: Bulgaria Has Worst Waste Management in Europe. (Online); 13.01.2011 (cited 2011 May 2). Available from: http://www.novinite.com/view_news.php?id=124128
57. Novinite Sofia News Agency (c). Novinite.com: Brussels Very Likely to Sue Bulgaria over Poor Waste Disposal. (Online); 15.06.2010 (cited 2011 May 2). Available from: http://www.novinite.com/view_news.php?id=117149
58. Novinite–Sofia news agency (d). Novinite.com: EC to Finance Initial Works at Sofia Waste Plant. (Online); 06.11.2010 (cited 2011 May 5). Available from: http://www.novinite.com/view_news.php?id=121888

59. Novinite–Sofia news agency (e). Novinite.com: Bulgarian Supreme Court Confirms Sofia Waste Plant Bid. (Online); 24.09.2010 (cited 2011 May 5). Available from: http://www.novinite.com/view_news.php?id=120463
60. Novinite–Sofia news agency (f). Novinite.com: Bulgarian Socialists: Sofia Waste Plant Project Flopped. (Online); 21.09.2010 (cited 2011 May 5). Available from: http://www.novinite.com/view_news.php?id=120371
61. Novinite–Sofia news agency (g). Novinite.com: Bulgarian Eco Minister Claims Waste Plant Project on Schedule. (Online); 17.09.2010 (cited 2011 May 5). Available from: http://www.novinite.com/view_news.php?id=120249
62. Novinite–Sofia news agency (h). Novinite.com: Bulgarian Watchdog Rejects Another Appeal over Sofia's Waste Plant. (Online); 19.07.2010 (cited 2011 May 5). Available from: http://www.novinite.com/view_news.php?id=118260
63. Novinite–Sofia news agency (i). Novinite.com: Bulgaria Watchdog Rejects Appeal over Sofia's Waste Plant. (Online); 02.07.2010 (cited 2011 May 5). Available from: http://www.novinite.com/view_news.php?id=117717
64. Novinite–Sofia news agency (j). Novinite.com: German, Greek Firms Appeal Sofia Waste Plant Tender. (Online); 30.06.2010 (cited 2011 May 5). Available from: http://www.novinite.com/view_news.php?id=117654
65. Novinite–Sofia news agency (k). Novinite.com: Sofia City Hall Announces Winning Waste Plant Bidder. (Online); 16.06.2010 (cited 2011 May 5). Available from: http://www.novinite.com/view_news.php?id=117205
66. Novinite–Sofia news agency (l) Novinite.com: Bulgaria to Set Strict Targets in Waste Management. (Online); 21.01.2011 (cited 2011 May 5). Available from: http://www.novinite.com/view_news.php?id=124396
67. Odpady Portal (a). Odpady-portal.sk: Skládka odpadov Považský Chlmec: Komisia žaluje Slovensko. (Online); 17.02.2011 (cited 2011 May 5). Available from: http://www.odpady-portal.sk/Dokument/100654/skladka-odpadov-povazsky-chlmec-komisia-zaluje-slovensko.aspx
68. Odpady Portal (b). Odpady-portal.sk: Európska komisia žaluje Slovensko za bioodpady. (Online); 19.07.2010 (cited 2011 May 5). Available from: http://www.odpady-portal.sk/Dokument/100460/europska-komisia-zaluje-slovensko-za-bioodpady.aspx
69. Olofsson M, Sundberg J, Sahlin J (2005) Evaluating waste incineration as treatment and energy recovery method from an environmental point of view. In Paper no. NAWTEC13-3168, 13th Annual North American Waste-to-Energy Conference (NAWTEC13); 2005; Orlando, Florida. pp 175–192
70. Orosz Z, Fazekas I (2008) Challenges of municipal waste management in Hungary. AGD Landsc Environ 2(1):78–85
71. Pajak T (2010) Abfallwirtschaftsplanung in Polen. ReSource. 2010 January: pp 14–15
72. Petrlik J, Havel M, Skalsky M (2006) The liberec municipal waste incinerator–a significant source of POPs. A report part of the International POPs Elimination Project. Prague, Czech Republic: Arnika–toxics and waste programme
73. Popovych O, Malovanyy M, Plaza E (2009) Hrybovychi municipal solid waste landfill reclamation and coherent pollution prevention in holistic approach. In research and application of new technologies in wastewater treatment and municipal solid waste disposal in Ukraine, Sweden and Poland. Proceedings of a Polish-Swedish seminar.; Stockholm, Sweden. pp 23–30
74. REC (2001) Waste management policies in central and Eastern European countries: current policies and trends. The regional environmental center for central and Eastern Europe, Szentendre
75. Recyclingportal. Recyclingportal.eu: Latvia: projects on waste collection stadions and requirements for dump sites finished. (Online); 07.05.2009 (cited 2011 May 2). Available from: http://www.recyclingportal.eu/artikel/22068.shtml

76. Recyclingportal. Recyclingportal.eu: Lithuanian ministry developing the national waste management system. (Online); 17.01.2011 (cited 2011 May 2). Available from: http://www.recyclingportal.eu/artikel/25884.shtml
77. REPORT (2010) (Unknown authors). Devisen für Deponien. Sekundär Rohstoffe. September: pp 20–22
78. ReTech-a (2009) Länderprofil Abfallwirtschaft Rumänien. Länderprofil Abfallwirtschaft. Bonn, Germany: ReTech–die BMU Exportinitiative Recycling und Effizienztechnik
79. ReTech-b (2009) Länderprofil Abfallwirtschaft Tschechische Republik. Länderprofil Abfallwirtschaft. Bonn, Germany: ReTech–die BMU Eportinitiative Recycling und Effizienztechnik
80. ReTech-c (2009) Länderprofil Abfallwirtschaft Ungarn. Landerprofil Abfallwirtschaft. Bonn, Germany: ReTech–die BMU Exportinitiative Recycling und Effizienztechnik
81. Romania Insider. Romania-insider.com: Romania pledges to spend EUR 1.2 billion on waste recycling until 2015. (Online); 10.03.2011 (cited 2011 May 5). Available from: http://www.romania-insider.com/romania-pledges-to-spend-eur-1.2-billion-on-waste-recycling-until-2015/20236/
82. SAKO Brno (2011) Sako.cz: Historie. (Online); (cited 2011 May 5). Available from: http://www.sako.cz/spalovna/historie/
83. Schäfer A, Kapsa K (2010) Das kommunale Abfallrecht in Polen (1. Teil). ReSource. March: pp 34–38
84. Scortar LM, Lazar I, Zagan Zelter D (2009) The condition of waste management in Romania. Manage Marketing J 7(1):29–35
85. Skovgaard M, Villanueva A, Andersen FM, Larsen H (2007) Environmental outlooks: municipal waste. ECT/RWM working paper. Copenhagen, Denmark: European topic centre on resource and waste management; Report No.: 2007/1
86. Stracansky P. Guardian.co.uk: Eastern Europe lagging on waste and recycling, NGOs warn. (Online); 01.04.2010 (cited 2011 May 10). Available from: http://www.guardian.co.uk/environment/2010/apr/01/waste-recycling
87. Tchobanoglous G, Kreith F (2002) Handbook of solid waste management, 2nd edn. edn. Irwin/Mcgraw Hill, USA
88. Tere J. The baltic course: Estonian Eesti Energia approves investments into WTE plant. (Online); 28.01.2010 (cited 2011 May 2). Available from: http://www.baltic-course.com/eng/investments/?doc=23000
89. The Sofia Echo (a). SofiaEcho.com: Sofia's waste treatment facility still pending EC clearance. (Online); 08.11.2010 (cited 2011 May 5). Available from: http://sofiaecho.com/2010/11/08/989658_sofias-waste-treatment-facility-still-pending-ec-clearance
90. The Sofia Echo (b). SofiaEcho.com: Bulgaria to unveil strategy for construction waste at the start of 2011. (Online); 10.12.2010 (cited 2011 May 5). Available from: http://sofiaecho.com/2010/12/10/1008623_bulgaria-to-unveil-strategy-for-construction-waste-at-the-start-of-2011
91. UK Without Incineration Network (UK WIN). ukwin.org: Sites where incinerators exist or are proposed. (Online); 2011 (cited 2011 March 3). Available from: http://ukwin.org.uk/knowledge-bank/incineration/sites-where-incinerators-exist-or-are-proposed/
92. Umweltbundesamt (2008) Waste incineration and waste prevention: not a contradiction in terms. Background paper. Dessau: Umweltbundesamt; Report No.: 052/2008
93. UN CSD- a (2010) Estonia UN CSD18: National Reporting on Waste management. National Reporting on Waste management. New York: United Nations Commission on Sustainable Development; Report No.: UN CSD18
94. UN CSD- b (2010) Romaina's National Report for the 18th Session of The United Nations Commission on Sustainable Development (CSD-18). National Report. New York, USA: United Nations Commission on Sustainable Development; Report No.: CSD-18
95. Vaccani A (2011) Überkapazitäten in der Abfallverbrennung in Europa. In Thomé-Kozmiensky KJ, editor. Planung und Umweltrecht–Band 5. TK Verlag Karl Thomé-Kozmiensky, Neuruppin pp 31–38

96. Waste Management World. WM World: Waste to Energy Project Expected to Pay Off in Romania. (Online); 16.11.2010 (cited 2011 May 5). Available from: http://www.waste-management-world.com/index/display/article-display.articles.waste-management-world.waste-to-energy.2010.11.Waste_to_Energy_Project_Expected_to_Pay_Off_in_Romania.html
97. Waste-to-Energy Research and Technology Council Germany (WtERT Germany)-a. WtERT.eu: Vysočina Region plans energy recovery from waste. (Online); 19.05.2010 (cited 2011 May 2). Available from: http://www.wtert.eu/Default.asp?Menue=18&NewsPPV=7938
98. Waste-to-Energy Research and Technology council Germany. (WtERT Germany)-b. Baltic's first WtE plant will be in Lithuania. (Online); 24.02.2010 (cited 2011 April 15). Available from: http://wtert.eu/default.asp?Menue=26&NewsPPV=7110
99. Waste-to-Energy Research and Technology Council Germany(WtERT Germany)-c. WtERT.eu: Lithuania: WtE could be the solution for an ineffective yet costly waste management. (Online); 26.03.2010 (cited 2011 April 14). Available from: http://wtert.eu/default.asp?Menue=26&NewsPPV=7436
100. Waste-to-Energy Research and Technology Council Germany. (WtERT Germany)-d. WtERT.eu: EIB mulls loans to waste-to-energy CHP plants in Sweden, Lithuania. (Online); 24.12.2010 (cited 2011 May 2). Available from: http://wtert.eu/default.asp?Menue=26&NewsPPV=9909
101. Waste-to-Energy Research and Technology Council Germany. (WtERT Germany)-e. WtERT.eu: New Vilnius waste incineration plant? Experts say Yes. (Online); 09.12.2010 (cited 2011 May 2). Available from: http://wtert.eu/default.asp?Menue=26&NewsPPV=9760
102. Waste-to-Energy Research and Technology Council Germany. (WtERT Germany)-f. WtERT.eu: CNIM to start building the first WtE plant in Estonia this summer. (Online); 07.02.2010 (cited 2011 May 2). Available from: http://wtert.eu/default.asp?Menue=26&NewsPPV=6938
103. Waste-to-Energy Research and Technology Council Germany. (WtERT Germany)-g. WtERT.eu: Veolia and the Tallinn Landfill starting the manufacturing of fuel from waste. (Online); 16.02.2010 (cited 2011 May 2). Available from: http://wtert.eu/default.asp?Menue=26&NewsPPV=7023
104. Waste-to-Energy Research and Technology Council Germany. (WtERT Germany)-h. WtERT.eu: Ragn-Sells launched construction of a waste incineration plant in Tallinn. (Online); 03.01.2011 (cited 2011 May 2). Available fromhttp://wtert.eu/default.asp?Menue=26&NewsPPV=9934
105. Waste-to-Energy Research and Technology Council Germany. (2009) (WtERT Germany)-i. WtERT.eu: Ukraine, Donetsk: EcoEnergy building its first of 20 billion € worth WtE investments. (Online); 27.12.2009 (cited 2011 May 5). Available from: http://wtert.eu/default.asp?Menue=26&NewsPPV=6565
106. Waste-to-Energy Research and Technology Council Germany. (WtERT Germany)-j. WtERT.eu: EU fines Poland for poor waste disposal. (Online); 23.08.2010 (cited 2011 May 5). Available from: http://wtert.eu/default.asp?Menue=26&NewsPPV=8769
107. Waste-to-Energy Research and Technology Council Germany. (WtERT Germany)-k. WtERT.eu: Nine Polish WtE projects submitted for EU funding. (Online); 07.07.2010 (cited 2011 May 5). Available from: http://wtert.eu/default.asp?Menue=26&NewsPPV=8453
108. Waste-to-Energy Research and Technology Council Germany (WtERT Germany)-l. WtERT.eu: Czech town Pilsen to get a new WtE plant in 2016. (Online).; 07.04.2010 (cited 2011 May 5). Available from: http://wtert.eu/default.asp?Menue=26&NewsPPV=7543
109. Waste-to-Energy Research and Technology Council Germany. (WtERT Germany)-m. WtERT.eu: Waste incinerator in the Czech Pilsen region to help reduce dependence on coal. (Online); 06.01.2011 (cited 2011 May 5). Available from: http://wtert.eu/default.asp?Menue=26&NewsPPV=9942

110. Waste-to-Energy Research and Technology Council Germany (WtERT Germany)-n. WtERT.eu: Czech Citizens against the Karviná incinerator. (Online); 06.05.2010 (cited 2011 May 5). Available from: http://wtert.eu/default.asp?Menue=26&NewsPPV=7816
111. Waste-to-Energy Research and Technology Council Germany. (WtERT Germany)-o. WtERT.eu: Vysočina a step closer to an incinerator. (Online); 10.06.2010 (cited 2011 May 5). Available from: http://wtert.eu/default.asp?Menue=26&NewsPPV=8146
112. Waste-to-Energy Research and Technology Council Germany (WtERT Germany)-p. WtERT.eu: Companies announce new waste facilities in Slovakia. (Online); 01.09.2010 (cited 2011 May 5). Available from: http://wtert.eu/default.asp?Menue=26&NewsPPV=8876
113. Waste-to-Energy Research and Technology Council Germany (WtERT Germany)-q. WtERT.eu: Bulgarian waste: EC starts it's first trial against EU newcomer. (Online); 19.04.2010 (cited 2011 May 2). Available from: http://wtert.eu/default.asp?Menue=26&NewsPPV=7645
114. Waste-to-Energy Research and Technology Council Germany (WtERT Germany)-r. WtERT.eu: Serbia holds 7 locations for WtE plants. (Online); 22.06.2010 (cited 2011 May 5). Available from: http://wtert.eu/default.asp?Menue=26&NewsPPV=8231
115. Wielenga K (2008) Waste-to-Energy and the revision of the waste framework directive. Waste-to-Energy's contribution to climate protection. Confederation of European Waste-to-Energy Plants, Braine l'Alleud
116. Wollmann R (2011) Unterstützung der Abfallwirtschaft in den Staaten Mittel- und Osteuropas, des Kaukasus und Zentralasiens. Wasser und Abfall. Jan, pp 26–30
117. Zeschmar-Lahl B (2010) Waste-to-energy compared to Mechanical Biological Treatment (MBT) And co-combustion of municipal waste. In ISWA World Congress; Hamburg pp 1–10
118. Zhechkov R, Viisimaa M (2008) Evaluation of waste policies related to the Landfill Directive–Estonia. ETC./RWM working paper. Copenhagen, Denmark: European topic centre on resource and waste management; Report No.: 3/2008

Energy from Biomass in Mauritius: Overview of Research and Applications

Romeela Mohee and Ackmez Mudhoo

Abstract Viewing the increase in energy demand and bearing in consideration the associated pollution issues, a shift in energy paradigm from a monopolized conventional energy regime of fossil fuels to a non-conventional renewable energy mix is necessary in the Mauritian context. Besides achieving considerable waste reduction, the incineration and fermentation of renewable biomass feedstock can also generate revenues from energy production, known as waste-to-energy (WtE) and anaerobic digestion technology (ADT), respectively. In Mauritius, municipal solid wastes (MSW) typically include food and garden wastes, other green wastes and organic residues, different grades of paper and plastics, textiles, rubber, glass, metals, wood, process sludges and inerts. This chapter provides a succinct review of research studies and large-scale applications of the WtE sector and ADT in Mauritius.

1 Biomass and Renewable Energy Needs

Demand for energy is increasing day by day due to the rapid growth of population and urbanization. Global energy needs are likely to continue to grow steadily for at least the next two-and-a-half decades [3]. According to Birol [3], if governments stick with current policies, it is much likely that the world's energy needs would be more than 50% higher in 2030 than today. More than 65% of the growth in world energy use will come from the developing countries, where economic and

R. Mohee (✉) · A. Mudhoo
Department of Chemical and Environmental Engineering,
Faculty of Engineering, University of Mauritius, Réduit, Mauritius
e-mail: rmh@uom.ac.mu; rkmohee@yahoo.com

A. Mudhoo
e-mail: a.mudhoo@uom.ac.mu; ackmezchem@yahoo.co.uk

A. Karagiannidis (ed.), *Waste to Energy*, Green Energy and Technology,
DOI: 10.1007/978-1-4471-2306-4_12, © Springer-Verlag London Limited 2012

population growths are highest. Fossil fuels continue to dominate energy supplies, meeting more than 80% of the projected increase in primary energy demand in this scenario, and with oil remaining the single largest fuel and having an projected demand of 115 mb/d in 2030 [3]. As the major conventional energy resources like coal, petroleum and natural gas are rapidly being depleted [13, 17], biomass is emerging as one of the promising environment friendly renewable energy option [8, 20]. The process of obtaining energy from the conventional energy sources causes atmospheric pollution, resulting in problems like global warming, acid rain and disturbed climatic patterns. Viewing the increase in the energy demands and bearing in consideration the pollution cascades, a shift over to non-conventional sources like wind, sunlight, water and biomass is inevitable [1, 13].

Different thermo-chemical conversion processes that include combustion, gasification, liquefaction, hydrogenation/pyrolysis and biochemical processes like anaerobic digestion biotechnology have been used to convert the biomass into various energy products [2, 13]. Biomass is being used from the ancient times as a combustion fuel for cooking and keeping warmth in houses. Biomass is available in abundance and is relatively cheaper and its better utilization is to convert it to energy rich products using suitable processes [13]. Biomass resources can be divided into two broad categories, i.e., natural and derived materials. Biomass resources include wood and wood wastes, agricultural crops and their waste by-products, municipal solid waste, animal wastes, waste from food processing and aquatic plants and algae [11]. Biomass resource can be further subdivided into the following categories according to Demirbas [7]: wastes: agricultural production wastes, agricultural processing wastes, crop residues, mill wood wastes, urban wood wastes and urban organic wastes.

The utility of biomass as feedstock for conversion depends upon the chemical constituents and physical properties, and in principle, biomass contains varying amounts of cellulose, hemi-cellulose and lignin [13]. According to Rabaey and Verstraete [29], biomass has an energetic value, whether considered as foodstuff, energy crop or waste. On average 1 kg of sugar, as a model biomass, contains a potential of 4.41 kWh of energy; and out of 1 kg carbohydrates, currently 0.5 L ethanol, 1.2 m^3 H_2 gas, 0.36 m^3 CH_4 gas or 0.5 m^3 biogas may be produced [29]. On average, these processes yield $\sim$1 kWh of useful energy. Biomass currently accounts for about 13–16% of world energy consumption and is a main source of energy for many developed and developing countries [18].

2 MSW Biomass and Waste-to-Energy

Different waste treatment options for municipal solid waste (MSW) have been studied to date and different combinations of incineration, materials recycling of separated plastic and cardboard containers, and biological treatment (anaerobic digestion and composting) of biodegradable waste have been extensively studies and reported in the literature and compared to landfilling [9]. The various options for the recovery of energy from biomass or biodegradable fractions from MSW may be subdivided into anaerobic digestion [19], production of biofuels, direct production of electricity microbial fuel cells [29], incineration with energy recovery [15], co-incineration in coal-fired power plants [22], gasification [26] and pyrolysis, supercritical wet oxidation of sewage sludge [23], and hydrothermal. Anaerobic digestion and incineration for energy recovery are discussed in more detail in the sections to follow.

2.1 MSW and WtE Incineration

MSW incineration in developing countries is generally limited by many factors, including significant capital and operating costs, potential environmental impacts, and technical difficulties of operating and maintaining an incinerator and its pollution control equipment [9, 16, 31]. One distinct characteristic of MSW in developing countries is its high moisture level (typically around 50% or more), which is much higher than that (20–30%) in the MSW in the U.S. and European countries. The high moisture content further lowers the MSW's energy content. Besides waste reduction, incineration can also generate revenues from energy production, known as waste-to-energy (WtE), which partially offsets the cost of incineration [5, 28]. The heat released from combustion of the MSW can be collected through steam generation, which is subsequently used for heating or power generation. MSW power plants are designed to dispose of MSW and to produce electricity as a byproduct of the incinerator operation. MSW is a source of biomass [4, 21]: food waste, yard trimmings and other green wastes are examples of biomass trash, while materials that are made out of glass, plastic, and metals are not [9].

2.2 Essentials of Anaerobic Biotechnology

The anaerobic digestion biotechnology has been considerably employed in the recovery of value-added products and biofuels from waste streams. Carbon, nitrogen, hydrogen, and sulphur from municipal, industrial, and agricultural solid and liquid wastes are converted into value-added resources. These include biofuels (hydrogen, butanol, and methane), electricity from microbial fuel cells (MFCs), fertilizers (biosolids), and useful chemicals (sulfur, organic acids, etc.).

The anaerobic digestion process consists of a hydrolysis step in which organic compounds from the biomass substrates, such as polysaccharides, proteins, and fat, are hydrolyzed by extracellular enzymes; an acidification step in which the products of the hydrolysis are converted into hydrogen, formate, acetate, and higher molecular-weight volatile fatty acids, and a third step in which biogas, a mixture of methane and carbon dioxide, is produced from hydrogen, formate, and acetate [30].

The high-molecular-weight volatile fatty acids have to break down to hydrogen, formate, and acetate before further conversion to methane and carbon dioxide is possible. The complete methanogenic conversion occurs by mixed microbiological communities, yielding methane as the sole reduced organic compound.

The biogas can be used as an energy source for the production of electricity and/or heat. The vast majority of the anaerobic processes applied in practice are mesophilic. Normally, the biogas production from a mixture of primary and secondary (biological) sludges roughly amounts to 1 m^3 of biogas/kg of organic solids biodegraded [30].

3 MSW Generation and WtE Potential in Mauritius

With rapid economic growth and massive urbanization equally happening in Mauritius, the management and reuse of MSW disposal has become a major economic, environmental and energy concern for decision makers, researchers, engineers and environmentalists. As observed from Fig. 1, over the last 9 years the national imports of diesel as an energy source, the corresponding thermal energy generation, peak energy demand, total amount of MSW landfilled and the average monthly income in Mauritian rupees (MUR) have kept on increasing steadily. These concomitant trends support a relatively steady rate of economic growth (average monthly income) accompanied by an increasing energy demand and the associated unhealthy dependency on diesel as an energy source. Also, Mauritius being a small island, landfilling selected as a most ‘straightforward’ option to dispose of MSW generatedmay not be a long-term and sustainable solution. MSW refers to the materials discarded in urban areas for which municipalities are usually held responsible for collection, transport, and final disposal. Additionally, with the price of light crude oil leaping over USD70 per barrel, increasing concern over global warming and the security of fossil fuel supplies, biofuel (mainly biodiesel) is rapidly gaining prominence as an alternative renewable energy source in line with the Kyoto Protocol.

3.1 Maurice Ile Durable Concept

The Republic of Mauritius is a group of islands in the South West of the Indian Ocean, consisting of the main island of Mauritius, Rodrigues and several outer islands located at distances greater than 350 km from the main island. The population of Mauritius is estimated at 1.3 million. Mauritius and Rodrigues, with a total area of 1,969 km^2, have an overall population density of 649 persons/km^2. About 43% of the area is allocated to agriculture, 25% is occupied by built-up areas and 2% by public roads; the remaining consists of abandoned canefields, forests, scrub land, grasslands and grazing lands, reservoirs and ponds, swamps and rocks (Source

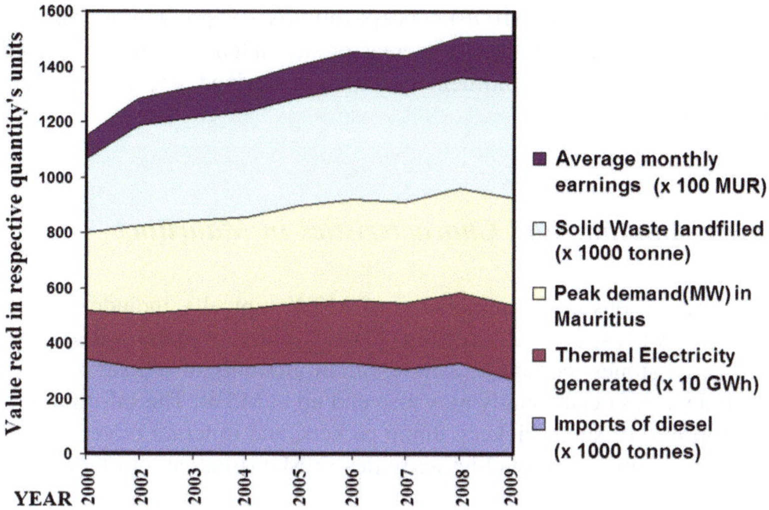

Fig. 1 Trends in national imports of diesel, thermal energy generation, peak energy demand, total amount of MSW landfilled and average monthly income in Mauritius (2000–2009). Source of numerical data: Central Statistics Office, Ministry of Finance & Economic Development. Individual annual reports (Mauritius in figures) were accessed on 17 September 2010 at http://www.gov.mu/portal/site/cso/menuitem.19621772f6bc90fe965c062ca0208a0c/

of data: 'Mauritius in Figures' annual reports available at the Central Statistics Office, Ministry of Finance & Economic Development website http://www.gov.mu/portal/site/cso/menuitem.19621772f6bc90fe965c062ca0208a0c/. Accessed on 16 September 2010).

During the past 35 years, the Mauritian economy has diversified from a sugar-cane monocrop economy in the 1970s to one based on sugar, manufacturing (mainly textiles and garments) and tourism in the 1980s. Global business (offshore) and freeport activities have also been growing continuously since the mid 1990s. The economy in 2009 grew by 3.1% and the Gross National Income per capita at market prices reached up to USD 8,455. While fossil fuels are being depleted, their prices are ever increasing and this weighs heavily on the Mauritian economy, too. Mauritius imported 406 tonnes of oil equivalent of primary energy during the year 2009 for the transportation sector and this amounted to nearly Rs 8,000 million (Source of data: 'Mauritius in Figures' annual reports available at the Central Statistics Office, Ministry of Finance & Economic Development website http://www.gov.mu/portal/site/cso/menuitem.19621772f6bc90fe965c062ca0208a0c/. Accessed on 16 September 2010).

During the combustion of these fossil fuels for the generation of electric power, many pollutants such as oxides of nitrogen (NO_x) and oxides of sulphur (SO_x) together with greenhouse gases like carbon dioxide and carbon monoxide are evolved. Therefore, the search for alternative renewable energy resources to alleviate the situation is equally important for the sustainability of the Mauritian economy and environment. Mauritius has recently adopted and embarked on the

Maurice Ile Durable Fund (MIDF) concept (http://www.gov.mu/portal/sites/midf/index.htm) which primarily aims at decreasing dependency on fossil fuels through the promotion and use of biofuels or biomass derived energy mainly in the transportation sector.

3.2 MSW Amounts and Characteristics in Mauritius

In Mauritius, the physical components of MSW typically include food wastes, paper, green wastes, textiles, rubber, plastic, glass, metals, wood, and inert materials (e.g., stones, ceramics, ashes). Small amounts of industrial wastes and construction wastes occasionally may also end up in MSW. The informal recycling sector, comprised of street pickers, dump pickers, and itinerant buyers, is involved in waste scavenging and recycling activities in Mauritius, as in other developing countries [9].

The current average and official MSW generation in Mauritius is 1,200 tons/day. Because of the lifestyle differences and current trash disposal practice, MSW in Mauritius shows some distinct compositional characteristics. Food and garden wastes make up the largest fraction (average of 42–65%) of MSW in most regions in Mauritius, while the contents of paper and metals are very low because of their high recycling levels. Tables 1 and 2 present the forecast of waste quantities without waste minimization initiatives and composition of household, tourist, commercial and non-hazardous healthcare wastes of Mauritius, respectively. Figure 2 depicts a summarized typical solid wastes composition in Mauritius, and this composition is similar to the situations in many other developing countries.

Incineration of MSW for energy recovery (i.e. WtE) in Mauritius presents some unique potential and challenge because of its relatively low calorific value (average of 5–17 MJ/kg on dry basis [25]) but relatively high water content (average of 45–55% [24]), respectively. Other reports mentioned in Mohee and Rughoonundun [25] have indicated the following calorific values and moistures for MSW. A report from SIDEC Arup in 1998 concluded that the net calorific value of MSW in Mauritius would likely be between 6.7 and 7.5 MJ/kg while Fichtner [10] reported net calorific values for MSW in Mauritius between 9.3 and 13.5 MJ/kg, with an average moisture content of 47%.

Recycling will normally affect the gross calorific value, and ash or residue content of the wastes to be incinerated in a WtE facility. Increased removal rates of metals and glass will reduce ash or inert residues and increase the bulk calorific value. Also, the removal of garden waste will improve the overall calorific value of the remaining wastes. The likely impact on gross calorific value is shown in the Table 3. The major benefit of recycling green waste, apart from obtaining a potentially useable commodity (compost) is that it improves the quality of the residual waste for combustion by raising the net calorific value and reducing moisture content. Removing paper, card and plastics will reduce the calorific value. However, given the nature of the wastes recovery market in Mauritius, it is

Table 1 Forecast of wastes generation without waste minimization in Mauritius (tons/year) [10]

Year	Household wastes	Commercial wastes	Industrial wastes	Construction & Demolition wastes	Household hazardous wastes	Industrial hazardous wastes	Total wastes
2011	340,078	70,314	71,159	64,938	1,701	28,134	576,324
2012	354,021	74,533	74,717	67,535	1,770	29,262	601,838
2013	368,536	79,005	78,453	70,237	1,843	30,437	628,511
2014	383,646	83,745	82,375	73,046	1,918	31,660	656,390
2015	399,375	88,770	86,494	75,968	1,997	32,934	685,538
2016	415,750	94,096	90,819	79,007	2,079	34,260	716,011
2017	432,796	99,742	95,360	82,167	2,164	35,641	747,870
2018	450,540	105,727	100,128	85,454	2,253	37,079	781,181
2019	469,012	112,070	105,134	88,872	2,345	38,577	816,010
2020	488,242	118,794	110,391	92,427	2,441	40,136	852,431

Table 2 Average composition (% wet mass of MSW) of household, tourist, commercial and non-hazardous healthcare waste in Mauritius [10]

Waste fraction	Household	Tourist	Commercial	Non-hazardous healthcare
Plastics & Plastic Bottles	7.9	8.0	9.1	8.6
Glass	2.6	3.3	4.2	1.0
Textiles	3.2	0.3	2.3	1.3
Ferrous metals	4.1	1.2	2.3	1.1
Non-ferrous metals	0.2	0.6	0.2	0.0
Paper & Cardboard	14.4	10.2	19.6	18.7
Packaging materials	3.7	2.6	2.6	4.6
Hazardous wastes	0.5	0.2	0.1	0.0
Vegetables & Organic matter	49.1	67.5	44.0	44.2
Wood & Tree branches	1.6	0.6	2.7	0.0
Leather & Rubber items	1.1	0.3	5.1	0.0
Composite materials	0.7	0.3	0.7	0.0
Fine wastes (<20 mm)	5.6	3.3	3.1	2.2
Inerts & Other wastes	4.6	1.7	4.2	18.3
Total	100	100	100	100

unlikely that there would be a major trend to recover non-source separated plastics and paper.

It is often quoted that there is a major conflict between incineration, which requires long term contracts on minimum waste inputs and efforts to recycle waste. However, besides the facts that wastes are not sorted in Mauritius and the amounts to be potentially recycled are small, it has already been considered that the opportunity for "recycling" in Mauritius is limited due to the lack of reprocessing industries and a market for the recycled or recovered products, certainly on an economic basis. The main exception is composting. If an initial target recovery of green waste is achievable at 30–35%, the theoretical calorific value of the waste

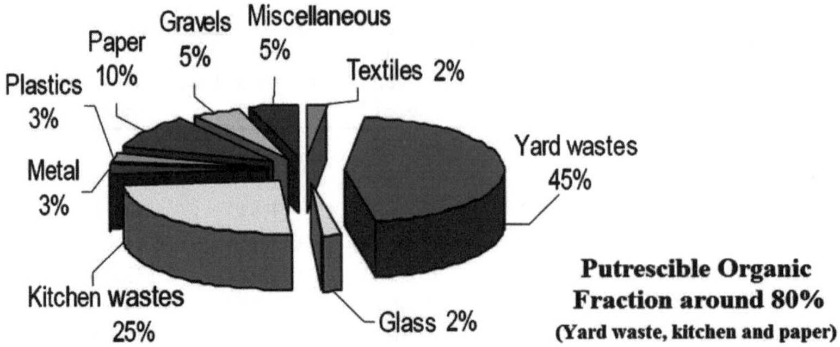

Fig. 2 Typical MSW composition in Mauritius [25]

Table 3 Trends predicted after removal of components from MSW streams

	Components being removed				
	Paper	Glass	Plastics	Metals	Organics
Moisture	↑	↑	↑	↑	↓
Ash content	↑	↓	↑	↓	↓
Gross calorific value	↓	↑	↓	↑	↑

↑: Concentration increases in residual waste as component is removed
↓: Concentration decreases in residual waste as component is removed

residue could increase to 7.2 and 8.1 MJ/kg for urban and rural wastes, respectively. With a maximum achievable target recovery of 75%, the calorific value could then improve to 7.8 and 8.4 MJ/kg, respectively. The amount of energy (electricity) that can be generated and then recovered from MSW fractions is dependent on the waste throughput, the calorific value of the wastes and the thermal efficiency of the combustion system. Normally, the thermal efficiency of larger WtE power plants (incorporating mass incineration units) is of the tune of 30% +. This efficiency is the efficiency of converting the thermal energy in the waste to electricity and/or heat. However, a smaller plant may be less efficient at 20–25%.

The Stoker diagram (Fig. 3) shows the relationship the heat input/electrical generation relationship for a small plant. Stoker diagrams are normally specific for each plant, and more particularly each grate design. Figure 3 also depicts the impact of calorific value on energy generation. Based on the findings of a study conducted by Brown & Root Environmental in 1998 for reviewing the National Solid Waste Management Plan of Mauritius, for a mean calorific value of 8.4 MJ/kg, it may be expected that the maximum gross output in a potential Mauritian plant be 0.473 MWh_e/ton, with a net export of 0.402 MWh_e/ton after parasitic loads (power used by the plant and normally of the order of 12–15%) have been deducted. For example, for a throughput of 53,000 tonnes/annum ($\approx$6 tonnes/h in

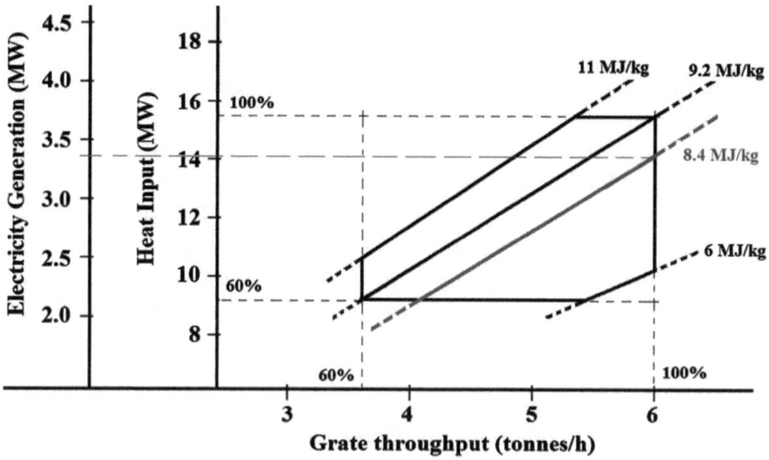

Fig. 3 Typical stoker diagram for an average calorific value of 8.4 MJ/kg for MSW fractions to be incinerated for a throughput of 6 tonnes/h in a WtE stoker incinerator

Fig. 3) in a small WtE incineration plant, the net power export might come down to 29 GWh_e/year for a 365 days/year and 24 h/day operation schedule.

4 Biomass-Derived Energy: Research in Mauritius

In Mauritius, research in the use of biomass sources to recover useful energy in the form of biofuels or electricity have been mostly conducted at the University of Mauritius (http://www.uom.ac.mu/) and through the Mauritius Research Council (MRC, http://www.mrc.org.mu/) supported and/or funded projects. The following sections substantiate on the latter.

4.1 Feedstocks Studied

Research at the University of Mauritius (UoM) on the utilization of biomass for the recovery of bioenergy has been conducted since the last 25 years in particular on the incineration and anaerobic digestion (and/or fermentation) themes using a variety of inorganic and organic substrates (Table 4).

- There has been a general shift in research from first generation energy recovery by incineration to third generation bioenergy recovery. This is evidenced from the fresh pool research project themes studied to explore the use of the anaerobic biotechnology for biogas and ethanol production from lignocellulosic substrates like bagasse, algae, vegetable wastes, kitchen wastes, some refractory MSW organic fractions and used oils.

- The wide majority of research so far conducted in the recovery of energy from wastes has been concentrated in the use of the calorimetric approach to determine the energy potential of one waste fraction or a blend of solid waste fractions. Such blends of non-biodegradable (plastics and textiles) and biodegradable (paper, cartons, sludges, MSW organic residues and used oils) have been tested for their net calorific values, volatile solids reduction and ash contents produced after combustion in the standard bomb calorimeter (Oxygen Bomb Calorimeter–PARR Instrument Company Inc., Moline, Illinois, Fig. 4) available in the chemical engineering laboratory of the Department of Chemical and Environmental Engineering.
- The results of the proximate analyses thus obtained have been used for the design of WtE power plant and in identifying potential Refused Derived Fuel mixes that could be harnessed for electrical energy in these power plants. In point of fact, in 2004 Lofur found that a mixture of bagasse (75% by mass), used cooking oil (12.5% by mass) and the balance being used oil had an average calorific value of 27.98 MJ/kg, and could be used to generate electricity in a medium scale sugar factory in substitution to coal.
- The next set of energy recovery projects from biomass sources have diversified to the application of the anaerobic digestion process to eventually produce biogas using substrates like organic MSW fractions, vegetables wastes, landfilled wastes, algal biomass and landfill leachate. Various reactors configurations (UASB), anaerobic digesters, SEBAC systems, biochemical methane potential (BMP) assays have been used and tested for this kind of research. A majority of research work on ADT have used the closed AD reactor configuration with the essential instrumentation and control accessories for pH, temperature and level control while a few have tested the feasibility of producing biogas using simple BMP assays and/or the UASB reactor mainly in the treatment and digestion of wastewaters with high organic pollution loads. In a work in 2006 (Table 4), Poinen designed a preliminary full scale SEBAC plant for the Municipality of Beau-Bassin/Rose-Hill to treat a throughput of 105 tonnes of MSW fractions. The plant consisted of 7 modules of 3 bioreactors and had a retention time of 30 days. The main conclusions from this design and its operation were that if offered an economical, effective and sustainable treatment of MSW with recovery of useful biogas for nearby communities, hotel resorts and some industries.

In another study in conducted in 2005 (Table 4), Ponnusawmy investigated the feasibility of using the SEBAC technology to treat a mixture of vegetable wastes and 10% chicken wastes. Experimental monitoring data from the two pilot scale mesophilic reactors and results indicated that both SEBAC set ups produced the required level of waste degradation and solubilization. An average VS reduction of 62.5% and an overall methane yield (57.2% of biogas generated) of 0.24 m^3/kg VS were recorded after 93 days of retention. All the more, the final residues from both

Table 4 Research work completed on the use of biomass and other materials for energy recovery at the University of Mauritius

Author	Year	Title of study	Biomass used	RT
Domun, P.	2002	Feasibility of using bagasse and used oil as an alternate source of energy	Bagasse and used oil	CS
Surroop, D.	2001	Investigating the potential of using Municipal solid waste with bagasse for energy generation	MSW and Bagasse	CS
Kokil, P.	2005	Energy generation through the use of solid municipal waste	MSW fractions	CS
Ramadhin, R.	2003	Characterisation of bagasse for energy generation	Bagasse	CS
Beesoo, K.	2004	Energy recovery alternatives from future mix of wastes combustibles in Mauritius	Mixed fractions of combustible wastes from MSW	CS
Lofur, V.	2004	Energy recovery from used oil and use cooking oil by combustion with bagasse	Used oil and bagasse	CS
Bhengeerothee, P.	2003	An investigation of using textile, paper and plastic waste as a source of renewable and sustainable energy for Mauritius	Textiles, paper and plastics wastes	CS
Lavenerable, J.C.S.	2009	Assessing the potential of emerging technologies for bagasse energy exploitation	Bagasse	CS
Domun, P.	2002	Feasibility of using bagasse and used oil as an alternate source of energy	Used oil and bagasse	CS
Victor, E.	2001	Assessment of the environmental and socio-economical impacts of energy from waste incineration as an appropriate technology for solid waste treatment	MSW fractions	CS
Bunjun, R.	1982	Recycling of municipal waste to produce electrical energy or compost	MSW fractions	CS
Jogeedoo, Y.	2009	Anaerobic digestion of algae and effect of thermal pretreatment on methane yield	Algal biomass	AD
Lobin, Y.	2000	Potential for the anaerobic co-digestion of municipal solid waste and wastewater derived sludge in Mauritius	MSW	AD
Vithilingum, D.	2002	Estimation of methane generation from Mare Chicose landfill using empirical models	Landfilled wastes	AD
Padiachy, K.	2002	The sequential batch anaerobic composting (SEBAC) technology for solid waste treatment a pilot plant investigation	MSW	AD

(continued)

Table 4 (continued)

Author	Year	Title of study	Biomass used	RT
Chadee, K.	2003	Anaerobic treatment of an organic wastewater on a labscale Upflow Anaerobic Sludge Blanket (UASB) reactor	Organic wastewaters	AD
Heerah, Sk. A.	2004	Steam pre–treatment of lignocellulosic wastes for biomethanogenesis: A preliminary study	Mixed vegetable wastes	AD
Ponnusawmy, S.	2005	A pilot plant investigation based on the sequential batch anaerobic composting (SEBAC) for the treatment of vegetables waste in Mauritius	Vegetable wastes	AD
Poinen, Y.	2006	Appropriateness of the sequential batch anaerobic composting (SEBAC) process for solid waste treatment	Solid wastes	AD
Lutchmanen, P.	2005	Treatability of leachate from Mare Chicose landfill by the UASB process	Landfill leachate	AD
Moholee, A.K.	2001	Oil extraction of *Jatropha curcas* L. for biodiesel production by solvent extraction	*Jatropha Curcas* L.	BDP
Rahiman, M.Z.	2003	The enhancing of ethanol production from molasses and cost benefit analysis	Sugarcane molasses	BEP
Khadun, B.N.	2005	A comparative life cycle assessment of the production of ethanol from sugar cane molasses in Mauritius	Sugarcane molasses	BEP
Rao, C.	1994	Studies on improved methods in ethanol fermentation using cane molasses as substrate	Sugarcane molasses	BEP
Sooknah, R.	1989	Low grade ethanol from molasses as a cooking fuel substitute for kerosene	Sugarcane molasses	BEP
Govind, N.	2010	Bioethanol production potential from organic municipal solid waste	MSW fractions	BEP
Noormamode, K.B.	2010	Potential production of ethanol from *Ulva lactura* algae	Algae	BEP

Research themes (RT) studied are classified as follows: *CS* for Calorimetric studies, *AD* for Anaerobic digestion, *BDP* for Biodiesel production, *BEP* for Bioethanol production

reactors had an average energy content of 14.82–15.29 MJ/kg on dry basis, indicating a further energy recovery potential from these final biosolids.

The use of appropriate pretreatment techniques to increase the bioavailability of biodegradable fractions from the lignocellulosic biomass for an enhanced biomethanogenesis has been scantily explored except for the work of Heerah et al. [14]. Heerah et al. [14] studied the steam pretreatment of lignocellulosic biomass (grass clippings, acacia branches, vegetable wastes and chicken wastes) at 95°C and 103 kPa for four consecutive steam cycles each lasting 45 min followed by

Fig. 4 Oxygen bomb calorimeter. Courtesy of Department of Chemical & Environmental Engineering, Faculty of Engineering, University of Mauritius

anaerobic digestion. The first set of experiments showed that steam pre-treatment had increased the final filtrate volatile fatty acids (VFA) level two-fold to 30.1 meq/L and that the filtrate chemical oxygen demand (COD) had increased from 10,500 to 16,000 mg/L. The anaerobic digestion produced a 49.6% decrease in COD from 38,450 mg/L, a 90.1% decrease in VFA from 42.5 meq/L and a 49.4% decrease in total solids from 8.40% for steam-pretreated biomass; and a 34.5% decrease in COD from 33,400 mg/L, a 93.0% decrease in VFA from 27.5 meq/L and 47.8% decrease in total solids from 11.5% for non steam-pretreated biomass. In terms of the bio-chemical methane generation performance, it was observed that after 36 days of anaerobic digestion, 12.72 and 12.76 L biogas from the non steam- pretreated (consisting of 50.3% CO_2 and 49.7% CH_4) and steam pretreated biomass (consisting of 35.3% CO_2 and 64.7% CH_4), respectively, had evolved.

Interestingly, since a couple of years the research themes have started to address the production of biofuels namely biodiesel and bioethanol using biomass such as algae and Jatropha Curcas L., and sugarcane molasses (Table 4, BDP and BEP), respectively. The results obtained so far are interesting and have been reasonably conclusive. Recently, [12] studied the conversion of organic MSW fractions to bioethanol by an acid catalyzed fermentation process. In this study, bioethanol was recovered using distillation and the highest ethanol concentration in the wort was 2.12% by volume. The ethanol yield was 185.5 L/tons dry organic MSW corresponding to a potential annual production of around 15.9 million cubic meters ethanol. The economic analysis showed that the production cost of 1 L of ethanol was lower at USD 0.45.

In March 2010, Noormamode investigated the possibility of producing of ethanol from *Ulva lactura* algae using a combined approach of acid hydrolysis, fermentation and distillationthis project. This studied showed that at an optimum acid concentration of 27% (w/w), the ethanol yield reached up to 3.87% purity. Further calculations demonstrated that around 3,173 m^3 of ethanol at 95.5% purity could be produced annually for a daily 60 tonnes throughput of *Ulva Lactura* algae at 300 days operation per year.

4.2 Research at MRC

The Mauritius Research Council (MRC) (http://www.mrc.org.mu/) was set up in May 1992 (Act no. 10 of 1992) as body to promote and coordinate Governments investment in research. The MRC acts as a vital body to advise Government on Science & Technology issues and to influence the direction of technological innovation by funding research projects in areas of national priority and encouraging strategic partnerships.

All projects funded by the Council are classified under one of the following areas: Biomedical and Pharmaceutical; Biotechnology; Energy Efficiency and Renewable Energy; Information Communication Technology; Land & Land Use; Manufacturing Technology; Ocean Technology and Marine Resources; Science & Technology Education; Social/Economic; Water Resources and Waste Management and Water Recycling. The energy efficiency and renewable energy projects undertaken at the MRC are outlined below and more details of these may be obtained from the website of MRC at http://www.mrc.org.mu/.

- CNO & WVO: MRC has completed a study on the use of Coconut Oil (CNO) and Waste Vegetable Oil (WVO which is used cooking oil, grease, frying oil probably including animal fats and/or fish oils from the cooking) as substitutes for diesel oil. Both short-term (i.e. emissions tests) and long-term (i.e. engine wear and tear tests) were carried out. The results were intended for use to further promote the recycling of WVO as a biofuels, with immediate target groups being hospitals (Ministry of Health & Quality of Life) and hotels.
- Islands of Agalega: MRC has also worked in close collaboration with Outer Islands Development Corporation to investigate the alternative use of CNO as a biofuel to generate electricity. Projects are in the pipeline to (1) generate electricity using various mixtures of CNO/diesel, and (2) production of biodiesel from CNO. These projects fall under the umbrella project ‘Energy self-sufficiency of Agalega—biofuels and solar energy’. MRC has already carried out a field trip to Agalega to investigate the prospects of using CNO and PV for electricity generation.
- Jatropha Biofuel Feasibility Study: MRC has carried out an economic feasibility study of Jatropha biofuel on behalf of the Ministry of Industry, Small & Medium Enterprises, Commerce & Cooperatives. The study investigated two scenarios: (1) the cultivation of Jatropha on marginal land in Mauritius for biodiesel production; and (2) the importation of raw materials vegetable oil and/or Jatropha seeds for biodiesel production in Mauritius.
- Biofuels Committee: MRC is the co-Chair of the Biofuels Committee (and Chair of the Technical Sub-Committee) set up by Cabinet to propose policies—covering technical, legislative and pricing mechanism(s) issues—for the introduction of biofuels in Mauritius. The findings of the committee are intended to complement the Energy Policy that is being worked out by EU consultants for Mauritius.

- Seaweed Biomass: This research is still at its infancy, and it aims at developing a seaweed industry in Mauritius. However, the thrust will be on the production of bio-fertilizer from the sap of the seaweed, and the generation of electricity from the combustion of the remaining biomass. A number of studies are underway to test the hypothesis and the initial findings are indeed promising.

5 WtE in Mauritius: Large-Scale Applications

In Mauritius, there are a number of large-scale installations where energy is recovered from the biomass feedstocks available in the country. The main feedstocks are bagasse produced as a by-product of the sugarcane milling process for sugar production, vinasse and molasses from the sugar producing operations in sugar factories and secondary sludge produced from wastewater/sewerage treatment processes. Vinasse is a byproduct of the sugar industry. Sugarcane is processed to produce crystalline sugar, pulp and molasses. The latter are further processed by fermentation to ethanol, ascorbic acid or other products. After the removal of the desired product (alcohol, ascorbic acid, etc.) the remaining material is called vinasse. Vinasse is normally sold after a partial dehydration and usually has a viscosity comparable to molasses.

The sugar cane crop has been occupying a prominent position in the Mauritian economy over the years since its introduction by the Dutch in the seventeenth century. After trials on a number of crops, it has been found to be the best crop suited for the agroclimatic conditions in the island which includes frequent visit of cyclones. Sugar production has increased over time to reach a plateau of around 600–650,000 tonnes, export markets and arable land area putting a limit to this production. The sugar industry is a major net foreign exchange earner but its relative contribution in the economy has been declining with the development of the tourism and manufacturing sectors. The essential features of the processes for bagasse cogeneration and biofuel production in Mauritius are now outlined with examples of actual large-scale installations.

5.1 Cogeneration of Bagasse in Sugar Factories

Mauritius has in the form cane biomass a very potent asset, which is not yet fully tapped. Of all the crops in Mauritius, sugarcane best assimilates solar energy both from the qualitative and quantitative perspectives. For every 100 tonnes of cane produced over each hectare, an average of 55 tonnes of carbon dioxide are fixed. In this way, every year 5 millions of environment friendly biomass is produced in the form of sugar cane.

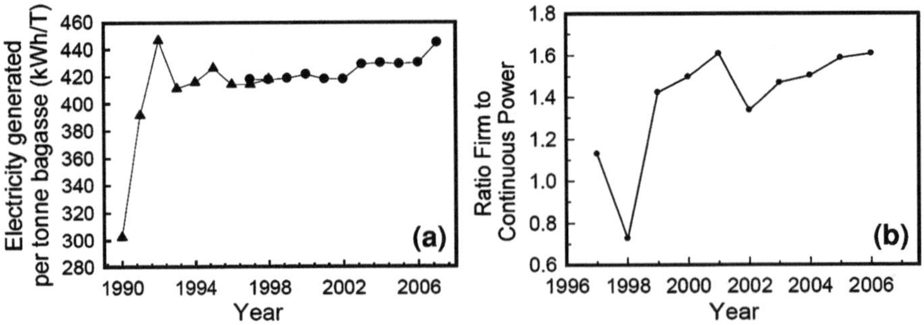

Fig. 5 **a** Electricity generated from bagasse (1990–2007). **b** Ratio of firm to continuous power. Source of two graphs is from Deenapanray [6] and original source of data for graphs is from CSO, Mauritius. Digest of Agricultural Statistics, http://www.gov.mu/portal/goc/cso/report/natacc/agri07/sugar.pdf (Accessed 19 September 2010)

The dependency of oil for electricity generation, excluding transportation and industry, has been reduced to some 45–48% today through the enhanced used of renewable energy sources such as the more efficient use of bagasse biomass and the use of coal as a complementary fuel to bagasse during the sugarcane off-crop season. The share of bagasse in electricity generation is around 13–17%.

Bagasse cogeneration was in many respects pioneered in Mauritius, and as early as 1926–1927, 26% of electricity generated in Mauritius was in sugar factories [6]. At this time, electricity was produced during only the sugar cane harvest season [6]. In this mode of operation, electricity generation is referred to as 'continuous power'. Although continuous power contributes positively to broaden the electricity mix of Mauritius, its seasonal character implies that there should be an equivalent power generation backup capacity held by the public utility [6]. Cogeneration is the simultaneous generation of electricity and steam (or heat) in a single power plant. When compared with the separate generation of electricity and steam, cogeneration is beneficial as it saves primary energy. The efficiency of a cogeneration plant is given as the sum of electric energy generated and heat energy in process stream divided by heat added to plant. There are two types of cogeneration. In the bottoming cycle cogeneration, primary heat is used at a high temperature, directly for process requirements. The low-grade waste heat is then used to generate electricity at low efficiency. It is generally of little thermodynamic or economic interest. In the topping cycle cogeneration, primary heat at the higher temperature end of the *Rankine* cycle is used to generate high-pressure and high-temperature steam and electricity. Depending on process requirements, process steam at low-pressure and low-temperature is extracted from the steam turbine. The topping cycle can provide true savings in primary energy.

Figure 5a shows the change in electricity generated per tonne of bagasse between 1990 and 2007. In general, there has been an increase in the electricity output per tonne of bagasse burnt, revealing an increase in the efficiency of

Table 5 Summary of power production from cogeneration of bagasse in sugar factories in Mauritius (annual estimates)

Sugar factory	Bagasse cogenerated (tonnes)	Cogeneration process conditions	Power production (MWh)
F.U.E.L.	171,780	High pressure boilers at 44 bars, 440°C with turbo alternator	4,990
CTSAV	100,000	Bagasse with coal cogeneration	19,355
CTBV	290,000	Bagasse.coal cogeneration using boilers operated at 82 bars, 520°C with turbo alternator	32,500

electricity output from bagasse cogeneration. In 2007, the electricity generated was around 445 kWh/tonne bagasse, which corresponds to ~110 kWh/tonne cane crushed. One of the main technological improvements leading to higher efficiency has been the use of high-pressure boilers. For instance, the two most recently built bagasse cogeneration power plants operate at a boiler pressure of 82 bars (producing superheated steam at 525°C), as opposed to older facilities operating at boiler pressures between 31 and 44 bars. Efficiency gains, leading to a surplus of electricity generation for export into the grid, have also been accomplished through the use of the turbo-alternator and the optimization of other process parameters, including process steam consumption, increasing fibre content of cane through genetic manipulation, lower moisture content of bagasse, and reducing the electricity consumption in the sugar mill and in the power plant. As will be discussed in the next section, several policy instruments have been put in place to increase the efficiency of cogeneration in Mauritius. Further, the development of bagasse cogeneration has been promoted by providing incentives for the cogenerator to export firm power onto the grid—i.e. electricity throughout the year and not just during the crop season. In an effort to shift from continuous to firm power generation, and in the absence of other renewable biomass in Mauritius, bagasse cogeneration projects co-firing with coal have been developed to ensure year-round operation (firm refers to bagasse during crop and coal during intercrop, and continuous refers to bagasse during crop season only). Figure 5b shows the ratio of firm to continuous electricity produced in Mauritius between 1997 and 2007. The trend reveals a clear shift towards the generation of firm power.

The sugar factories in Mauritius supply about 10–12% of the island's energy. Their expansion into electricity generation is partly a consequence of the Sugar Energy Development Project, part of which was World Bank funded. These facilities have sometimes been promoted as WtE plants but this is in fact rather misleading. Whilst waste bagasse biomass is used to generate electricity, this is generally required for internal uses on the plants and has to be imported to supplement the power generation for export. However, as internal power requirements are generally limited to a short season (July–December), the generation plant is available to produce electricity for export for the remainder of the year. A number of factories, which operate as the only Independent Power Producers (IPPs) in the

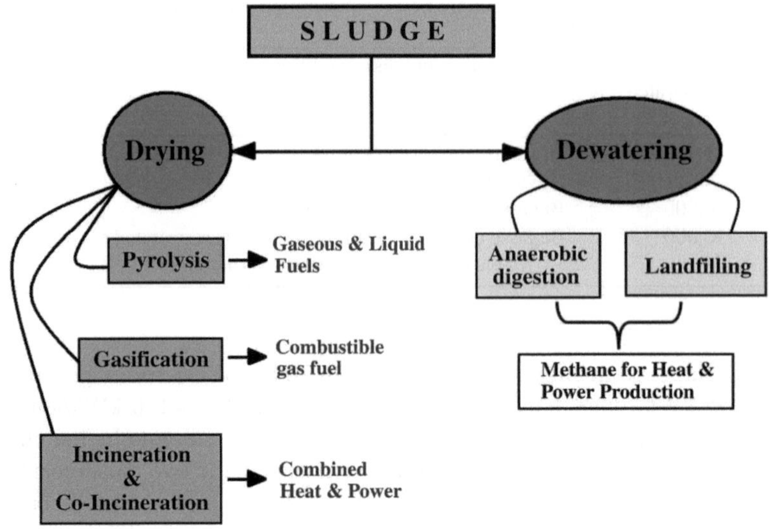

Fig. 6 Routes for recovery of energy from sludges

island, have expanded their capacity (and plan to expand further) to enable the generation of power for export throughout the year to the National Grid system (Table 5). In 2002, 299.1 GWh of electricity was sold to the National Grid while this figure increased to 317.4 GW h in 2004. This figure has since then been also increasing slowly but steadily with an average of 310.3 GWh per year over the last 5 years.

As an example, the Flacq United Estates Limited (F.U.E.L) sugar factory is expanding to provide 26 MW guaranteed export capacity during cropping, with 30 MW during the rest of the year. At present, the F.U.E.L. power plant operates for 24 h from Monday to Sunday. The technologies used are high pressure boilers of about 44 bars producing superheated steam at around 440°C and turbo-alternator to export electricity to the grid for public utility. During the crop season of 2010 (June–September 2010), around 10,737 tonnes of bagasse per week has been burned and co-fired with 1,031 tonnes of coal to produce 4.99 million kWh of electricity, of which 3.499 million kWh has been sold to the national grid managed by the Central Electricity Board for public utility.

The Central Thermique de Savannah (CTSAV) situated in the southern part of Mauritius has one on bagasse–cum–coal at La Baraque of 2 × 45 MW. It is the largest installation of its kind in this region of the world, and the unit at La Baraque produced 22.4% (equivalent to 510,261 MWh of electricity) of the national electric output in 2008. This power plant is operational for 4,000 h per year during crop season and from the records, it was observed that 19,355 MWh of electricity was generated from bagasse for a power plant which has an annual requirement of 100,000 tonnes of bagasse.

Table 6 Estimates of sludge generation at 25% dry solids at wastewater treatment plants in Mauritius

Treatment plant	Phase 1 (2005)		Phase 2 (2013/2015)	
	Tonnes of sludge	Dry solids (t/year)	Tonnes of sludge	Dry solids (t/year)
St Martin	26,412	6,603	61,640	15,410
Montagne Jacquot	24,820	6,205	51,820	12,955
Grand Baie	8,395	2,086	8,400	2,100
Central Housing Estates	400	98	400	100
Total	60,027	14,992	122,260	30,565

(Adapted from Mohee and Rughoonundun [25]. Original source of data: Environmental Solid Waste Management Programme, Feasibility report, [10])

Sample calculations for electricity from biogas	
Amount of dry sludge in 2013	≈30,565 tonnes (or 3,489 kg/h)
Volatile solids (VS) content	≈80%
Amount of VS destroyed	≈45% on average of VS available
Average biogas yield	≈1.0 Nm^3/kg VS destroyed
Motor electric yield	≈35% of Lower Calorific Value
Lower calorific value of biogas	≈6.0 kWh/Nm^3 @ 60% CH_4 in biogas
Energy generation	≈2,638 KWh

The Centrale Thermique de Belle Vue (CTBV) plant was commissioned in the year 2000. This plant is considered to correspond to the state-of-the-art technology in the field of bagasse advanced cogeneration. The Belle Vue sugarcane mill connected to CTBV has an annual cane crushing capacity of 1.0–1.2 million tones of canes and the crushing rate is 350 tonnes cane per hour. The CTBV power plant marked a step further in the improvement of technology with bagasse/coal boilers operating at 82 bars and 520°C, and a steam output of 140 tonnes/h. In this set–up, in an average year, the CTBV power plant produces a minimum of 325 million kWh of electricity for the grid (including 105 million kWh from bagasse), 26 Million kWh for the factory as well as 340,000 tonnes of low pressure steam, by burning 290,000 tonnes of bagasse and 130,000 tonnes of coal. This is extremely important for a small country like Mauritius, where bagasse is the only fuel to be produced locally, and dependency on oil or other imported fuels may severely affect the economy.

5.2 *Power Generation from Sludge-Derived Biogas*

Sludges may be handled and treated conventionally by land application, composting and ocean dumping; and there are also several energy recovery routes for this biomass source as illustrated in Fig. 6. Sludges generated from the wastewater treatment plants operational in Mauritius are also used to generate electricity through the combined heat and power combustion of biogas produced from anaerobic digestion processes. Table 6 gives a summary of the average amounts of sludges produced at the wastewater treatment plants in Mauritius. A sample set of calculations immediately after gives an indication of the biogas generation potential from sludge as a biomass source for energy recovery with St Martin as case study.

The Wastewater Management Authority (WMA) has been established as a body corporate under the Wastewater Management Authority Act to be responsible for all matters relating to the collection, treatment and disposal of wastewater. It operates as an autonomous organisation under the aegis of the Ministry of Renewable Energy and Public Utilities. The WMA thus plays a vital role in the protection of the environment and in ensuring the country's sustainable development by the provision of appropriate water pollution standards, wastewater control systems and management services to the entire population of Mauritius. *Berlinwasser International* has recently won a 7 year Operation & Maintenance contract for the 70,000 m^3/day St Martin Sewerage Treatment Plant in Mauritius (http://www.berlinwasser.com/content/language1/html/1378.php).

The St–Martin Sewerage Treatment Plant is located in the west of Mauritius, and is rectangular site of approximately 30 hectares. The expansion of St Martin Sewerage Treatment Plant has involved an upgrading in the system of oxidation ponds to the introduction of the more complex low rate activated sludge process with biological nutrient removal followed by tertiary treatment with rapid gravity sand filters followed by UV disinfection. More importantly, and with regards to the recovery of bioenergy from the process wastes, sludge treatment is carried out by an efficient anaerobic digestion process in digesters followed by centrifugal dewatering. With an average sludge feed of usually varying between 250 and 285 m^3/day (with a sludge dry solid content varying from 22.0 to 25.5% and volatile solids ranging between 68.3 and 73.5%) to the anaerobic digester, the biogas production yield typically turns around 2,500–2,600 Nm^3/day. A combined heat and power plant then generates electricity from the surplus biogas generated.

5.3 Bioethanol Production

Omnicane was launched in June 2009 through a strategic re-branding of Mon-Trésor–Mon-Désert Ltd, a long established sugar cane group in Mauritius as sugar was produced on its very industrial site as early as 1818. Omnicane is engaged in sugar cane production and processing, electricity production, food crop, flower and venison production. With a view to mitigate the pollution associated with the use of fossil fuels, Omnicane is currently working towards establishing an ethanol distillery of 22.5 million litres per year capacity at La Baraque. A Memorandum of Understanding (MoU) has already been signed between Alcodis and Omnicane for the transfer of the distillery unit to the site identified. Energy requirements for the distillery will be optimized with the biomass available and the vinasse produced will be used to produce liquid bio fertilizer for use in sugar cane production. The production of ethanol as a biofuel can cater for the 10% blend with gasoline (E10) consumed in Mauritius.

6 Landfill Gas Management at Mare Chicose Landfill

The Ministry of Local Government and Solid Waste Management is responsible for the sanitary landfill site at Mare Chicose in Mauritius, and which is the only official landfill on the island. The Mare Chicose landfill is situated in a very wet area, prone to cyclones with a high annual rainfall in excess of 2,600 mm per annum. The original design and construction of the landfill infrastructure was undertaken by Scott Wilson Kirkpatrick Consulting Engineers, and Grinaker Construction Ltd, respectively. For the past 15 years since its inception, the landfill has been operated for Cells 1, 2 and 3, and more recently by the end of 2006 for Cell 4. The construction of a new cell (Cell 5) was started in September 2004 for an estimated capacity in excess of 250,000 m^3.

Although based on limited surveys, it is generally clear (as indicated in Table 2) that the largest fractions of the wastes reaching the landfill are putrescible kitchen and yard wastes. The Mare Chicose landfill in Mauritius has been in operation since 1997 and has received more than 1.3 million tons of wastes up to now. The wastes being highly biodegradable in nature with more than 80% of organics, a large amount of biogas is being generated. The latter have a significantly higher yard waste component. On a country average basis, the fractions of yard and kitchen wastes outweigh the remaining categories. The existing flares at the Mare Chicose landfill are enclosed flares. Enclosed flares are devices where the residual (landfill) gas is burned in a cylindrical or rectilinear enclosure that includes a burning system and a damper where air for the combustion reaction is admitted. The residual gas is the gas stream containing methane that is to be flared as part of a potential project activity.

Current landfill gas collection and flaring at the Mare Chicose landfill only manage some 15–25% of the total gas yield. A higher recovery of methane through a Clean Development Mechanism (CDM) landfill gas utilization project will potentially realize the collection and destruction of in excess of 50% of the landfill gas from the site. The CDM, established under the Kyoto Protocol, allows the funding of measures that can decrease the emission of greenhouse gases. This mechanism is designed to assist developing countries in achieving sustainable development and developed countries to meet greenhouse gas emission reduction targets. A study by Mudhoo and Mohee [27] predicted the annual generation of landfill gas (LFG) and methane using the *Scholl–Canyon* model for 1998–2030 for five scenarios of landfill management at Mare Chicose. Using the Approved Large Scale Methodology AM003 and Approved Baseline Methodology AM0010 of the Approved Methodologies for CDM activities, the carbon dioxide emissions

Table 7 Predicted total electricity generations (GWh) from LFG under a potential CDM project. Data compiled from Mudhoo and Mohee [27]

EI_{gd} (tCO_{2e}/MWh)	Scenario 1	Scenario 2	Scenario 3	Scenario 4	Scenario 5
0.451	95.32	391.52	628.25	937.01	186.62
0.708	60.41	248.14	398.19	593.88	118.28
0.961	183.52	294.49	439.23	87.48	183.52

reduction and electricity generation capacities for three CO_2 emissions intensities from flaring of methane were actually estimated for 2008–2030. The total LFG generation was found to amount to an expected 319.8–2,371.2 Mm^3 for 1998–2030, and at 50% (v/v) methane content and 60% recovery, the maximum LFG generation and recoverable volume of methane amounted to 1,869.5 and 560.9 Mm^3 for 2008–2030, respectively. This corresponded to an emissions reduction of up to 421,657.3tCO_{2e}. These very large volumes of untapped landfill gas were eventually shown to correspond to potential power generations varying from 44.6 to 937 GWh for the three emissions intensities considered at the grid system (EI_{gd}) (Table 7).

In October 2010, Sotravic Ltée and the Bilfinger Berger Group which are the present joint venture managing the operation and maintenance of the Mare Chicose landfill, have come into an agreement with the Central Electricity Board (CEB) of Mauritius to implant an electric power station under a 'Gas-to-Energy' CDM project. This power station shall burn LFG captured at the Mare Chicose landfill (40–60% CH_4) for producing electricity. The construction of this power station started in October 2010 and it shall be operational as from June 2011 for the generation and subsequent sale of electricity. The design shall comprise 3 generators each of a capacity of 1 MW and the promoters of this Gas-to-Energy project have evaluated the project value to the tune of USD 8,000,000 and expect to achieve an emissions reduction of 300,000 tCO_{2e} by 2016. The total 3 MW of 'green' electric power thus produced shall correspond approximately to 1% of the daily energy production for the National Grid of the CEB and will be able to meet the energy demand of around 20,000 households. In order to further enhance the capture of LFG and optimize on the generation of LFG-derived green power, it is planned to install a further 50 vertical tubes to collect more LFG by 2016.

7 Concluding Remarks

The pool of research work and applications of WtE from biomass sources in Mauritius have been focused on incineration and anaerobic digestion processes. Although the fundamentals of these WtE technologies are fairly well understood,

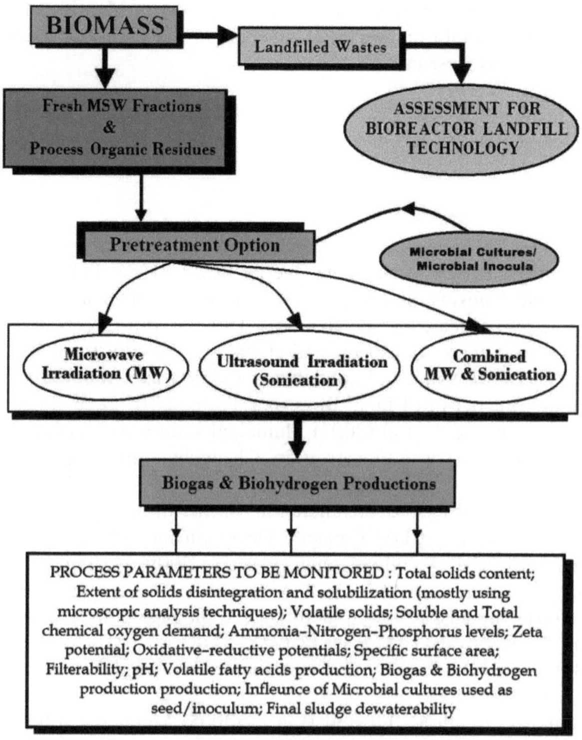

Fig. 7 Multilevel approach to research for pretreatment of biomass for energy recovery

the need for further research to optimize the exhaustibility of biomass for energy recovery should become a major research area of concern in Mauritius.

All the more, it is also important to gear this research towards potential process applications which fit with the concept of Green Chemistry and Green Engineering (Fig. 7). Based on a survey of specific literature on novel and green technologies, the areas of res earch on energy recovery from biomass which may be planned, initiated and worked on in Mauritius could comprise (1) the study of ultrasonic energy in its effects and potential application as pretreatment (i.e. sonication) of biomass; (2) the study of the effects of microwave irradiation on biomass in combination with other pretreatment conditions for biogas production; (3) developing research themes and projects on biohydrogen generation from the biomass feedstocks and at a later stage also investigate the effects of sonication and microwave irradiation on the production of biohydrogen, and (4) an assessment of the potential of integrating a bioreactor landfill system at Mare Chicose landfill.

Acknowledgments The authors express their gratitude to all those persons whose valuable data, inferences and views have been of significance in adding substance to this chapter. The authors also acknowledge the institutional support from the University of Mauritius Library, Department of Chemical and Environmental Engineering (University of Mauritius). A special words of thanks is also due to the Mauritius Research Council.

References

1. Asif M, Muneer T (2007) Energy supply, its demand and security issues for developed and emerging economies. Renew Sustainable Energy Rev 11:1388–1413
2. Balat M (2008) Mechanisms of thermochemical biomass conversion processes part 3: react liquefaction. Energy Sources, Part A: Recovery Util Environ Effects 30:649–659
3. Birol F (2006) World energy prospects and challenges. Austral Econ Rev 39:190–195
4. Cheng H, Hu Y (2010) Municipal solid waste (MSW) as a renewable source of energy: current and future practices in China. Bioresour Technol 101:3816–3824
5. Cheng H, Zhang Y, Meng A et al (2007) Municipal solid waste fueled power generation in China: a case study of waste-to-energy in Changchun city. Environ Sci Technol 41: 7509–7515
6. Deenapanray P (2009) Bagasse cogeneration in Mauritius: policy lessons for African countries. UNDP Mauritius, CDM Capacity Development in Eastern and Southern Africa. http://un.intnet.mu/UNDP/downloads/energy_sector/2009-04_Bagasse_Cogeneration_In_Mauritius-Policy_Lessons_For_Africa.pdf. Accessed on 14 September 2010
7. Demirbas A (2001) Biomass resource facilities and biomass conversion processing for fuels and chemicals. Energy Convers Manag 42:1357–1378
8. Dincer I (2000) Renewable energy and sustainable development: a crucial review. Renew Sustainable Energy Rev 4:157–175
9. Eriksson O, Carlsson Reich M, Frostell B et al (2005) Municipal solid waste management from a systems perspective. J Cleaner Prod 13:241–252
10. Fichtner Baseline and Concept Report (2000) Available at the ministry of local government in Mauritius
11. Gerbens Leenes PW, Hoekstra AY, van der Meer Th (2009) The water footprint of energy from biomass: a quantitative assessment and consequences of an increasing share of bio-energy in energy supply. Ecol Econ 68:1052–1060
12. Govind N (2010) Bioethanol production potential from organic municipal solid waste. Bachelor Dissertation, University of Mauritius
13. Goyal HB, Seal D, Saxena RC (2008) Bio-fuels from thermochemical conversion of renewable resources: a review. Renew Sustainable Energy Rev 12:504–517
14. Heerah ASk, Mudhoo A, Mohee R, Sharma SK (2008) Steam pre-treatment of lignocellulosic wastes for biomethanogenesis: a preliminary study. RASĀYAN J Chem 1:503–514
15. Holmgren K, Henning D (2004) Comparison between material and energy recovery of municipal waste from an energy perspective: a study of two Swedish municipalities. Res Conserv Recycl 43:51–73
16. Kalogo Y, Habibi S, MacLean HL et al (2007) Environmental implications of municipal solid waste–derived ethanol. Environ Sci Technol 41:35–41
17. Kavouridis K, Koukouzas N (2008) Coal and sustainable energy supply challenges and barriers. Energy Policy 36:693–703
18. Kaygusuz K, Türker MF (2002) Biomass energy potential in Turkey. Renew Energy 26: 661–678
19. Khanal S (2008) Overview of Anaerobic Biotechnology. In: Khanal S (ed) Anaerobic biotechnology for bioenergy production: principles and applications. Wiley–Blackwell Publishing, New York http://www.wiley.com/WileyCDA/WileyTitle/productCd0813823463.html
20. Keoleian GA, Volk TA (2005) Renewable energy from willow biomass crops: life cycle energy, environmental and economic performance. Crit Rev Plant Sci 24:385–406

21. Li A, Khraisheh M (2010) Bioenergy II: Bio-ethanol from municipal solid waste (MSW): the role of biomass properties and structures during the ethanol conversion process. Int J Chem Reactor Eng 8:85
22. Luts D, Devoldere K, Laethem B et al (2000) Co-incineration of dried sewage sludge in coal-fired power plants: a case study. Wat Sci Technol 42:259–268
23. Mizuno T, Goto M, Kodama A et al (2000) Supercritical water oxidation of a model municipal solid waste. Ind Eng Chem Res 39:2807–2810
24. Mohee R (2002) Assessing the recovery potential of solid waste in Mauritius. Res Conserv Recycl 36:33–43
25. Mohee R, Rughoonundun H (2006) Solid wastes as a potential energy source. Renewable Energy Workshop, Mauritius Research Council, 25–26 July 2006
26. Morris M, Waldheim L (1998) Energy recovery from solid waste fuels using advanced gasification technology. Waste Manag 18:557–564
27. Mudhoo A, Mohee R (2009) Estimates of CO_2 emissions reduction and potential power generation from biogas at Mare Chicose landfill. Int J Global Environ Issues 9:169–192
28. Psomopoulos CS, Bourka A, Themelis NJ (2009) Waste-to-energy: a review of the status and benefits in USA. Waste Manag 29:1718–1724
29. Rabaey K, Verstraete W (2005) Microbial fuel cells: novel biotechnology for energy generateon. Trends Biotechnol 23:291–298
30. Rulkens W (2008) Sewage sludge as a biomass resource for the production of energy: overview and assessment of the various options. Energy Fuels 22:9–15
31. Tyskeng S, Finnveden G (2010) Comparing energy use and environmental impacts of recycling and waste incineration. J Environ Eng 136:744–748

Potential of Municipal Solid Waste in Hanoi for Energy Utilisation

Trang Nguyen thi Diem, Giang T. H. Nguyen, Sven Schulenburg and Bernd Bilitewski

Abstract The integration of Renewable Energy Sources (RES) within the contextual framework of existing thermal energy production systems has emerged as a promising and sustainable policy towards addressing the growing global energy demand. Especially for developing countries, as they are characterized by decentralized energy systems, locally available RES are a viable option for generating thermal energy. In this chapter, we provide a methodological framework for integrating waste biomass into a portfolio of supply chains for thermal energy production, by presenting the relevant drivers for waste biomass usage making especially the case for developing countries, the associated systems and the supply chain operations. A generic strategic optimization model is proposed for determining the optimal mixture of energy sources for a specific region. This model could be employed by a system's regulator to conduct various 'what-if' analyses, in order to develop comprehensive effective policies that also integrate waste biomass into the existing energy system. Finally, a real-world case study is presented, and interesting managerial insights are discussed.

Abbreviation

MoC	Vietnam Ministry of Construction
WB	World Bank
ISTEAC	Integration of Solid Waste Management Tool into specific settings of European and Asian Communities

T. N. t. Diem (✉)
Chemistry, Hanoi University of Science, 19 Le Thanh Tong, Hanoi, Vietnam
e-mail: trang_ntd@yahoo.com

G. T. H. Nguyen · S. Schulenburg · B. Bilitewski
Institute of Waste Management and Contaminated Site Treatment,
Technical Univarsität Dresden (TU Dresden), Pratzschwitzer Strasse 15,
D-01796 Pirna, Germany
e-mail: abfall@mail.zih.tu-dresden.de

A. Karagiannidis (ed.), *Waste to Energy*, Green Energy and Technology,
DOI: 10.1007/978-1-4471-2306-4_13,

JICA	Japan International Cooperation Agency
URENCO	Urban Environment Company
VEM	Vietnam Environment Monitoring
VEPA	Vietnam Environment Protection Agency
VS	Volatile substance
WtE	Waste to Energy
RDF	Refused Derived Fuel
FS	Fresh waste (wet waste)
GHGs	Green house gas
HV	Heating value
CHP	Combined heatpower plant
MBT	Mechanical biological treatment
Cbio.	biogenic carbon/regenerative carbon
Cfoss.	fossil carbon
LCV	Lower calorific value
MSW	Municipal solid waste
NCV	Net caloric value
oTS	organic dry substance (organic matter)
Rbio.	biogenic part
TC	Total carbon
TS	Total solid

1 Introduction

Waste generation in urban areas in Vietnam is increasing parallel with the rapid increasing of citizens, the economic growth and the improvement of living-standard. Between 1997 and 1999, waste generation increased by 30%; in which municipal solid waste (MSW) accounted for three-quarters, followed by industrial and medical waste (11%) [6]. In 2000, waste generation ratio for large cities in Vietnam ranged between 0.5–0.8 kg/cap/d while it was 0.3–0.4 kg/cap/d for rural areas. In 2003, around 50% of the whole solid waste was produced in urban areas although only 25.8% of the country's population was living there. With the increasing of generation ratio over 10%/cap/year, it is expected that the average waste generation ration in urban areas in Vietnam would be 1–1.6 kg/cap/d in 2010 (Table 1).[1]

[1] Survey data from 2006–2007 (VEPA [17]), Hanoi and HoChiMinh are belong to centrally-controlled municipalities.

Table 1 Solid waste generation in urban areas, divided by city level. Source: [17]

City level		Urban MSW generation			Contribution of MSW generation in urban areas in Vietnam 2007
		kg/cap./d	t/d	t/a	
Centrally-controlled municipalities	DB (2)	0.84	8,000	2,920,000	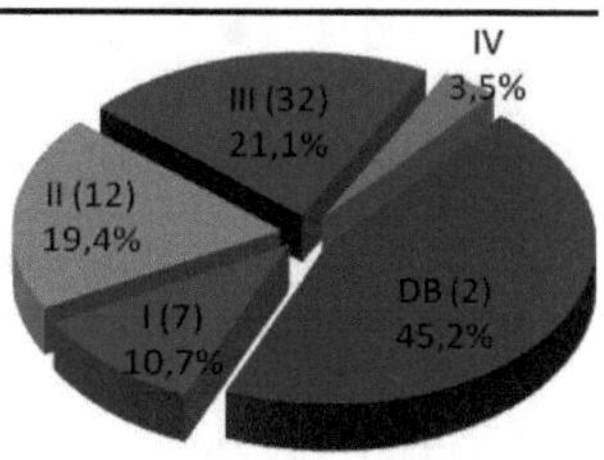
Provincial municipalities	I (7)	0.96	1,885	688,025	
	II (12)	0.72	3,433	1,253,045	
	III (32)	0.73	3,738	1,364,370	
	IV (-)	0.65	626	228,490	
Sum			17,682	6,453,930	

Table 2 Percentage of household in urban areas by methods of garbage disposal. Source: statistic of world bank (WB) [14]

Location	Garbage truck	Burning	Burying	Throwing to river	Throwing to animal closure	Other
Rural	6.8	63.0	23.0	15.0	16.7	18.9
Urban	71.0	20.0	7.5	6.3	4.1	2.8
Total	21.9	52.9	19.4	12.4	13.7	15.1

Landfill is a dominant form of solid waste disposal in Vietnam currently. According to VEM [16], there are 91 disposal sites in the country, only 17 are sanitary landfills which located in only 12 out of 61 cities and provincial capitals. The remains are open-dumps which do not satisfy the basis requirement of landfill (i.e. ground linings; adequate top covers) and located next to residential areas. These poorly operated dumps cause a multitude problem for environment and surrounding residents, such as ground-/surface water pollution by leachate; bad odours etc. In rural areas where it hardly access to environmental services, self-disposal is a common. The waste is often dumped in rivers or lakes, or at the sites nearby their homes.

MSW in Vietnam is composed much of organic, between 50.27 and 62.22%, and high content of construction materials such as soil, sand, stone and broken bricks [10] which reduces the average calorific value (NCV) to 3,800 kJ/kg MSW. In the big cities like Hanoi and Hochiminh, NCV of mixed MSW is approximately 6,000–7,000 kJ/kg depending on the season. Although MSW is used for energy utilization in European under form of Refuse Derived Fuels (RDF) either directly or co-combustion in power plants, it is still new in Vietnam (Table 2).

Regarding to the potential for energy utilization of MSW, a case study focuses on analysis of parameters influenced to energy utilization was undertaken in the laboratory of Hanoi University of Science (HUS), Vietnam in cooperation with Institute of Waste Management and Contaminated Site Treatment (IAA), TU Dresden, Germany. Results focus on net calorific value (NCV); biogenic part

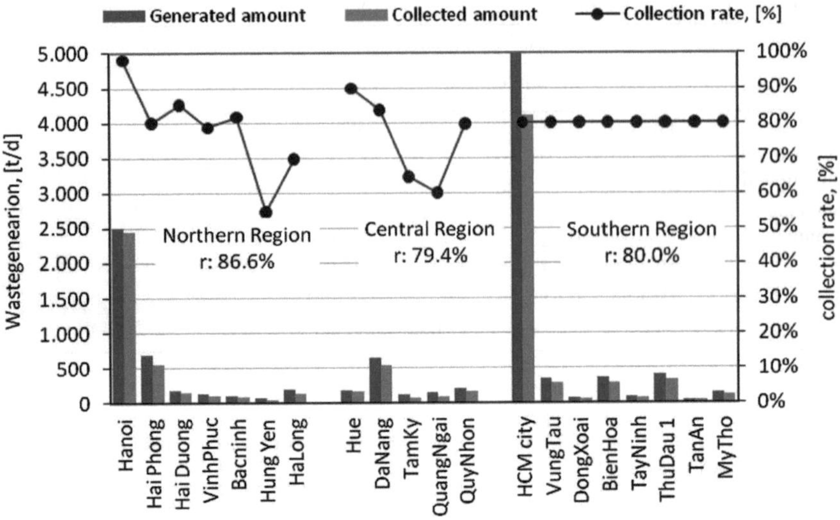

Fig. 1 Domestic waste generation and collection in urban Vietnam, divided by cities and regions (2005–2006)

(Rbio); total carbon content (TC) and regenerative carbon. Moreover, a RDF production process of Vietnam has been also investigated.

2 Municipal Solid Waste in Hanoi

2.1 Quantities and Characteristics

To be the capital of Vietnam with over 3 million citizens (2008), Hanoi is the major waste producer in the North with average ratio of 1.3 kg/cap/d and reaches to the second place of the whole country (after HCM city). In northern Vietnam, Hanoi also takes the lead with 54.3% of the total urban solid waste discharged by the whole region (Hanoi URENCO 2006). The average waste collected rate is estimated 86.6% of the whole region, where it is up to 98% in Hanoi [14] and 2006. In the South, there are 6,566 tons MSW produced daily, of which 80% are collected. HCM city produces the largest volume, about 81.83% of the total region's MSW. Urban in Central region is the least waste-generating area with 0.7–0.85 kg/cap/d; whereby Da Nang tops the list with 0.83 kg/cap/d, followed by Hue of 0.76 kg/cap/d (Fig. 1).[2] Waste generation in Hanoi in comparison with urban areas over Vietnam would be seen in picture 13.1 (MoC-2008).

[2] Sources: URENCO (2005-2006); local Department of Natural Resources and Environment (2005); Institute on Rural and Urban Planning, Ministry of Construction; Centre for Research and Urban Environment Planning.

Table 3 Prognosis of waste generation in Hanoi, Vietnam from 2001 to 2020

Year	Waste growth rate yearly (%)	Generated MSW (t/d)	(t/a)
2001	5.04	1,545	563,925
2002		1,623	592,347
2003		1,705	622,201
2004		1,791	653,560
2005		1,881	686,499
2006	4.86	1,972	719,863
2007		2,068	754,849
2008		2,169	791,534
2009		2,274	830,003
2010		2,384	870,341
2011		2,472	902,108
2012		2,562	935,035
2013		2,655	969,164
2014	3.65	2,752	1,004,539
2015		2,853	1,041,204
2016		2,957	1,079,208
2017		3,065	1,118,599
2018		3,177	1,159,428
2019		3,292	1,201,747
2020		3,413	1,245,611

Calculation based on the growth rate of GDP and waste generation per year
Source: (ICA and Hanoi URENCO Report 2006)

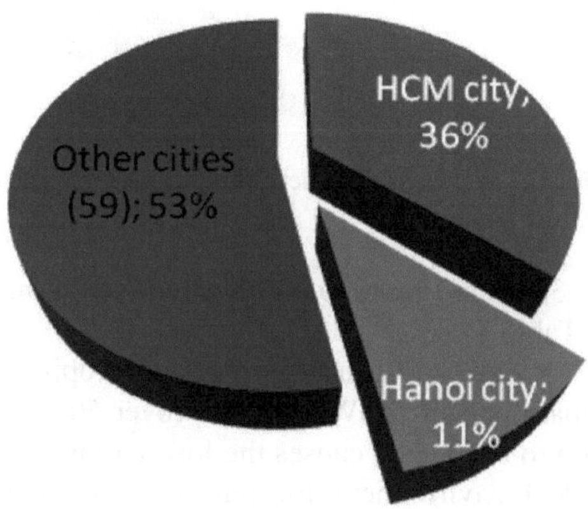

Fig. 2 Waste generation in urban 2009

The population growth, urbanization and increasing of consumption are reasons of rapid waste generation, which also increase the portion of hazardous waste (batteries, household solvents etc.) and non-degradable waste (plastics, metal) in MSW [16]. According to (JICA 2006), MSW quantity in Hanoi would increase to

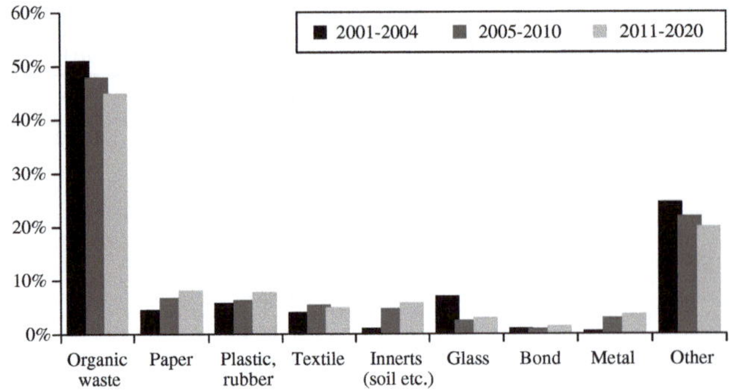

Fig. 3 Prognosis of MSW composition in Hanoi by period of time [7, 14]

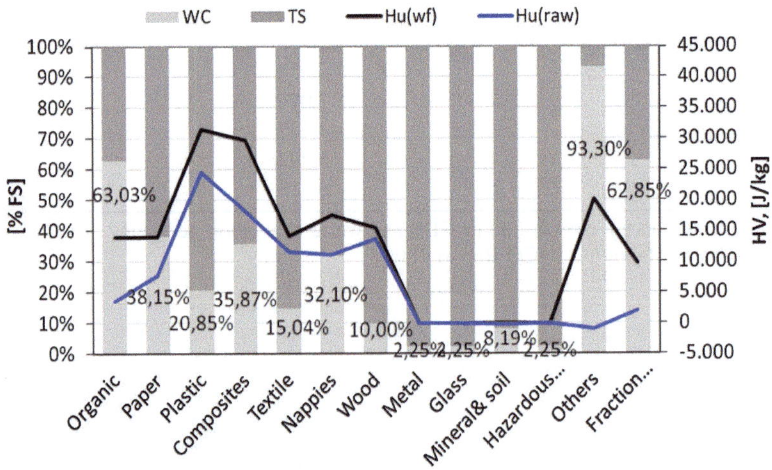

Fig. 4 Water content, net calorific value by waste fractions [8, 12]

2,384 ton/d in 2010 and continue to be the top of the North in the incoming years (Table 3) (Fig. 2).

Likely with waste in other developing countries, organic fraction constitutes a main part of MSW in Hanoi (over 50%) [11]. The high water content of organic, up to 63% [11], causes the low average calorific value of the waste and creates an ideal environment for micro-degradation. According to prognosis reported by JICA [7], the constitution of organic fraction in Hanoi waste would be reduced in the future; while other high-calorific fractions such as: plastics, packaging, paper, textile etc., would be increased (see Fig. 3).

An experiment on MSW in Hanoi to provide a basic data on characteristics of different waste fractions for forecasting the potential treatment of MSW later on was taken by Schulenburg [12] and Nguyen [8].

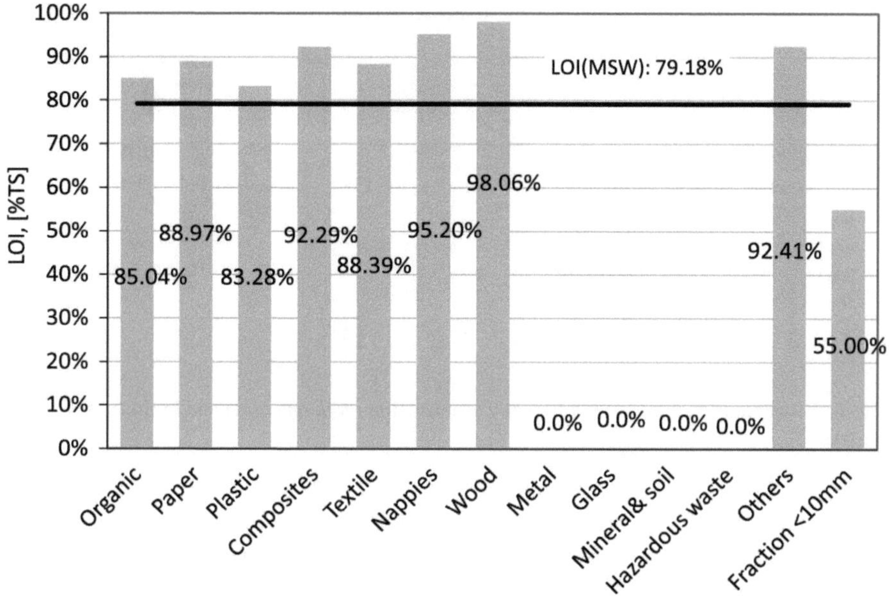

Fig. 5 Loss on ignition (*LOI*) in dry waste by waste fractions [8]

The experiment focused on determination of water content, biogenic content, total carbon, calorific value and the concentration of some heavy metals (Pb, Cd, Cr, Zn, Cu, Ni) (detail data see Schulenburg [12]). Within this paper, some main characteristics by waste fraction (e.g. Rbio, TC, Cbio, LCV etc.) will be presented.

The first results on water content (WC) and net calorific value (NCV) are provided in Fig. 4, of which the highest water content is found in "others" fraction. The composition of this fraction, however, is variable during year and constitutes normally less than 1% in the total waste; its effect on the whole MSW is negligible. The highest contribution in MSW is "Organic" with more than 50% mass content, as a result, its water content would be strongly influent on the average WC of MSW (WC MSW: 49.41%).

Net calorific value (NCV, kJ/kg) is presented in Fig. 4 in two values, with and without water. Similarly with other study, paper, plastic, composites, nappies and wood have highest calorific values with up to 24,491 kJ/kg (LCV, plastic) while metal, glass, mineral & soil get minus values for LCV. The minus values could be understood as a required energy to evaporate available water in theses fractions during combustion process. Average lower calorific value (LCV) of MSW in Hanoi in 2010 is calculated as 6,948 kJ/kg.

Investigation of volatile substance (loss on ignition, LOI) in MSW, Hanoi 2010 was carried out at Institute of Waste Management and Contaminated Site Treatment (IAA), TU Dresden, Germany, using the same waste samples used by Schulenburg [12]. LOI not only presents the behaviour of waste fraction in incineration process (burnable capacity) but also gives an estimation of organic matter content for degradable

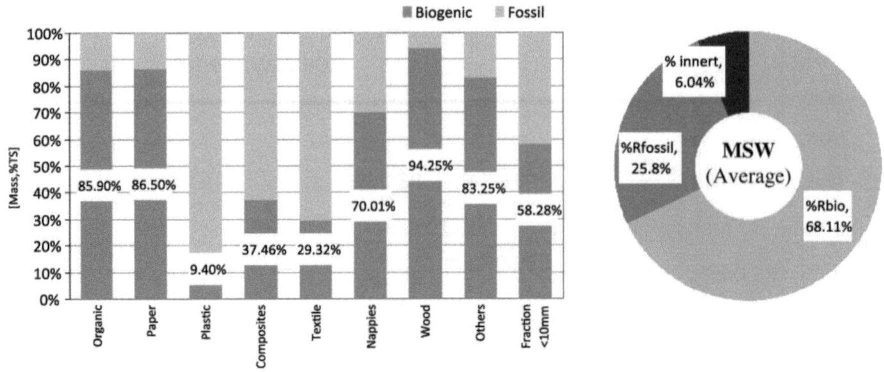

Fig. 6 Biogenic contribution by different waste fraction (*column chart*) and in mixed waste (*pie chart*) [12]

fractions (e.g. biowaste, paper, textile etc.). The result is presented by waste fraction in Fig. 5. In general, LOI of waste fraction is high, maximum value is found at wood with 98.06% and minimum is 55% for fraction <10 mm. Other fraction, such as: "metal", "glass", "mineral & soil" and "hazardous waste", are not experimented and be considered as inert (LOI = 0). In average, loss on ignition of MSW in Hanoi, 2010 is 79.18%.

Nine waste fractions were experimented on biogenic content, excluding: metal, glass, mineral & soil and hazardous waste, which are considered as inert in MSW (incombustible). Of all investigated fractions, organic, paper and wood have the highest biogenic content with 85.9, 86.5 and 94.25% respectively. Plastic and composites, in contrast, contain only 9.4 and 37.5% of biogenic for each fraction. With regard to mass contribution, the average biogenic percentage in MSW, Hanoi 2010 is calculated by 68.11% (respected to the calorific value).

However, it should be recognized that the biogenic data of some specific factions (e.g. organic, paper and wood) given in literature are almost 100% while experimental results in these cases pointed out a values of less than 100%. In essence, this caused by the inhomogeneity of waste fractions during sampling; that means, some fine plastics/composites or other inert fraction were mixed with the organic/paper and hard to separate by manual sorting. The mixing of foreigner fraction in these fractions, on the other hand, also increases their mass percentage in MSW. Respecting to the mass contribution, the result is thus acceptable. The same explanation also uses for plastic (Fig. 6).

Total carbon (TC) by waste fractions (regarding to wet waste) is experimented at IAA, TU Dresden [8], of which regenerative carbon (Cbio) is determined basing on the result of specific biogenic content. Plastic and composites contain a high content of carbon (50.2 and 37.6% respectively) but originated mainly from fossil carbon (Cfoss) which contributes to GHGs emission during combustion process. A summary of TC result is displayed in Fig. 7, whereby average carbon content of MSW is 20.48% (wet waste) and regenerative carbon constitutes 68.15% in total carbon.

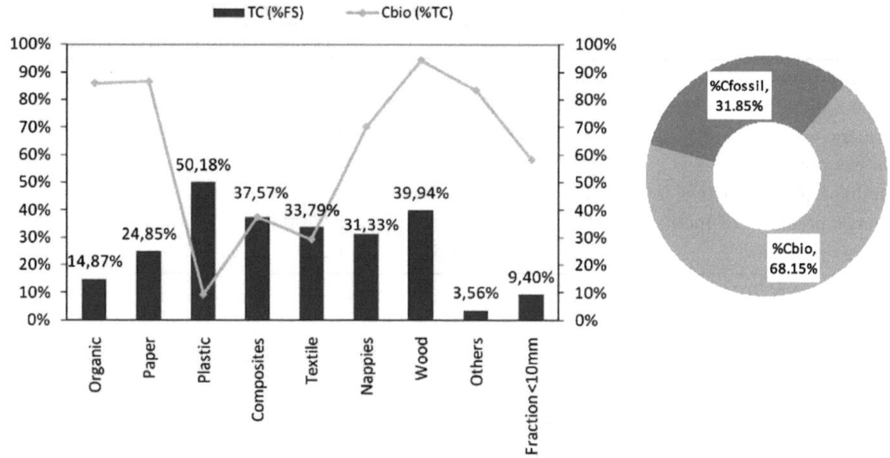

Fig. 7 Contribution of regenerative carbon (Cbio.) in total carbon content (TC) [8, 12]

2.2 *Potential of Energy Utilization*

When considering the potential of energy utilization from waste, biogas production and thermal utilization are mentioned as popular methods. Within this paper, an overview on energy potential by the utilization on MSW, Hanoi 2010 will be presented, the loss of energy/energy demand during utilization process is not involved.

2.2.1 Production of Biogas

Biogas, in term of energy utilization, can only be operated with CHP plants (Combined Heat-Power plant). Biogas which is produced during anaerobic digestion can be converted into electricity by gas-engine (e.g. gas-Otto motor). In order to get optimum conditions, in which biogas yield is 100–130 m^3/Mg input and methane content in biogas up to 65%, only organic waste is digested by anaerobic process. Data for this process is showed in Table 4.

Energy potential (E, kJ/kg) per ton of organic waste by anaerobic digestion is calculated by: [2, 3, 5]

$$E[kJ/kg] = oTS \times TS \times V_{\text{biogas}} \times V_{CH4} \times NCV_{CH4} \quad (1)$$

where:

oTS: organic substances, ($\%TS$)
TS: total solid (dry substance), ($\%FS$)
V_{biogas}: biogas yield, (m^3/Mg FS)

Table 4 Key figures for biogas production of organic waste in Hanoi 2010

	Unit	Organic fraction, MSW, Hanoi 2010
Water content, WC	(%FS)	63.03%
Total solid, TS	(%FS)	36.97%
Organic solid, oTS	(%TS)	85.0%
Biogas potential	(m^3/Mg FS)	170–200 (185)[a]
Methane (CH4) content in biogas	(vol.%)	65.0%[a]
Energy potential of methane	(kJ/m^3)	32,700
Energy potential	(kJ/kg FS)	1,235.7

[a] (UFOPLAN 2010)

V_{CH4}: methane content in biogas, (vol%)
NCV_{CH4}: energy potential of CH4, (kJ/m^3)
Note: the loss of energy/energy demand/emission during anaerobic process is not involved.

2.2.2 Thermal Utilization

A short summary for thermal utilization of Hanoi, Vietnam MSW in 2010 is given in Table 5 with a comparison to German household waste in 2005 and 2010 (data from IAA). Energy potential in this case is illustrated through lower calorific value (LCV, kJ/kg); in case of MSW, Hanoi 2010 the average energy potential is 6,948 kJ/kg.

A comparison between emission factor (EF, kgCO_2-eq/MJ) of MSW and fossil fuels per unit energy is computed as formula: [1, 4]

$$EF = \frac{(TC \times C_{\text{foss.}} \times 44/12)}{Hu}$$

where:

TC: total carbon, kgC/Mg *MSW*
$C_{foss.}$: percentage of fossil carbon in total carbon, %
44/12: convert from C to CO_2
Hu: lower calorific value (*LCV*), MJ/Mg
Note: energy demand during process is not computed in the formula.

An interesting result observed from the Fig. 8 is that, the total CO_2 emission per unit energy (kg CO_2/MJ) by combustion of Hanoi MSW (Hcomb. = 100%) is 0.1073 kg CO_2/MJ but only 0.034 kg CO_2/MJ is accounted as GHGs emission. The left amount (0.074 kg CO_2) is considered as neutral to environment; thus the specific CO_2 emission of MSW is much lower per unit energy (MJ) in comparison to conventional fossil fuels (IPCC, UNFCC).

Table 5 Key figures for thermal utilisation of MSW in Hanoi 2010 (compare with MSW Germany) [4, 13]

	Hanoi 2010	Germany 2005[a]	Germany 2010[a]
Total carbon, TC (Mg C/Mg MSW)	0.205	0.209	0.230
Biogenic carbon, Cbio. (%TC)	68.15%	66.6%	61.9%
Fossil carbon, Cfoss. (%TC)	31.85%	33.4%	38.1%
Biogenic carbon content, (MgC_{bio}./MgMSW)	0.1396	0.139	0.1423
Fossil carbon content, (MgC_{foss}./MgMSW)	0.0652	0.070	0.0876
Lower calorific value, LCV (kJ/kg)	6,948	7,240	8,844

[a] Household waste

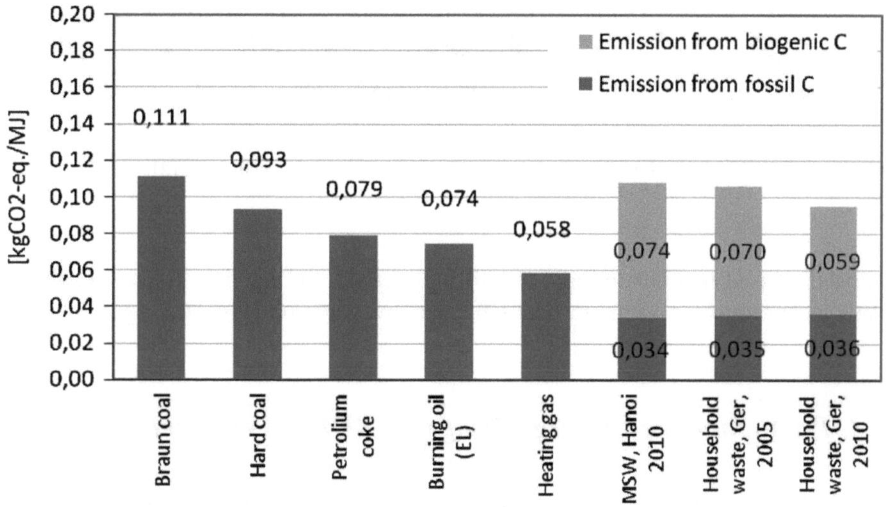

Fig. 8 CO_2 emission from combustion of MSW (Hanoi, Vietnam) in compared with fossil fuels

It is obvious that the energy content of MSW is lower than NCV of fossil fuels; especially comparing with oil and gas.

So far, there are some researches/projects concerning to waste quantities-/composition in urban areas in Vietnam; but a detail study on characteristics of waste fraction (e.g. total carbon, biogenic carbon, heating value, heavy metal, ash content etc.) has not investigated yet. Although MSW in Hanoi is not representative for MSW of the whole country (HCM city would be a better sample in the South), the data given in this paper would provided a necessary information for more understanding about energy utilization of MSW in urban areas in Vietnam. The investigation on fluctuation of waste quantities/composition during season and the behavior of waste fractions by different waste management activities, however, should be carried out for a fully information. Theoretically, energy utilization of MSW has multiple positivenesses; e.g. replacing partly for conventional fossil fuels, reducing the emission of GHGs, decreasing the potential of environmental

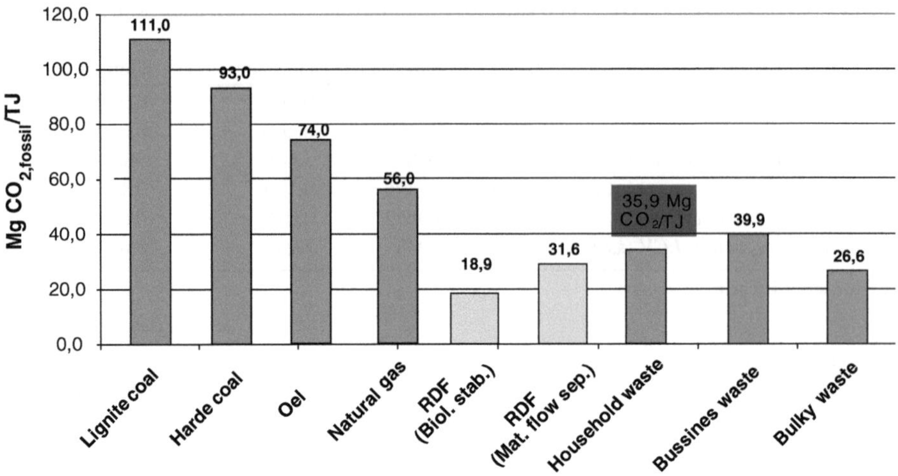

Fig. 9 Fossil CO_2-emission factors for several waste types and fossil fuels [2]

impacts caused by conventional landfilling method; but the selection of a suitable WtE technology for Vietnam needs further investigation (Fig. 9).

2.3 Production and Energy Usage of RDF in Vietnam

Refuse Derived Fuel (RDF) is defined differently across countries. According to its origin, three sources of waste relevant for the production of RDF can be distinguished: Residential waste from private households and small enterprises, commercial waste and industrial waste. Commercial and industrial waste usually has a homogeneous but very producer specific composition. Recyclable, combustible and non-combustible fractions are often segregated already at source.

2.3.1 Background

For fuel processing, the main components for production of RDF are paper, plastic and wood. The content of inert material and food or garden waste is negligible. In Europe, due to the segregation at source, no initial segregation of high calorific fractions is necessary in the beginning of the process.

RDF from residential waste, which consists of many different material fractions, is produced in MBT plants according to two main concepts:

- MBT type 1—Separation which seeks to split the waste into 'biodegradable' (that may be composted and afterwards landfilled) and 'high calorific' fractions

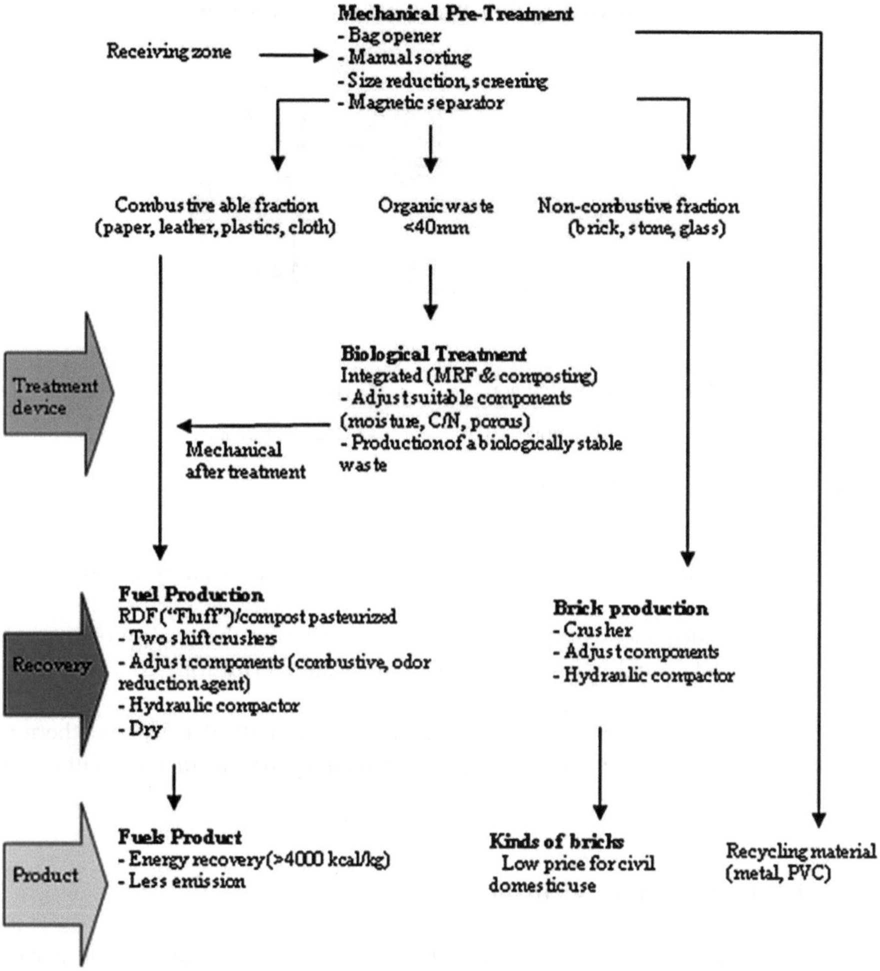

Fig. 10 Schematic diagram of the MBT-CD.08 [9]

- MBT type 2—Dry stabilization which is less concerned with the splitting into fractions, and more focuses on the use of heat from a 'composting' process to dry the waste (bio-drying) and increase its calorific value, thereby making it suitable for use as a fuel as well as facilitating the separation of fractions. As alternative to the bio-drying, drying with natural gas, landfill gas or biogas can be used (physical drying).

Objectives of the RDF production process are:

- Removal of water and inert components in order to improve calorific value
- Removal of chlorine, aluminum and zinc in order to avoid corrosion and other fuel related technical difficulties in furnace and boiler

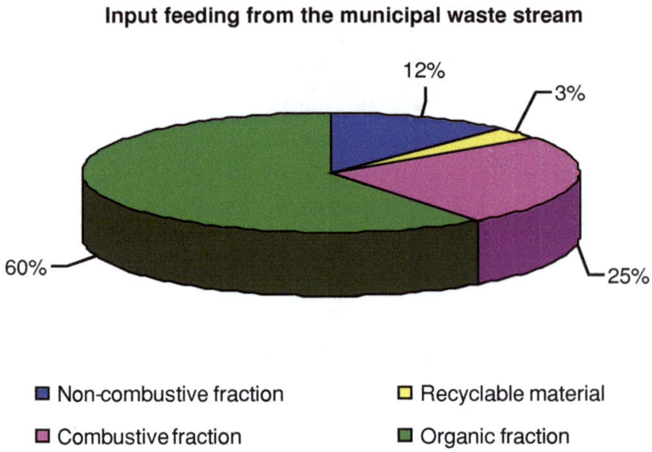

Fig. 11 Input feeding from the municipal waste stream [9]

- Reduction of volatile substances that have a negative environmental impact when combusted
- Reduction of substances that have a negative impact on the quality of byproducts such as ashes and gypsum from flue gas desulphurization

Furthermore, mechanical properties like grain size, bulk density and thermal stability of the fuel are important for the fuel feeding system of the combustion plant [15].

2.3.2 RDF Production in Vietnam

In general waste can be processed to a RDF in order to improve the fuel properties but this technology in Vietnam is not yet high developed. "Technology for Treatment of Solid Waste into Fuels" which called MBT-CD.08 developed by a Vietnamese private enterprise can be seen as the first one. This technology fits to MBT type 2 categories.

The flow chart of MBT-CD.08 is summarized in Fig. 10, whereas the input feeding from municipal waste stream looks like in Fig. 11, material flow shows in Fig. 12 and the heat value presented in Fig. 13 respectively.

The MBT-CD.08 is used for a capacity of 15 tons/day, which has been tested successfully in Duy Tien district, Ha Nam province since June 2006; this treatment model is therefore suitable for small town or district. However the system is constructed with module, the capacity can reach to 100 tons/day if more modules are connected, it can be therefore applied also for cities.

In this plant waste of the commune was collected and treated right away in a day. Fuels product was distributed for domestic use (industrial plant, heat, stream

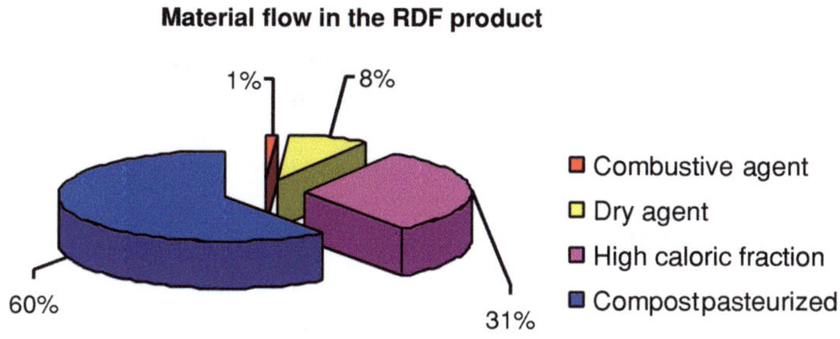

Fig. 12 Material flow in the RDF product [9]

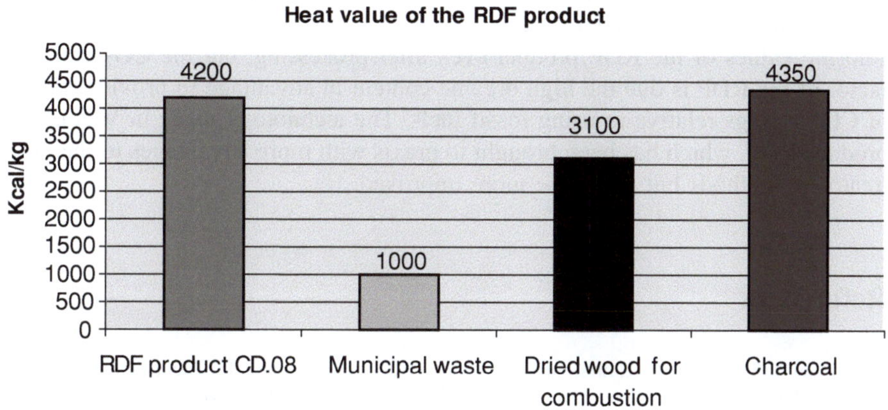

Fig. 13 Heat value of the RDF product CD.08 and others [9]

Table 6 Emission from compactor bricks of residual from RDF process

Parameter	Test No.[a]	Unit	Result	Standard of Vietnam (TCCP867/1998/QD-BYT)
Lead	867/1998/QD-BYT & VA231/1	ppm	0,098	2
Cadmium	867/1998/QD-BYT & VA231/1	ppm	<0,1	0,2
Arsen	867/1998/QD-BYT & ICP-MS	ppm	0,032	0,2
Antimon	867/1998/QD-BYT & ICP-MS	ppm	0,019	0,2

[a] Quatest 1—2/7/2008

and power). Ash remains 30–40% after burn, which is used for construction purpose. Table 6 presented the emission from compactor bricks of residual.

This technology was introduced to apply over the country by the Vietnam Minister of Construction in Decision No.925/QD-BXD dated from 18/7/2008.

This technology has been appreciated in Vietnam as it has advantages prominent to the others. However, most important parameters for the conversion of RDF into energy as heating value, water content, ash content, ash melting point, sulphur and chlorine have been not yet examined. The emission needs carefully tested and study on hazardous chemicals distribution in sorting process needs also to be done.

3 Concluding Remarks

On the way looking for a new source of fuel this case study has contributed some initial results from Vietnam. This case study has clarified that the energy utilization of MSW has multiple positivenesses; e.g. replacing partly for conventional fossil fuels, reducing the emission of GHGs, decreasing the potential of environmental impacts caused by conventional landfilling. Though the waste composition in Vietnam and other Asian countries lead to a reduced fuel yield and lower calorific values of the RDF product even after processing, but the CO_2 emission factor of the RDF is due the high organic content in advantage to provide options of CO_2 savings relative to using fossil fuels. The technology made in Vietnam to produce RDF, which has been brought to praxis with more advantages to the other treatment methods but it will be more improved.

References

1. Bilitewski B (2007) CO_2 Emissionminderung durch Müllverbrennung. In: Tagungsband für Abfall wirtschaft und Klimaschutz: Emissionshandel- Emissionsminderung- Klimaschutzprojekte, TU Dresden
2. Bilitewski B (2008) Production and energy usage from RDF in germany. In: Workshop Mechanical-Biological Treatment in Hanoi, Vietnam, Hanoi 9-10/04/2008
3. Bilitewski B, Wünsch C, Hoffmann G (2008) CO_2 Reduktion in Abfallverbrennungsanlagen durch Energieeffizienz. In: Waste to Energy- Internationale Fachmesse und Konferenz für Energie aus Abfall in Biomasse, Bremen, Dezember
4. EdDE (2005) Ökologiesche Effekte der Müllverbrennung durch Energienutzung, Entsorgergemeinschaft der Deutschen Entsorgungswirtschaft e.V. -EdDE-
5. Kern M, Raussen T (2006) Energiepotenzial von Bio- und Grünabfall. In: Anaerobe biologische Abfallbehandlung, Entwicklungen-Nutzen und Risiken der Biogastechnilogie, Tagungsband, IAA, TUD
6. Nippon Kei Co. Ltd (2003) Environment sector study for Japanese ODA in the socialist republic of Vietnam
7. Nguyen BT (2007) CO_2 balance for landfill sites_ comparison of gas emission balances for a chosen LF site with and without final capping with landfill gas prognosis, MSc thesis, co-operation between TU Dresden, Germany and Hanoi University of Science (HUS), Vietnam
8. Nguyen TH (2010) Giang, possibilities and limitation of energy recovery from MSW in Vietnam, Draft report, IAA-TU Dresden
9. Nguyen Thi Diem Trang, Nguyen Binh, Nguyen Gia Long (2009) In: 3rd International symposium MBT and MRF, Hanover, Germany 12–14 May 2009, p 99
10. Nguyen Thi Diem Trang (ISTEAC) (2004) Project "Integration of solid waste management tool into specific settings of European and Asian communities", Hanoi University of Science
11. Result from Practical training of Master Course at HUS 2010

12. Schulenburg S (2010) Analyze of RDF production in Vietnam, Master thesis, cooperation between TU Dresden, Germany and Hanoi University of Science (HUS), Vietnam
13. UFOPLAN (2010) Nutzung der potenziale des biogenen Anteils im Abfall zur Energieerzeugung IAA/INTECUS GmbH
14. URENCO report at local area from 2005 to 2006
15. Vera Susanne Rotter (2010) Waste-to-Energy in the City of tomorrow International Workshop Hanoi, 25th–27th October 2010, p 22
16. Vietnam Environment Monitor (VEM) 2004
17. Vietnam Environment Protection Agency (VEPA) (2008) Project report "Model design and deploy at pilot scale of the separation, collection and treatment of MSW for new urban zone"

Waste to Energy in Brazil

Luciano Basto Oliveira, Claudio Fernando Mahler and Luiz Pinguelli Rosa

Abstract This chapter discusses the current status of waste treatment in Brazil, initially presenting the projects for mitigating greenhouse gas emissions in landfills already approved by the Ministry of Science and Technology. It addresses the issue of consumerism and the "extraction-production-consumption-post wasteful consumption" that makes a more modern technical solution unfeasible. It also discusses the regulatory framework and issue of conservation and energy supply. The laws of the National Climate Change and National Solid Waste policies signed in 2009 and 2010, respectively, are also discussed. It finally stresses the importance of these laws and not just waste disposal in landfills, but does not make them a problem for future generations, always alert to the possibilities of waste recycling and energy use, which will certainly be significant for future societies, considering the increasing production of waste worldwide and the problems caused by different forms of energy production, such as nuclear power, evidenced recently in Japan.

1 Introduction

Although energy potential in Brazil is significant, only a few cases of energy generated from urban solid waste are in operation; namely, two landfills in the São Paulo capital generating the equivalent of 44 MW electric power, the Horizonte Asja 6 MW plant in Minas Gerais, the recently inaugurated Minas de Leão plant of

L. B. Oliveira · C. F. Mahler (✉) · L. P. Rosa
Coppe/UFRJ – Alberto Luiz Coimbra Institute,
Federal University of Rio De Janeiro, Rio De Janeiro, Brazil
e-mail: cfmahler@acd.ufrj.br

A. Karagiannidis (ed.), *Waste to Energy*, Green Energy and Technology,
DOI: 10.1007/978-1-4471-2306-4_14,

Table 1 Projects to mitigate gas emissions responsible for the greenhouse effect in landfills in Brazil (MCT [13])

Landfill	Year of approval	Landfill	Year of approval
Nova Gerar	2004	Itapevi	2006
Vega Bahia		Quitaúna	
MARCA		Pedreira	2007
Lara	2005	SANTECH	
Tremembé		PROBIOGÁS-JP	
ESTRE		Tijuquinha	
Caieiras		URBAM	
Bandeirantes		Irani	
Anaconda		Vila Velha	
São João		Feira de Santana	
Canabrava	2006	Gramacho	2008
Aurá		TECIPAR	2009
Bragança		CTRS	
SIL		Natal	
Manaus		Corpus/Araúna	2010
Alto Tietê		Manaus	
Terrestre Ambiental		Itaoca	

6.5 MW in Rio Grande do Sul, and Gramacho landfill in Rio de Janeiro about to provide treated biogas for the Duque de Caxias Refinery.

Table 1 shows which landfills in Brazil have submitted MDL projects to the Ministry of Science and Technology.

In general, waste consists of food leftovers, electro-electronic material, packaging, clothes, animal remains, paper, cardboard, leather, contaminant construction and inert material. Its disposal in landfills produces gas emissions responsible for the greenhouse effect, producing leachate, loss of area for other activities and a tendency to real estate devaluation in the region. When the disposal is in refuse dumps, the damage is even worse in terms of real estate and traffic in the region, in addition to much environmental damage to the soil, surface and ground water in the vicinity and the air, in addition to a variety of social and health hazards, by occupying surrounding regions and sometimes in the actual waste disposal site.

Questions regarding waste production can lead to a discussion on consumerism, very often justified by planned obsolescence—a strategy adopted by the global production sector to reduce costs of goods and make them accessible to most of the population so that it can maintain the growing demand and therefore keep jobs and "turnover" of the economy. But this capitalist strategy has failed to specify the importance of recycling waste and demonstrate its unfeasibility. Inertia seems to be mostly responsible for continuing with the current "extravagant extractivism-production-consumption-post-consumption" pattern than any technical unfeasibility. In fact, it is possible to substitute some of the inputs from extractivism by waste from the process, preventing depletion of natural resources and damages from the end disposal of still useful materials.

2 The Regulatory Framework

European regulations provided disposal restrictions on food leftovers in landfills from 2014 on in order to prevent emissions. Since this is a costly procedure, although being a good practice for society as a whole, it is to be expected to be required as a minimum parameter against environmental dumping in international negotiations. In Germany, this procedure has been adopted since 2005 with excellent results and has been increasingly adopting procedures of waste treatment for energy use.

Brazil's National Climate Change Policy [11] is committed to a goal to reduce the emissions of various sectors, including waste by 2020. However, since the disposal of a large part of Brazilian waste is in refuse dumps and controlled landfill, the sanitary solution provided by the National Solid Waste Policy [12]—which considers the sanitary landfill as its minimum standard -, tends to increase emissions [3]. It will, therefore, be necessary to have more feasible alternatives to this method, including compliance with the prerogative of this legal instrument, which only permits refuse disposal in sanitary landfills, the concept of which is "waste without economic, environmental and technical feasibility" for reuse.

Since world technologies are, by their very existence, technically feasible and also environmentally feasible based on the granting of environmental licences, there is still the need to demonstrate economic feasibility, since disposal charges currently paid by local governments in Brazil are around half or less than the European charges. It is worth recalling that, unfortunately in developing countries, even when in comfortable economic phases, the view of waste costs is in the short term, and very often long term monitoring costs, environmental damage, traffic losses, damaged public thoroughfares, the health of the neighbouring population, real estate devaluation, and so on are not considered, and which will be paid by future societies.

Accordingly, the question to be asked from this viewpoint is how this issue should be addressed: socially or privately? Socially, it would be sufficient for the project to prove that it is more profitable than the country's discount practice, but this would require investment and operating the system by public authorities and would, therefore, require technical specialization, also involving risk, so it is better if there is interest in using private enterprise.

And this has been the trend in Brazil since around the 1970s in São Paulo and since the 1990s in the rest of the country and large towns with a population of over a million. More recently, there have been attempts to create the figure of multi-municipal consortia in order to share the implementation and operating costs and create more solid opportunities for selling carbon credits and for power generation. It should be noted that, in the case of private enterprise, feasibility depends, of course, on profitability competing with the other applications and existing return rates—including consideration of the risks, in the last case, of innovation and payment of part of the revenue by the public authorities. This, then, is one of the challenges of this generation: to make waste utilization feasible in Brazil.

Table 2 Conservation of energy from recycling, according to various references

Material	Brazil	Canada	USA	Australia
Reference:	[5]	[14]	[7]	[22]
Paper and cardboard	3.51	1.75	2.95	1.37
Plastic (including PET)	5.06	5.55	15.39	5.91
Glass	0.64	0.08	0.62	1.25
Metals (aluminium, steel)	5.3	3.25	5.85	2.67

It is also worth mentioning that the first spark of the many environmental accidents since the 1960s is caused by inadequate waste disposal on hillsides, associated in the sequence with the presence of unsuitable waste disposable in valleys, close to rivers, creeks and culverts. If these costs are added up, certainly proper solutions for waste would be considered important and costs would not be addressed just as short-term factors.

One of the potential aids for the waste issue involves the fact that packaging and a number of electro-electronic items can be recycled by processes that prevent extractivism of natural resources, while at the same time, increase their working life and that of the existing sanitary landfills.

3 Energy Conservation

One potential aid is the fact that packaging can be recycled; a process against extractivism of natural resources, which at the same time increases the working life of natural resources and sanitary landfills.

One of the benefits of waste recycling is saving energy. Although there are bibliographic references on the topic, which suggests considerable values, as shown in Table 2, there is very little international exploitation.

In Brazil, considering the production of 60 Mt/a (IBGE [8]) and composition of the waste [10], as shown in Table 3, it is possible to estimate potential energy conservation from the [14].

Accordingly, the potential of 87.5 TWh electricity equivalent, or 22 Mtep, is achieved, enough to attend the consumption of the national residential sector (BEN) and similar to all electric power generated from waste in the 2,000 or so thermopower plants in the world (CEWEP [6]).

This is, in fact, a greater quantity of energy since the Brazilian energy matrix consists first and foremost of hydropower plants on sites appropriate for using this source, and the majority are far from consumer centres, which requires transmission over 2,000 km. This activity incurs losses, today in the 15% range of the generation (ANEEL [1]). This is why recycling prevents not only electricity consumption but also loss in transmission. It is worth mentioning that this supply is

Table 3 Composition of Brazilian waste and potential energy conservation

Material	Composition (%)	Conservation factor (MWh/t)	Potential Conservation (MWh/a)
Paper and cardboard	12	3.51	54.6
Plastic (including PET)	18	5.06	25.3
Glass	3	0.64	1.15
Metals (aluminium, steel)	2	5.3	6.36

equivalent to that of the Itaipu hydropower plant and more than the entire nuclear complex in existence and planned [2], [20].

In the Brazilian case, where there is a fund for mandatory investment in activities with this profile—the Energy Efficiency Programme (PEE), controlled by the regulatory national electricity agency (ANEEL [1]) and adopted by the distribution and generation concessionaires –, of R$ 400 million a year (2010), and must benefit less favoured classes with 60% of the total, it is possible to establish a public policy for recycling.

Moreover, one of these international references on conservation from recycling founded the UN-approved AMS-III.AJ methodology on reducing greenhouse gas emissions based on recycling plastics (IPCC [9]), corroborating the proposal made by Pimenteira et al. [19].

The allocation of resources available in the PEE to encourage recycling can, therefore, be used to drastically reduce the need for private investment in waste utilization projects, which increases their profitability and directly their feasibility, in response to the legal precept and against burying the recyclable material.

4 Energy Supply

Consequently, the food leftovers to be segregated may undergo anaerobic composting to generate electricity or sell the resulting gas, since the investment will have been reduced and it will then also be feasible—although the amount of available electricity is below 40% of the conserved. Lastly, in the case of selling electricity, the exhaust heat should be used to dry the organic compost and consume it in a boiler to generate a little more electricity, reducing the disposal of material in a landfill by around 10%.

As a result, the waste is now supported by the energy sector as a feasible proposition and, in counterpart, helps it to postpone building new plants, optimising financial, human and natural resources of society.

Thus, the energy potential of urban solid waste is now a function of the minimum scale of feasibility of the projects, which was based on international experiences in which biodigestors that treat 150 t/d (OWS [18]) of organic waste, successfully supply a thermopower plant of 850 kW. Making this restriction compatible with the available waste per Brazilian citizen, there are around 300 cities that can provide plants to treat their own waste, which totals around 60% of all waste produced in the country. This achieves the power of 460 MW—or the equivalent in use of the gas now made available, including for vehicles as a substitute for diesel fuel (BOSCH [4])—in approximately 550 plants, as stated by [15]. Possibly these values are higher when implementing multi-municipal consortia, but this requires assessments regarding transport, energy balance, emissions and finance still to be undertaken. It should be noted that Brazil is a country with more than 5,500 towns, many of them in densely populated regions, where joint solutions for solid waste would certainly be extremely welcome.

If the segregation of recyclables or consumer market of this packaging is not enough to absorb them only by recycling, it is possible that incineration technologies take over some of the space—even if a larger scale of materials is required, it represents fewer plants and greater supply of electricity.

In any case, it is a good opportunity for different technologies in energy use from landfill waste, since the Brazilian market proves able to increase by more than half the number of plants currently in operation in the world market.

5 Incentives and Sanctions

In this case, the National Solid Waste Policy itself, by its regulatory decree, provides economic incentives for feasible projects in this sector, such as lower interest rates, financeable portion and longer terms, plus lower taxation on products and incentive for its procurement by public authorities.

On the other hand, failure to meet the requirements incurs fines between R$ 50 and R$ 500 per relapse (which may be charged on a daily rate), to discourage non-action.

Since this alternative also attends the National Climate Change Policy, it is desirable to use the concept of mitigating cost to hierarchise the alternatives and choose the cheapest. In this sense [17] show that energy utilisation of urban solid waste is cheaper than alternative wind energy, which has been given incentives based on this environmental criterion.

And, lastly, because of the transversality of the topic, which also covers the social and strategic issue—especially to prevent public health hazards and help increase energy security as the supply is being decentralised -, it is suitable to apply the concept of sustainability to this source, so that the characteristics of the economic, environmental, operating, social and strategic dimensions can be analysed jointly, which was done by Oliveira et al. [16] and Rovere et al. [21].

Solid waste solutions that enable financial gain for a large part of the population are welcome and must be considered in the future. The sanitary landfill, however good its disposal, is a solution that in essence benefits the owners or the companies operating the landfills. There is room in Brazil to implement processes that consists of more recycling, composting, anaerobic fermentation and incineration with energy use, which are processes that will involve many more new companies and jobs, and which will certainly be implemented in our country in the next few years.

References

1. ANEEL (Agência Nacional de Energia Elétrica—National Agency of Electric Energy) (2011) Programa de Eficiência Energética—Energetic Efficiency Program. Available at http://www.aneel.gov.br. Accessed 14 May 2011
2. BEN (Balanço Energético Nacional) (2011).(National Energetic Balance). Empresa de Pesquisa Energética (Energetic Research Company). Available at http://www.epe.gov.br. Accessed 12 May 2011
3. Bogner JE, Oliveira LB (2007) Reduction of global landfill methane emissions and using energy for waste. In: Ribeiro SK, Araújo MSM. (Org.). IPCC outreach in Rio de Janeiro. Rio de Janeiro: Rio 360 Comunicação, pp 26–33
4. BOSCH (2010) (Robert Bosch Gmbh) Diesel natural gas flex system
5. Calderoni S (1996) Os Bilhões Perdidos no Lixo. In: São Paulo (ed) USP (3rd ed.)
6. CEWEP (Confederation of European Waste-to-Energy Plants) (2011) Available at http://www.cewep.eu. Accessed 14/05/2011
7. EPA (US Environmental Protection Agency) (2007) Waste reduction model. Washington, EPA, 2007. (Available at http://www.epa.gov). Accessed December 2007
8. IBGE (Instituto Brasileiro de Geografia e Estatística) (2008) (Brazilian Institute of Geographie and Statistics). III Pesquisa Nacional de Saneamento Básico. Rio de Janeiro, IBGE
9. IPCC (Intergovernmental Panel on Climate Change) (2010) Meth panel. MAS—III.AJ
10. IPT (1998) (Instituto de Pesquisas Tecnológicas)/CEMPRE. Lixo Municipal: Manual de Gerenciamento Integrado. (Municial Waste—Integrated Management Manual) (2nd reprint). São Paulo, IPT/CEMPRE
11. Law (2009) 12,187, which institutes the National Climate Change Policy. Brazil
12. Law (2010) 12,305, which institutes the National Solid Waste Policy
13. MCT (2011) (Science and Technology Minister). CDM Projects. Available at http://www.mct.gov.br. Accessed 20 May 2011
14. Morris J (1996) Recycling versus incineration: an energy conservation analysis. J Hazard Mater 47:277–293 (Elsevier, Amsterdam)
15. Oliveira LB, Rosa LP (2003) Brazilian waste potential: energy, environmental, social and economic benefits. Eng Policy 31:1481–1491 (Elsevier)
16. Oliveira LB, Araújo MSM, Rosa LP, Barata M, Rovere EL (2008) Analysis of the sustainability of using waste in the Brazilian power industry. Renew Sustain Eng Rev 12:883–890 (Elsevier)
17. Oliveira LB, Henriques RM, Pereira AO Jr (2010) Use of wastes as option for the mitigation of CO_2 emissions in the Brazilian power sector. Renew Sustain Eng Rev 14(9):3247–3251 (Elsevier)
18. OWS (Organic Waste System) (2010). Available at http://www.ows.be. Accessed 16 Apr 2010
19. Pimenteira CAP, Pereira AS, Oliveira LB, Rosa LP, Reis MM, Henriques RM (2004) Energy conservation and CO_2 emission reductions due to recycling in Brazil. Waste Manage 24:889–897 (Elsevier)

20. PNE (2030) (Plano Nacional de Energia 2030) (National Energy Plan 2030). Empresa de Pesquisa Energética (Energetic Research Company). Available at http://www.epe.gov.br. Accessed 12 May 2011
21. Rovere EL, Soares JB, Oliveira LB, Lauria T (2010) Sustainable expansion of electricity sector: Sustainability indicators as an instrument to support decision making. Renew Sustain Eng Rev 14(1):422–429 (Elsevier)
22. Warnken ISE (2007) Potential for greenhouse gas abatement from waste management and resource recovery activities in Australia. Final report, Sydney, Warnken ISE/SITA p 54

The Ambiguous Relation Between Waste Incineration and Waste Prevention

Henning Wilts

Abstract Waste incineration technologies are crucial elements of many modern waste management concepts. But from a sustainable resource management point of view especially the qualitative and quantitative prevention of waste will gain importance in the future as the generation of waste is more and more shifted into developing and emerging countries. The German case study highlights the risk of path dependencies and load logics caused by billion investments in waste incineration plants and their contradicting effects on the recovery of secondary raw materials. Conclusions can be drawn on the importance of integrating the thermal recovery of waste in a much more systematic way in existing socio-technical systems of waste management.

1 Introduction

The amended EU Waste Framework Directive 2008/98/EC (WFD) which came into effect in December 2008 defines for the first time the goals of resource conservation and efficient resource management as elements of sustainable waste management policy:

H. Wilts (✉)
Research Group Material Flows and Resource Management,
Wuppertal Institute, Döppersberg 19, 42103 Wuppertal, Germany
e-mail: henningwi@wupperinst.org

H. Wilts
Institute IWAR, TU Darmstadt, Petersenstreet 13, 64287 Darmstadt, Germany

A. Karagiannidis (ed.), *Waste to Energy*, Green Energy and Technology,
DOI: 10.1007/978-1-4471-2306-4_15, © Springer-Verlag London Limited 2012

> This Directive lays down measures to protect the environment and human health by preventing or reducing the adverse impacts of the generation and management of waste and by reducing overall impacts of resource use and improving the efficiency of search use.

This is implemented by the new five-step waste hierarchy in Article 4 WFD. Accordingly the following order shall apply to the treatment of waste, deviations are only possible based on environmental aspects in a life cycle perspective:

- prevention;
- preparing for re-use;
- recycling;
- other recovery, e.g. energy recovery; and
- disposal.

However, the different hierarchical levels cannot be viewed as isolated from each other, as it might be the case at first glance. This is especially relevant for the relation between waste prevention and energy recovery.

Incineration technologies are crucial elements of many modern waste management concepts: incinerators reduce the volume of solid waste, thus reducing the volume of landfill waste (quantitative waste prevention). Simultaneously, the incineration results in an inertization of especially hazardous waste making ingredients dangerous for humans hazard-free. The deposal of in such a way pre-treated waste leads to a significantly reduced development of landfill gas whose high methane content would represent a significant burden for the global climate (qualitative waste prevention). In particular, the various waste-to-energy technologies reduce the use of fossil fuels, whose production is associated with enormous consumption of resources along the product chain. So incineration makes a substantial contribution for the purpose of waste prevention.

On the other hand the example of Germany, were since the 1970s a lot of emphasis has been put to incineration technologies and today more than half of the municipal waste is burned, shows very illustratively how this very capital-intensive technology produces path dependencies and a logic of utilization rates that are counterproductive to the prevention of waste.

This paper therefore aims to analyze under which circumstances waste incineration in developing and emerging countries can contribute to a sustainable resource management and what aspects should be considered from a socio-technical perspective beyond the mere technical artifacts. The construction of waste management infrastructures in these countries offers the opportunity to avoid mistakes that have been and are made in Germany.

The text is structured as follows: Sect. 2 considers the need and framework of a global sustainable resource management, Sect. 3 the possible transition of the waste sector in this direction. Section 4 describes very briefly the theoretical framework of socio-technical analysis which is applied to German case study in Sect. 5. The final Sect. 6 draws some preliminary conclusions and describes further need for research.

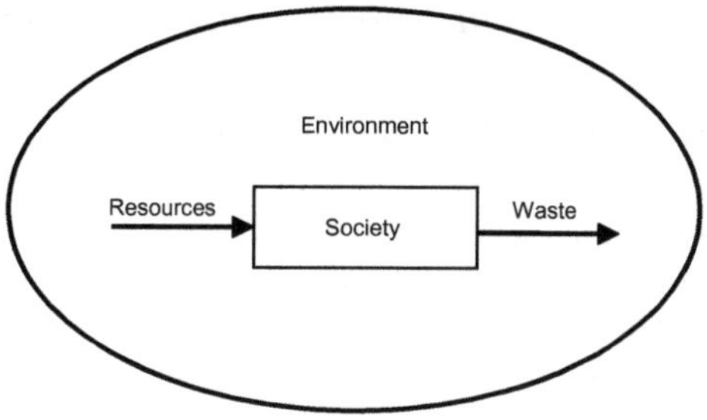

Fig. 1 The concept of the socio-industrial metabolism according to Hofmeister [13]

2 Necessity of Global Approach to Sustainable Resource Management

Following the discussions since the 1970s about the "Limits to Growth" [25] it is clear that our entire social system is based on the use of natural resources (cf. [3], p. 1). An economic development based on an unbridled physical expansion can therefore in any case not be sustainable in the long run. This is the starting point of ecological economics, according to which the resources in the closed system of the earth are not necessarily scarce, but their finiteness is the central challenge of sustainability [24]. The discussions since the UN conference in Rio in 1992 how to achieve sustainable development also refer to the natural basis of all social processes.

This juxtaposition of the socio-economic and biophysical effect relationships and the various exchanges between the two spheres is described by the concept of the socio-industrial metabolism. This biological term goes back to Robert Ayres [2] and indicates that our society as any other organism needs resources for its reproduction and also excretes them at the end: as waste. Figure 1 illustrates this context in a very simplified form: The socio-economic system is embedded in its natural environment and from this it takes resources and the energy used for production and consumption processes. Following the principle of mass conservation, energy and mass can neither be created nor destroyed, so any of these inputs ends up again after its use phase in a more or less transformed way as output, which is released into the natural environment.

The approach is thus based on a systemic understanding of sustainability, which differs from an isolated consideration of economics, ecology and social aspects as in the "three-pillar model" of sustainability and instead tries to consider the relationships between these subsystems in an integrated analysis. These relationships between society and nature are structured by physical infrastructures which decisively influence the sustainability of material flows. Nature and society interact in this way, but according to their own rationalities and rules of causality.

As Hofmeister ([13], p. 44f) points out, however, precisely this separation between inside and outside is no longer possible in the industrial society. A striking example of this process is the human induced climate change that impacts heavily on all ecosystems. These processes are also distinguished by the fact that they are irreversible – contrary to the assumptions of the neo-classical economics.

2.1 Problem Shifting into Developing Countries

By trade, each country has the opportunity to use the resources of other countries, so that both the importing and the exporting country can profit according to the principle of comparative advantage. But these imported products are again linked to the consumption of resources in the form of ecological rucksacks occurring together with the caused environmental burden in the exporting country, including for example the contaminated overburden volumes at the mine of raw materials and energy consumption in the production of semi-finished products (cf. [3], p. 59). The analysis of the physical material flows shows that the EU has intensively replaced domestic resource extraction by the import of products since the start of globalization, which has led to a reduction of the local environment burden and added additional damages especially to developing and emerging countries. In the period 1976–2000 the economic value of net imports to the EU increased about one-third, however, the physical net balance of the related products including the hidden flows has more than doubled (cf. [32]). This concerns in particular the pollution-intensive industries in the field of metal extraction and refining of oil, whereas the import industries benefit from lower environmental standards in developing countries.

2.2 The Case Study of PGM

This phenomenon of global relocation of environmental burdens caused by the use of resources can be demonstrated particularly well on the example of platinum group metals (PGM): PGM are mostly used in automotive catalysts because of their special chemical properties. The PGM content per catalytic converter depends on type and size of the engine and fluctuates between 1 and 15 g with an average of about 4.5 g (cf. [23], p. 3). Following the U.S. and Japan, where the use of catalytic converters was obligated nationwide since the 70s, the trend in Europe was mainly driven by the introduction of the EURO standard, which limits emissions for carbon monoxide (CO), nitrogen oxides (NO_x), hydrocarbons Set (HC) and particulate matter (PM) since 1992 and which were exacerbated regularly since then. But also in many emerging markets tightened emission standards cause a higher demand for platinum and palladium. For example in People's Republic of China in July 2008 exhaust limits came into force which can only be achieved by the use of catalytic converters. The use of catalytic converters in vehicles has contributed to a significant improvement in the environmental performance of motor vehicles and reduced their emissions by up to 95%.

However, at the same time the recovery of PGM taking place almost exclusively in Russia and South Africa causes enormous environmental damages: Norilsk Nickel alone, one of the world's largest producer of palladium and located in Russia, emits NO_2 eq-emissions of 156.000 t every year just by the recovery of

PGM, which corresponds about to the total annual emissions of Belgium. Considering the total resource consumption along the supply chain of the mining of PGM, one ton of PGM causes a resource consumption of approximately 388.000 t, whereof 86% are incurred in the surrounding of the mine (cf. [30]). Hence the catalytic converter as an end-of-pipe technology leads to a massive shift of environmental impacts: While the air quality in Europe has increasingly improved, whole regions of Siberia are now considered virtually uninhabitable because of the mining industry. Just to compensate the sulfur dioxide emissions of the production, a diesel vehicle with a catalytic converter has to drive about 95,000 km (cf. ibid., p. 25).

2.3 Waste Exports

Another increasingly important application of PGM is the field of electronic devices where palladium is used primarily on printed circuit boards. The UN estimates that worldwide about 20–50 million tons (depending on the definition) of electronic scrap arises. On the global level one must assume that only about 10% of these amounts are recycled properly (see [21]). Used electrical appliances leave to a great extent the European Economic Area, often as illegal shipments of electronic waste. According to a study published by the United Nations University [35] only about 2.1 million tons of the annual total 8.7 million tons of e-waste are collected and recovered in the EU meaning that over 75% are exported or disposed of in other illegal ways. According to another study by Sander and Schilling [29], approximately 155,000 t used electrical appliances and electrical equipment have been exported from Germany in 2008, including about 2 million monitors.

The contained about 150 kg PGM are almost completely lost because of the lack of adequate recycling infrastructures in the target countries. Not only the unregulated dumping of WEEE leads to environmental problems in these countries, but also the improper recycling. This is a particular problem for the recycling of PGM containing printed circuit boards (cf. [33]). Studies conducted in WEEE recycling spots show that at all stages of the WEEE processing toxic heavy metals and organic pollutants (e.g. lead, cadmium and PBDEs) are emitted in high concentrations to the environment and harm the workers.

2.4 Conclusion

The examples show that from the perspective of sustainable resource management national environmental policies are increasingly limited. Material flows become more and more international and attempts to solve the environmental impacts associated with them at the national level increasing only lead to their shifting.

Thus developing and emerging countries have to be much more in the focus of resource management:

- The largely resource-poor countries of the European Union depend on imports of raw material from those countries. This is particularly true for environmental technologies, which are often crucially dependent on critical materials such as PGM or rare earth elements. The resulting wastes, however, remain at the production sites. At the same time energy and resource-intensive processes are shifted into these countries.
- Also used and waste products are increasingly exported after their use phase from Europe into developing and emerging countries.

One reason for this development are the increasingly stringent European environmental standards in the waste legislation: for example, the prescribed removal standards in the end-of-life vehicle or WEEE directive lead to significant costs attached, so that it is often cheaper to export and dispose products to Asia or Africa. Therefore waste to energy concepts in these countries cannot be evaluated without taking into consideration infrastructure systems in developed countries. In addition to the environmentally friendly disposal of waste in these countries also the establishment of global redistribution systems for nonrenewable resources particularly critical metals gains importance.

3 Functional Transformation of Waste Management

For a long time it has been denied that waste management could and should have a function and responsibility for sustainable material flow management beyond disposal and recovery. Of course the global raw material consumption per head is by about a factor of 100 higher than the municipal waste generation. But nevertheless, the technical infrastructure systems of waste management establish a framework affecting not only the waste treatment but also the potentials of sustainable consumption and production patterns [19]. Such an active role of the waste management sector would, however, mean a paradigm shift from a reactive to a proactive self-image.

For the waste management sector this approach of a material flow management would mean in particular "a waste treatment optimized under aspects of recovery and prevention" [10]: Waste should be primarily seen as a source for secondary resources aiming to manage material streams as efficiently as possible throughout their life cycle. The basic function of the waste sector as a material manager is to set the course for the design of redistribution systems and recycling facilities:

- On the one hand these systems should be designed in a way so that at the end of the use phase of products the contained materials can be re-included into technical and natural cycles. Waste management would become a "basic industry backwards" [14]. Of course pollutants must still be discharged and supplied to appropriate sinks.

- Additionally the waste management sector should also contribute to an ecologically and economically efficient use of resources, by providing necessary information about the arising waste material flows in a transparent combination of qualitative and quantitative criteria. The different kinds of relevant information are described in the following.

3.1 Innovative Fields of Waste Management

The aim is to overcome the current "broken system" [9] between supply and disposal structures, which so far represents a major obstacle in the implementation of sustainable resource management. For the waste management sector a new task arises from this: to intensively cooperate with different steps in the value chain and especially to establish an improved coupling of these processes:

- The most relevant decisions concerning the environmental effects of production, use and disposal of a product are made in the phase of product design. Although already a large number of conceptual proposals for environment- or disposal-friendly design exist (e.g. IPP), an actual thinking from the end does almost not take place at all in reality. But for Flatz [9] this is a crucial starting point: "This neglect is the actual main reason why the organization of an efficient material flow management is an almost impossible task nowadays (…)". A design for recycling is complicated by conflicting targets, e.g. between the durability of materials on the one hand and the recyclability them on the other hand, which can only be resolved with precise knowledge of the conditions at the end of the product's use phase. Such information about the disposal of products in the specific context can generally only be provided by the actors in the waste management sector.
- In the production phase integrated environmental protection concepts aim etc. on the prevention of waste, instead of dispose or recycle it afterwards with end-of-pipe technologies. Especially the integrated management of recycling networks between different companies could generate considerable potentials.
- An increase of use time and intensity of products could also make an important contribution to the sustainability of resource use. If products are used for a longer time, ceteris paribus, this lowers the demand for new products, reduces waste in production as well as the generation of waste products. The waste management sector could contribute to the establishment of redistribution systems for products, which could be further used as complete products by remanufacturing or re-used by repairing individual product parts.
- The phases of redistribution, dismantling, recycling and disposal are already in the responsibility of the waste management sector and of course this will be the core task also in the context of sustainable resource management. However, as described a better feedback to the product design is needed in order to eject pollutants or enable the optimal disassembling of products for the recycling. For

this purpose also the actors in the waste management should formulate their requirements for needed information e.g. in form of hints for dismantlement of products.

Such a functional transformation of the waste sector could be fostered by national waste prevention programs which have been established against the opposition of the waste regime — but the actors slowly start to see the market potentials in this business.

3.2 Waste Prevention Programmes

In a new waste hierarchy the amended EU Waste Framework Directive (WFD) (2008/98/EG) confirmed the prevention of waste as a priority measure to protect the environment with regard to the production and handling of waste. Amongst others the Member States are requested to promote waste prevention. According to article 29 par. 1 WFD the prevention measures have to be planned in terms of waste prevention programmes to be created by the Member States until December 12th 2013. These prevention programmes have to describe existing waste prevention measures and set waste prevention goals. The progress has to be monitored and assessed by implementing appropriate, specific qualitative or quantitative benchmarks for adopted waste prevention measures. By the objectives and measures of prevention programmes the environmental impacts associated with generation of waste shall be decoupled from economic growth.

In a project funded by the German Ministry of Environment and the German Federal Environmental Agency existing national waste prevention measures which do already help to reduce waste on the national, state, regional and local level in Germany and abroad have been recorded and structured by Wuppertal Institute and Öko-Institut [12] considering voluntary, regulatory, economic, and eco-specific instruments:

- Measures that could affect the framework conditions related to the generation of waste;
- measures that could affect the design and production and distribution phase; and
- measures that could affect the consumption and use phase.

Altogether the project has identified and described 296 different public waste prevention measures which cover a very wide range referring to the actors involved as well as to the different fields of action. To give some examples:

- *Promotion of eco-design*: *Second Life*. Within this UBA-project quality criteria for the reuse of used electronic devices are analyzed and defined in order to create the basis for the introduction of a quality label.
- *Sensitization: Information and advice on prevention and recycling of waste requiring special monitoring* (*BIVA*). Programme to advise and to provide information for small and medium enterprises and cottage industries in Hessian

funded by a levy on hazardous waste with a focus on free consultation (1050 initial consultations), implementation advice, the provision of information through seminars, workshops, guidance documents and leaflets.
- *Promoting measures for reuse and repair:* Use of reusable dishes and mobile automatic dishwashers. Mandating the use of reusable dishes at events by regulatory law. Switching to reusable dishes has reduced the waste generation at the Munich Oktoberfest from 11,000 to 400 tonnes.

3.3 Conclusion

New requirements regarding the waste industry can therefore be derived from the objective of sustainable resource management that go far beyond the safe disposal of waste and focus on the prevention of waste, high-quality recycling and conservation of resources. And in fact, a variety of innovative projects shows that such a sustainable waste management is not so much a technical, but a management challenge. That raises the central question of the obstacles to such a development: Why do these approaches normally not go beyond pilot and research projects and fail to make the necessary structural change? In developing and emerging countries the possibilities of cheap disposal in more or less controlled landfills are still the central barriers. But why can even in such countries as Germany, where this unregulated disposal should be largely prevented by a very differentiated waste law, be seen no significantly different results? The focus should therefore be extended from individual technologies to their social embedding as an infrastructure system in order to understand the conditions under which change takes place in such a socio-technical system.

4 Theoretical Framework: Innovations in Socio-Technical Systems

4.1 Eco-Innovations

Environmental-friendly innovations are assumed to be the decisive approach to changing the socio-industrial metabolism towards sustainability: The limits to growth should be expanded by an ecological modernization and allow a decoupling of economic growth from resource use. The normative goal of sustainable innovation research was associated with twofold increase in the consideration, as well in terms of problem understanding as of the analytical framework (cf. [34]):

- Until the early 1990s, hope has been still set on the development of end of pipe technologies to "clean technologies" in each company. It is now clear that structures outside the company must be considered to understand when and under what conditions an ecological innovation can become successful in the market.

- The analysis must therefore be extended beyond neoclassical, demand-pull price effects. In a more evolutionary-economic understanding also non-market barriers such as routines, regimes and knowledge capacities have to be taken into account.

Only in such a comprehensive understanding of the innovation context it can be possible to understand and promote the radical change necessary for a sustainable development instead of only substituting old techniques with new ones.

4.2 Infrastructure Theory

Hence this paper thus does not try to analyze waste incineration as single technology, but the waste infrastructure system in its entirety. Especially Hughes [15] has pointed out that so called "large technological systems" (LTS) cover more than just their technical artifacts. According to Kemp they include also organizational, scientific, legal and natural components. All these system components, of course including the technical artifacts interact with each other in these systems; a change of an individual artifact will affect all of them. Thus the specific properties of the individual parts do not result from themselves but from their system context. As a social subsystem LTS are on the one hand socially constructed, on the other hand, they also make considerable impact on society.

According to Konrad et al. [18] the starting point of such socio-technical regimes for infrastructures is not the used technology, but their social function or the problem to be solved. Especially infrastructure systems are characterized by path dependencies due to their long-term planning dimension and capital intensity, giving them on the one hand a particular stability, which represents on the other hand a special challenge for the processes of transition management. Their regimes are usually so highly institutionalized and specialized based on the division of labor that changes are only possible in the form of incremental innovations.

Socio-technical regimes are based as shown not only on technologies and their artefacts, but also on rule systems which are accepted by all regime members: the continuous interaction of the actors on the basis of these rules leads to a continuous reproduction of the regime (cf. [34]). Kemp et al. [17] point out that the constitutive rules of a regime mutually strengthen themselves and in this way impede their change. The economic actors on the supply side of technology development are threatened by radical innovations because they mean the devaluation of their existing core competencies, at the same time the side of technology demand shows a high level of risk aversion: As yet there are no experiences with the new technology, despite high presumed benefits they usually confide in the established products. In addition to economic considerations also legal factors increase the stability of a regime, e.g. in the form of long-term contracts as they are commonly used in the disposal sector.

Path dependencies thus have the consequence that not necessarily the best system succeeds in the market competition. Hughes [15] describes very impressively the "momentum" which an infrastructure system develops throughout its consolidation. This is based partly on the people working within the regime, who are interested in its preservation and growth due to their specific virtue and the specific skills regarding the used technology. Secondly infrastructures are characterized by a high degree of sunk costs: Cost of technical artifacts that need to be charged off as losses when the dominant regime technique changes, because they cannot be used otherwise.

4.3 Conclusion

The consideration of complete socio-technical systems focuses on the specific social context in which a technology such as waste incineration is used. Technical systems that work well in one country may lead in other countries to a variety of unwanted side effects. Particularly in the light of the very export-oriented German industry for recycling plant technologies there have been a number of projects in recent years in developing and emerging countries where the technology used was completely oversized and clearly not adapted to the existing systems (cf. [7]).

Moreover, not only the current situation should be taken into consideration, but also from the beginning path dependencies triggered by necessary investments for a waste incinerator in order to enable long term sustainable solutions for waste management. In the following it shall be illustrated by the example of Germany, what kind of problems can be caused if these aspects are not sufficiently taken into account and what negative effects on prevention and high-quality recycling of waste may be associated in the individual cases.

5 Waste Management in Germany

The German waste industry in the 1950s after the 2nd World War had to face the task of building an entirely new infrastructure system whose basic structures still continue to have effects on today's waste management. The rapidly rising of consumer spending and the increasing popularity especially of plastic containers led very quickly to bottlenecks in the landfill and the still very low incineration capacities. In addition, the emerging environmental consciousness of the population was bothered about the consequences of the unregulated land filling and the first waste scandals. The first environmental report of the German federal government in 1970 issued the figure of 50,000 unauthorized dumps in Germany, which at this time neither required approval nor have been subject to monitoring, but of which many today are still dealt with as contaminated sites (cf. [31]).

With the Waste Disposal Act of 1972 a national waste disposal monopoly was constituted with a centralization of legal responsibilities at the federal level, while the local government units were defined as responsible for the implementation of the waste legislation, citizens and companies were obligated to leave their wastes to the public institutions (cf. [22]). By waste disposal plans and mandatory plan approval procedures technical disposal paths were defined. But especially the establishment of a technical infrastructure in the form of waste incinerators developed rapidly into a "prosperous business sector", however, without that any incentives to a prevention of waste would have been given.

On the contrary, the plants resulted in a load logic as the economic viability of the system depended on a secure input of waste (cf. [31]). However, in the late 1980s there was a lack especially of burning capacities because of the resistance in the population against new waste incineration plants in their neighborhoods out of fear of harmful substances such as dioxins and furans. The federal government responded with the 17th Federal emission control act of 1990, which lead to a drastic tightening of the limit values for waste incinerators.

Although the Waste Disposal Act of 1972 had stated that waste treatment facilities require planning permission, there was an urgent need to build new incineration plants after the German reunification, especially in the former GDR. Therefore, as part of the Investment Facilitation and Housing Land Act of 1993, the licensing of waste disposal facilities was revised: Since then thermal, biological and mechanical waste treatment facilities only call for an emission-related permit and an environmental impact assessment instead for a plan approval procedure. The main difference between these two procedures is the fact that in a licensing procedure according to the German Federal Control of Pollution Act the applicant is eligible for the approval if he meets all the required environmental conditions, in a plan approval procedure such an entitlement is not provided (cf. [20], p. 484f). Due to this change there is now no way for a national demand planning of waste treatment capacities in Germany, this is also not provided in the new draft of the German waste legislation necessitated by the revised WFD (cf. [26], p. 3).

5.1 Development of Incineration Capacities

The technical infrastructure of waste management in Germany is mainly characterized by a variety of waste incineration plants, whose total capacity is generally regarded as clearly exceeding of the actual needs. This is mainly due to two related recent developments: significant capacity constraints at the end of the transition period of the "technical instructions for municipal waste" (TASi) in 2005 which prohibits the disposal of untreated wastes and the dynamic development of extra capacities in the field of substitute fuels (RDF).

5.1.1 Ending of Untreated Disposal

With the expiry of a twelve-year transitional period in Germany landfilling of waste without treatment in a suitable facility is no longer possible since 1 June 2005. The local operators therefore tried to fill their landfills before the closure by the lowest possible prices. As a result waste incinerators lost heavily on inputs (cf. [28], p. 7). Already quite early it was predictable that after the TASi even with an intensified use of co-incineration only insufficient incineration capacity would be available: Alwast [1] estimated the additional needs at least 4.5 million Mg. The forecasts were very difficult because with the end of cheap landfilling much commercial waste that was previously "recycled" in self-responsibility of the producers, now again was declared as waste to dispose and left to the municipalities.

In fact, the burning rates increased dramatically for the remaining contingents in waste incineration plants since June 2005 and were charged for about three times the prices to be paid today. In order to avoid a chaos many authorities approved the formation of interim storage and at the same time invested huge amounts of money in new treatment facilities. In this way they wanted to close the projected gap in waste disposal in the medium and long term. But obvious parallel plannings for conventional incinerators and new substitute fuel power plants were drawn up and carried out. It should also be mentioned that in this time significant amounts of industrial waste were dumped illegally in clay pits. Given the high prices in incineration plants a lot of money could be earned very quickly in this grey market.

5.1.2 RDF

Compared to conventional waste incineration facilities RDF plants are relatively small steam power plants with about 50–220 MW, in which medium-to high-calorie waste is incinerated and the resulting energy is used by combined heat and as process steam or district heating for industrial facilities. The input material mainly consists on the one hand of commercial and industrial wastes which are characterized by uniformity and consistent burning properties such as residues from the paper industry, slugs, rubber and plastics, waste oils and on the other hand of heterogeneous high calorific fractions in the household waste (e.g. plastic packaging). RDF power plants have dramatically gained in importance since 2005: For one thing because of high prices in the waste incinerators at the end of TASi transition period, but for another thing also because of the significantly increased costs for primary energy sources. The RDF business model is thus based on three pillars: The operators accept deliveries of waste against payment of fees, they burn this and their own waste and save energy costs. In contrary to the German waste incineration market, RDF plants are almost exclusively run by private operators, mostly industries with a very high energy demand and suitable waste to burn, e.g. the paper industry (cf. [26], p. 15). Several environmental NGOs complain that the economic profitability of these plants is also caused by the fact that they only very

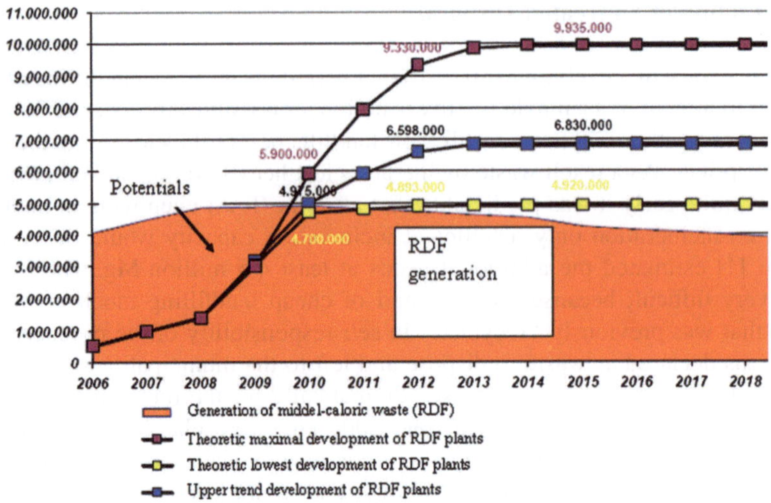

Fig. 2 Current and projected capacities and waste generation in the RDF market, cf. [26]

barley comply with the limit values of the German Federal Control of Pollution Act, even though the high calorific wastes contain higher pollutant loads (cf. [6]). Figure 2 illustrates the dynamics of the development in the RDF market since 2005.

In a very short time licensing procedures for RDF power plants with capacity of nearly 5 million Mg have been granted. The top line ("theoretical maximum capacity") will quite certainly not be reached due to project cancellations in the world economic crisis, but already today more plants are in operation or construction than RDF input materials will be available. The forecasts assume that the generation of high-calorific industrial wastes will soon decrease considerably in Germany (see [1], p. 153). It should be noted that waste incineration and RDF are no completely separate markets: These amounts of waste are now missing in the incinerators. This competition will intensify when both types of plants will be considered as recovery methods by the new Waste Framework Directive.

5.1.3 Forecast for 2020

According to current forecasts the residual waste (including bulky waste) in Germany will fall from 16.1 million Mg in 2006 to 13.9 million Mg in 2020 (cf. [26], p. 1). This decline is caused by efforts to increase resource efficiency in the industry and the expected population decline by about 2%, but also by the intensified separate collection of recyclable materials from households and the industry. But looking at the waste incineration capacities, contrary trends are apparent. The private recycling industry already today complains about enormous overcapacities for the incineration of waste, caused by massive failure investments

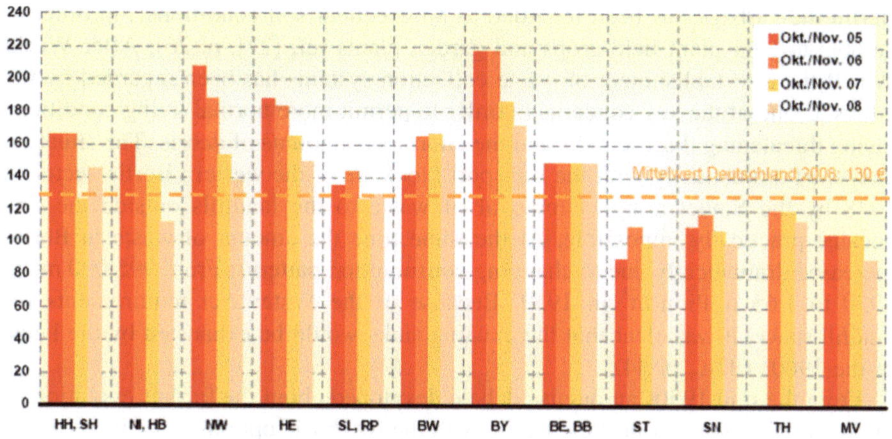

Fig. 3 Development of waste incineration fees, cf. [26]

of the municipal waste management sector in the amount of approximately € 4 billion (see [8], p. 7). In addition to the German over-capacity also expansions in the Netherlands and Poland will affect the German market.

In sum—considering waste incineration, RDF, co-incineration and other forms of waste treatment–results in 2015, an excess capacity of about 7% results, even if up to than expiring capacity would not be renewed (if the local authorities do not abandon the decommissioning of their facilities the deficit would be about 15%, cf. [26]). Therefore it can be anticipated that the falling of prices for energy recovery as illustrated in Fig. 3 will continue in the future. This could lead to a spate of insolvencies in the medium-sized recycling industry, since at lower incineration prices more materials will be thermally recovered rather than fed to material recovery facilities. The graph also shows the significant regional price differences within Germany: For household waste the transport costs cause a significant share of the price and thus the price level cannot be balanced by supply and demand, thus causing different regional playing fields for the waste management sector.

Two specific case studies of German cities shall illustrate this relationship between incineration capacity and innovation structures to avoid waste by high-quality recycling or reuse. Both examples also show how the institutional framework and economic circumstances differ even from city to city, thus emphasizing the relevance of the urban-specific context for sustainable planning of waste management infrastructures.

5.2 *Case Study Berlin*

Until 1936, the municipal waste of Berlin had been shipped to large parts with barges into one of the first nature reserves in Germany located in the hinterland of Berlin. There it has been "scenic recycled": for the many bogs and swamps around

Berlin the garbage has been regarded as an excellent soil conditioner, in order to be able to later plant field crops and garden plants (cf. [27], p. 26). After World War II the re-establishment of waste collection systems has been regarded by the Allies as one of the key concerns in order to permit more normal living conditions facing threatening disease risks and the enormous amounts of debris. The situation worsened dramatically with the transport blockade of the western Berlin sector in 1948 by the Soviet army, a removal of waste to the landfills outside the city became practically impossible. At the same time the amount of waste in Berlin increased dramatically due to changing consumption patterns: From 995.000 m^3 in 1950 to 1.65 million m^3 in 1959. Because of the Water Act requirements for landfill space, it was clear that the existing sinks would be exhausted by the latest in the 1960s (ibid., p. 44).

From the mid 1950s an intense debate developed about the further evolution of the waste management in West Berlin with the two basic options of incineration or composting of biogenic components. Against the composting on the one hand it has been argued that larger areas would have been needed for, which were not available in Berlin. As even more serious it has been considered, that sales market for the resulting compost would be missing because of the insularity of Berlin. Against the option of incineration it has been proffered, that first incineration experiments in the 1920s in Berlin had failed because of the specific waste composition- in contrast to the successful approaches to incineration in England Berlin households heated only with brown coal, so that the waste did not have a sufficient caloric value. But with the increasing amounts of paper and plastic in the 1950s the calorific value increased significantly and allowed an incineration of waste (cf. [27], p. 49). In 1963 the Berlin Senate published a study "The transformation from waste disposal towards waste incineration". For the financial viability of an incineration plant it was crucial that as well the waste heat could be used as the resulting waste products could be transformed by a sintering process in aggregate, which suggested a location in the immediate vicinity to a power plant. In 1973 the construction of at this time the largest facility of its kind in Europe was completed. Even then, however, it was clear that the original plan to use the slag as concrete aggregate would fail, because due to quality problems the sintered material could only be used as bedding or as a sport court surface. Overall, the building had cost about 52 million €.

In addition, the Senate of Berlin West started negotiations with East Germany concerning the disposal of West Berlin garbage in landfills near the city. Because the GDR was very interested in foreign exchange earnings, a treaty with a 20-year period was than agreed in 1974. It stipulated that until 1994 a total of 18 million m^3 of excavation, 38 million m^3 of construction waste and 35 million tons of municipal waste from West Berlin could be shipped to the GDR, in turn, the GDR would have taken profits of about 750 Mio. Euros over the entire term of the contract. Three landfills were especially constructed on the territory of the GDR for the garbage from West Berlin. The discussions about the composting of organic waste were thus finally terminated, the waste disposal problem was transformed into a pure problem of waste-transport. In 1979 78.1% of the total waste generation was brought into the GDR and deposited there (cf. [5]).

From the mid-1970s the recycling in Germany had a new and increasing social relevance; additionally private companies discovered the recovery of secondary raw materials as a new and lucrative business. In Berlin, the BSR tried to pursue this trend e.g. by establishing a working group on recycling in 1973 and tentatively invested in the separate collection of paper (but because of a collapse on the market for waste paper this project was stopped quite soon, cf. [27], p. 108). In 1975, a private company developed a new redistribution system, in which 20,000 households collected paper, glass and later also metal scrap in separate tons, what later became known as the "Berlin Model". It was financed by fees, but which were cheaper than the municipal garbage fees resulting in savings for the citizens ([36], p. 257). However, the amount of waste did not develop as assumed in the planning of the incinerator capacity and the waste shipment contract. This was due to a population decline in Berlin, to the increasing environmental awareness among the population but also due to the separate collection of recyclable materials and their recovery. Considering an utilization rate of the waste incineration plant of only 40% it does not surprise that no real interest existed in the reduction of waste by recycling. Given the focus on waste incineration in Berlin there were only little incentives to avoid harmful substances in products or to collect and dispose them separately. This pollution problem presented a major problem for the composting of organic waste: Facing the high pollutant loads in the domestic waste a recycling of the compost was almost impossible (cf. [4], p. 181). Nevertheless, facilities for the composting of organic wastes mainly from gardens and parks and the establishment of a "compost stock exchange" had been planed, but the project failed, however, because the Senate did not grant the necessary areas (cf. [36], p. 262).

5.3 Case Study Frankfurt

The city of Frankfurt has built a waste incineration plant in 1965 which is still in operation and has provided in combination with a thermal power plant district heating to a satellite city in the Northwest district. Facing the coming end of untreated waste disposal due to the TASi, in 2002 it was decided to fundamentally renovate and expand the waste incineration plant. The operator has been obligated to the extra condition that only waste collected in a radius of 70 km should be burned in order to prevent "waste tourism". The waste is collected primarily in the city of Frankfurt and the surrounding Rhine-Main area. The six-year renovation cost about € 279 million, since its reopening in September 2009, the plant is one of the most modern incinerators in Europe. The app roved annual capacity is about 525.000 t, the technical capacity would be even significantly higher. However, the amount of residual waste per capita decreased in Frankfurt in the last ten years from about 320 kg to less than 250 kg, so that the plant had already to deal with workload issues in the past.

In April 2005, the company Infraserv, operator of an industrial park in the north of Frankfurt, informed the responsible regional government office about plans for a refuse-derived fuel (RDF) plant located in the industrial park for about 300 million € with an annual capacity of 675.000 t/a in order to supply the resident companies with 70 MW energy or 250 t steam/h at bargain prices. This is one of the largest facilities of its kind in Europe. Ninety companies with about 22,000 employees are located in the industrial park, since 2000 about 3.7 billion euros have been invested in the site. From an economic policy perspective, it thus forms the industrial backbone of the region. The business competition towards the waste incineration plant in Frankfurt has been mentioned, but in the end the facility has been authorized as it met all environmental regulations. According to the operator all the materials to be burned are purchased from neighbouring federal states and countries.

Nevertheless, the consequences of these regional incineration overcapacities for the recycling of reusable materials are clearly visible. While the RDF power plant accepts processed waste for just 30 € per ton in order to guarantee full capacities, the price in the surrounding municipal solid waste incinerators is up to 200 €/t, causing massive under-utilization and economic losses to their owning cities. To minimize these deficiencies, the city of Frankfurt attempts to deflect as many waste streams into its incineration plant as possible: For example, a sorting plant for commercial waste has been closed. Also a non-profit company in Frankfurt, which deals with re-use and disassembly of electronic equipment, complains about lack of input, because the devices are often only shredded and burned, with the result that the contained precious metals are distributed dissipative in the slag and thus are ultimately lost for a recycling.

6 Conclusions

The generation of waste in developing and emerging countries is assumed to grow rapidly in the next few years. E.g. the Environmental Outlook of the EEA estimates an increasing adaptation of the per capita amount of waste in the new EU member states to the European average, for many emerging countries a strong population growth has to be added. For this waste a safe disposal has to be ensured, in the long term landfilling will only be an insignificant issue, so that waste incineration will be a key component of sustainable waste management.

The perspective of resource management highlights, however, that in addition to an optimized energy recovery the recycling of raw materials should be considered in greater extent. Where landfilling is increasingly restricted, one should focus on path dependencies caused by waste incineration plants. Especially for waste management systems in developing and emerging countries incinerators are enormous capital investments, constituting key elements for material flow management for a very long period. In view of this temporal dimension, especially the capacity planning for these systems is becoming enormously important. Inter alia the economic development (on local, national and international level), the spatial population development, modifications of the social and employment structure, measures of economic development and the development of industrial production techniques have to be considered.

Also expectations about the change of the legal framework can be of highly significant importance, as they can cause a redirection of entire waste streams, e.g. the TASi in Germany (cf. [20], p. 466), but also assumptions have to be made about the price development for recycled materials. Given this variety of uncertainties that make an accurate prediction of future amounts of waste virtually impossible, there is a clear trade-off between the objective of ensuring disposal guaranty on the one hand and the imperative of waste prevention on the other hand: For reasons of safe disposal it would be appropriate to be orientated towards scenarios with very high figures of waste generation, in contrast a most restrictive planning of waste disposal and recycling facilities would set much stronger incentives for waste prevention.

As the case studies have shown the capacity requirements in the waste infrastructure planning have often been overestimated in the past. This was partly due to financial incentive structures combined with corruption and bribery, on the other hand often due to a simple extrapolation of trends in the generation of waste, based on a simple linear relationship between GDP and amounts of waste. It should be analyzed in much more detail whether the described trends of waste shifting into developing countries enable a similar decoupling of GDP and waste, as empirically observable in Germany since about 10 years. By all means qualitative aspects of flexibility, compatibility with socio-economic conditions and market conditions for secondary raw materials should be taken into account in infrastructure planning from the beginning.

6.1 Impacts on Redistribution Systems for Recyclables

When establishing thermal recovery infrastructures a special focus should be put on not to endanger existing systems of material recovery. As illustrated in Fig. 4 only a small share of waste incineration plants in Germany is able to recover metals from the ash beyond iron and steel. But from a resource management point of view especially closing the loop for these non-renewable resources is of greatest importance:

Metallic raw materials are important for a variety of technical applications. With advances in technology in many areas, the use of metals expanded rapidly in recent decades (more applications, more metals). Accordingly, today the majority of the 60 metals is technically used routinely. Besides the well known ferrous and nonferrous metals, which dominate based on the used amount, a variety of metals have emerged, which are mainly used for specific functions in small amounts.

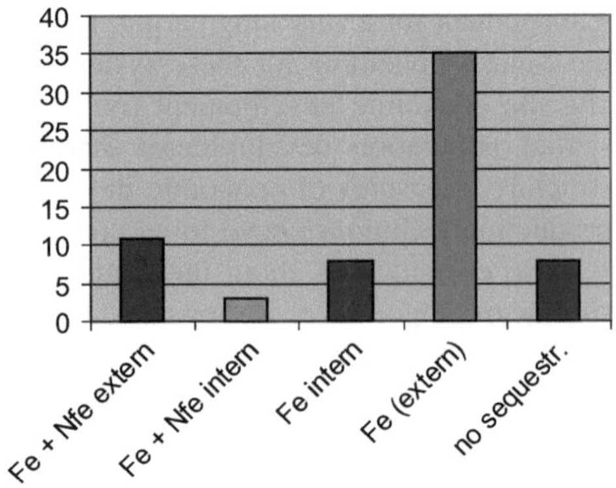

Fig. 4 Sequestration of metals in waste incineration plants, cf. [16]

Table 1 Environmental impacts associated with PGM production [30]

Impact indicator	Platinum
Primary production, TMR in t/t	683,564.91
Secondary production, TMR in t/t	8,738,82
Ratio primary/sceondary	78.22

In this sense they can be described as rare. Furthermore, some of these metals are discussed because of their limited availability–that's why they are also called critical metals, including for example the precious metals silver (Ag), gold (Au), palladium (Pd), the steel refiners manganese (Mn) and nickel (Ni), the heavy metals tin (Sn), zinc (Zn) as well as the "specialty metals" gallium (Ga), indium (In) and titanium (Ti).

Typical fields of application with strong growth rates are electrical and electronic equipment, medical technology and nanotechnology. Referring to the described case study on PGM Table 1 shows the differences in resource consumption between primary (mining) and secondary recovery by recycling:

Both case studies have highlighted the risk that incineration excess capacities can result in diverting additional material flows into waste incinerators. The contained metals in these products, e.g. shredded electronic products are dissipatively distributed in the ash and thus are irretrievable lost for high quality recycling. It is therefore necessary to establish as early as possible arrangements to sort out these products and to supply them to appropriate recycling methods.

In order to secure uniform disposal standards and a control of waste flows from a sustainable resource management point of view it must be prevented that the waste is directed only due to price differences between the different waste treatment options (cf.[28], p. 3). Thus in the future strategies are needed for a better resource management both for commodity markets which increasingly depend on secondary raw materials from recycling and for the energetic use of waste as RDF in energy-efficient facilities with an optimal power-heat coupling for energy-intensive industries (cf. [26], p. 46).

The efficient development of new raw materials potentials out of complex waste streams, arising distributed in space and time requires the establishment of networked recycling facilities in a multi-stage treatment process (cf. [11], p. 251). Prerequisites for an efficient recycling of secondary raw materials are suitable sales channels, which maximally exploit the material properties and have the appropriate capacity and flexibility to compensate these temporal variations. This requires an optimal size and location of each link in the value chain.

The example of German over-capacities in the market for thermal recovery particularly highlights that especially waste incineration plants must be integrated very carefully into the overall system of waste management. Thus the total capacity should not mainly depend only on short-term economic aspects if the goal is to enable a long term sustainable development—both in ecologic and economic terms.

References

1. Alwast H (2009) Ersatzbrennstoffmarkt in Deutschland. In: Recycling Almanach 2009, Eltville
2. Ayres RU (1989) Industrial metabolism and global change. Int Soc Sci J 121:363–373
3. Bringezud S, Bleischwitz R (2009) Sustainable resource management trends, visions and policies for europe and the world. Greenleaf Publisher, Austin
4. Brüggemann T (1985) Kompostierung in Berlin-Ein ziel im rahmen des abfallwirtschaftspro gramms Berlin. In: ANS info-band heft 7. Berlin
5. BSR (1979) Moderne Stadreinigung und Abfallbeseitigung in einem Ballungsgebiet. Berlin
6. BUND NRW (2010) Umweltdumping durch EBS-Anlagen? Avaiable online: http://www.bund-nrw.de/themen_und_projekte/abfallpolitik/ersatzbrennstoffe
7. Dilewski G (2010) Abfallverwertung und kreislaufwirtschaft in entwicklungs- und schwellenländern—potentiale und grenzen. IWAR-lecture neues aus der umwelttechnik und raumplanung, Darmstad
8. EUWID (2010) Europäischer wirtschaftsdienst recycling. vol 9. Springer, Berlin, p 7
9. Flatz A (1996) Von der abfallbewirtschaftung zum stoffstrommanagement. Organisation sansätze am beispiel elektrotechnischer produkte. signum, Wien
10. Friege K (2005) Infrastruktur für eine moderne Abfallwirtschaft. In: Loske R, Roland Schaeffer R (eds) Die Zukunft der Infrastrukturen. Marburg, Metropolis
11. Goldmann D (2009) Erschließung neuer Rohstoffpotenziale aus Abfallströmen. In: Thomé-Kozmiensky KJ, Goldmann, D (eds) Recycling und Rohstoffe, Vol 2. Neuruppin
12. Günter Dehoust G, Küppers P, Bringezu S, Wilts H (2010) Erarbeitung der wissenschaftlichen grundlagen für die erstellung eines bundesweiten abfallvermeidungs programms. UBA-Texte Nr. 59/2010, Dessau
13. Hofmeister S (1998) Von der Abfallwirtschaft zur ökologischen Stoffwirtschaft Wege zu einer Ökonomie der Reproduktion. Westdt Verl, Opladen
14. Huber J (2000) Industrielle Ökologie Konsistenz, Effizienz und Suffizienz in zyklusanalytischer Betrachtung. In: Kreibich R, Simonis UE (eds) Global change. Verlag Arno Spitz, Berlin, pp 109–126
15. Hughes TP (1989) The evolution of large technological systems. In: Bijker WE, Hughes TP, Pinch T (eds) The social construction of technological systems. New directions in the sociology and history of technology. MIT Press, Cambridge, pp 51–82

16. IFEU (2007) Beispielhafte Darstellung einer vollständigen, hochwertigen Verwertung in einer MVA unter besonderer Berücksichtigung der Klimarelevanz. Heidelberg
17. Kemp R, Schot J, Hoogma R (1998) Regime shifts to sustainability through processes of niche formation: the approach of strategic niche management. Technol Anal Strateg Manag 10(2):75–195
18. Konrad K, Voß J-P, Truffer B, Bauknecht D (2004) Transformationsprozesse in netzgebundenen versorgungssystemen. Ein integratives analysekonzept auf basis der theorie technologischer transitionen. CIRUS/EAWAG und Öko-Institut
19. Kopytziok N (2009) Das Ganze im Blick behalten. Müllmagazin 2:24–29
20. Kranert M, Cord-Landwehr K (eds) (2010) Einführung in die Abfallwirtschaft. Vieweg and Teubner, Wiesbaden
21. LaDou J et al (2007) Export of electronics equipment waste. Int J Occup Environ Health 14:1–10
22. Lamping W (1998) Kommunale Abfallpolitik. Ökologischer Strukturwandel und politisches Lernen. Dissertationsschrift an der Fakultät für Geistes- und Sozialwissenschaften der Universität Hannover, Hannover
23. Lucas R, Wilts H (2009) Roadmap–towards a more efficient recycling of PGM from catalytic converters on an international level. Wuppertal
24. Luks Fred (2005) Innovationen Wachstum und Nachhaltigkeit Eine ökologisch-ökonomische Betrachtung. In: Rennings K (ed) Innovationen und Nachhaltigkeit. Metropolis Verl, Marburg, pp 41–62 (Jahrbuch ökologische Ökonomik, 4)
25. Meadows DH (1974) The limits to growth. A report for the club of Rome's project on the predicament of mankind, 2nd edn. Universe Books, New York
26. NABU (2009) Müllverbrennung in Deutschland wächst unkontrolliert–recycling ist gefährdet. Müllimport wird attraktiver, Berlin
27. Park J (2004) Von der Müllkippe zur Abfallwirtschaft–Die entwicklung der hausmüllentsorgung in Berlin (West) von 1945 bis 1990. TU Berlin, Berlin
28. Rahmeyer F (2004) Abfallwirtschaft zwischen Entsorgungsnotstand und Überkapazitäten. Volkswirtschaftliche Diskussionsreihe, Beitrag Nr. 266, Augsburg
29. Sander K,Schilling S (2010) Optimierung der Steuerung und Kontrolle grenzüberschreitender Stoffströme bei Elektroaltgeräten/Elektroschrott. UBA-Texte Nr. 11/2010, Dessau
30. Saurat M, Bringezu St (2008) Platinum group metal flows of Europe–Part I. Global supply, use in industries and the shift of environmental impacts. J Ind Ecol 12(5–6):754–767
31. Schenkel W (2003) Zur Geschichte der Abfallwirtschaft in Deutschland. Müll und Abfall 35(12):620–625
32. Schütz H, Moll, Bringezu S (2003) Globalisation and the Shifting of Environmental Burdens: Material Trade Flows of the European Union. Wuppertal Paper Nr. 134, Wuppertal
33. Shinkuma T, Minh Huong NT (2008) The flow of E-waste material in the Asian region and a reconsideration of international trade policies on E-waste. Environ Impact Assess Rev 29(1):25–31
34. Smith A, Voß J-P, J Grin (2010) Innovation studies and sustainability transitions: the allure of the multi-level perspective and its challenge. Res Policy 39:435–448
35. UNU (2008) Review of Directive 2002/96 on Waste Eletrical and Electronic Equipment. Final Report, Bonn
36. Vondracek HJ (1985): Informationsstrategien zur Förderung von Kompostierungsaktivitäten in Berlin. In: ANS Info-Band Heft 7, Berlin

MIX
Papier aus verantwortungsvollen Quellen
Paper from responsible sources
FSC® C105338

If you have any concerns about our products,
you can contact us on
ProductSafety@springernature.com

In case Publisher is established outside the EU,
the EU authorized representative is:
Springer Nature Customer Service Center GmbH
Europaplatz 3, 69115 Heidelberg, Germany

Printed by Libri Plureos GmbH
in Hamburg, Germany